155
Advances in Biochemical Engineering/Biotechnology

Aims and Scope

This book series reviews current trends in modern biotechnology and biochemical engineering. Its aim is to cover all aspects of these interdisciplinary disciplines, where knowledge, methods and expertise are required from chemistry, biochemistry, microbiology, molecular biology, chemical engineering and computer science.

Volumes are organized topically and provide a comprehensive discussion of developments in the field over the past 3–5 years. The series also discusses new discoveries and applications. Special volumes are dedicated to selected topics which focus on new biotechnological products and new processes for their synthesis and purification.

In general, volumes are edited by well-known guest editors. The series editor and publisher will, however, always be pleased to receive suggestions and supplementary information. Manuscripts are accepted in English.

In references, Advances in Biochemical Engineering/Biotechnology is abbreviated as *Adv. Biochem. Engin./Biotechnol.* and cited as a journal.

More information about this series at http://www.springer.com/series/10

Qin Ye · Jie Bao · Jian-Jiang Zhong
Editors

Bioreactor Engineering Research and Industrial Applications I

Cell Factories

With contributions by

Hongjun Dong · C. Groeger · Yin Li · Zhao Lin
Pingping Liu · Bao-Di Ma · Yanhe Ma · Jiang Pan
Hao Qin · W. Sabra · Kazuyuki Shimizu · Zaigao Tan
Ya-Jie Tang · Han Xiao · Jian-Hui Xiao · Jian-He Xu
Jun-Wei Xu · An-Ping Zeng · Tianrui Zhang
Xueli Zhang · Yanping Zhang · Zhi-Jun Zhang
Chunhua Zhao · Jian-Jiang Zhong · Xinna Zhu

Springer

Editors
Qin Ye
State Key Laboratory of Bioreactor Engineering, School of Biotechnology
East China University of Science and Technology
Shanghai
China

Jie Bao
State Key Laboratory of Bioreactor Engineering, School of Biotechnology
East China University of Science and Technology
Shanghai
China

Jian-Jiang Zhong
School of Life Sciences and Biotechnology
Shanghai Jiao Tong University
Shanghai
China

ISSN 0724-6145 ISSN 1616-8542 (electronic)
Advances in Biochemical Engineering/Biotechnology
ISBN 978-3-662-49159-1 ISBN 978-3-662-49161-4 (eBook)
DOI 10.1007/978-3-662-49161-4

Library of Congress Control Number: 2015958844

Printed on acid-free paper

This Springer imprint is published by SpringerNature
The registered company is Springer-Verlag GmbH Berlin Heidelberg

Preface

Biotechnology industrialization is the result of basic life science progress plus engineering science contribution. Without doubt, without the good integration of biology and engineering, it is unbelievable that today's biotechnology becomes so popular and important to our human beings and the whole society of the world. The two special volumes entitled "Advances in Bioreactor Engineering Research and Application: I. Cell Factories" and "Advances in Bioreactor Engineering Research and Application: II. Bioreactors" are striving to reflect the recent advances in biology and engineering related researches with their significant impact on academic and industrial R&D.

Microbial cell factories are the basis to establishment of economical biomanufacturing processes, which provide various kinds of antibiotics, enzymes, vitamins, amino acids, pharmaceutical proteins, etc. With the progresses in metabolic engineering, metabolism in wild-type microbial strains can be well altered and metabolic flux can be effectively directed to target products, to meet the requirements for efficient production of interested metabolites. This special volume on Cell Factories is dedicated to establishment of bioconversion systems for efficient production of chiral chemicals, organic acids, biofuels, and other useful metabolites.

The complicated metabolic network in microbial cells is highly ordered and precisely regulated to adapt the changing environment and to survive under unfavorable conditions. In the review by Shimizu [1], various regulations in cells are presented. Understanding of these regulatory characteristics is very important to construction of more efficient cell factories for metabolite production.

Chiral chemicals are important building blocks for the synthesis of many pharmaceuticals, pesticides and food additives. Compared with chemical synthesis, biochemical processes have the advantages of better selectivity, higher productivity, and less environmental impact. The chapter by Zhang et al. [2] reviews the efficient synthesis of chiral chemical blocks by enzyme-mediated reactions. With the development of microbial genomics, efficient discovery of enzymes with special stereoselectivity and robust performance can be realized by genome mining which is much more efficient than traditional strain selection and breeding from soil samples. Many interesting case studies are presented.

With the constantly increasing consumption of huge amounts of fossil feedstock, the supply of fossil resources will become limited, while continuous use of such resources has caused serious environmental pollution. For sustainable development, researchers all over the world have been exploring production of fuels and bulk chemicals using microorganisms with renewable, alternative feedstocks as substrates. Some organic acids and alcohols such as citrate, lactate, gluconate and ethanol are commercially produced by fermentation, but some organic acids and alcohols cannot be produced economically by microorganisms. The chapter by Liu et al. [3] summarizes the principles of constructing efficient cell factories and reviews the progresses of *Escherichia coli* cell factories for the production of organic acids and alcohols. *n*-Butanol is an excellent biofuel whose performance is better than ethanol as a transpotation fuel. Traditionally *n*-butanol is commercially produced by *Clostridium* in acetone-butanol-ethanol (ABE) fermentation, but the titer is low and genetic manupilation is difficult. The chapter by Dong et al. [4] reviews the researches on construction of *E. coli* cell factories for production of *n*-butanol and methods of theoretical prediction. Diols have wide applications and microbial production of such diols has attracted the interest of many researchers. The chapter by Sabra et al. [5] shows the recent progress in construction of microbial cell factories for production of low molecular weight diols including 1,3-propanediol, 1,2-propanediol, 2,3-butanediol, 1,3-butanediol and 1,4-butanediol.

Higher fungi are important sources of many secondary metabolites which have excellent pharmaceutical and physiological properties or can be potential lead compounds for new drug development. However, the production levels of such metabolites are usually low due to the low content in cells. In the chapter by Qin et al. [6], the methodology of genetic manipulation in higher fungi and omics analysis are described, and various useful metabolites produced by higher fungi are also summarized.

We hope this volume can provide some basic principles and current status in construction of efficient microbial cell factories, especially in the field of industrial biotechnology. Here, we would like to thank all the contributing authors and referees for their superior collaboration, the Managing Editor Prof. Dr. Thomas Scheper, and the publisher and the book-series editorial staffs at Springer for their constructive suggestions, constant support and kind help during the entire process for this special volume.

Shanghai
November 2015

Qin Ye
Jie Bao
Jian-Jiang Zhong

References

1. Shimizu K (2015) Metabolic regulation and coordination of the metabolism in bacteria in response to a variety of growth conditions. ABE, doi: 10.1007/10_2015_320
2. Zhang Z-J, Pan J, Ma B-D, Xu J-H (2015) Efficient biocatalytic synthesis of chiral chemicals. ABE, doi: 10.1007/10_2014_291
3. Liu P, Zhu X, Tan Z, Zhang X, Ma Y (2015) Construction of *Escherichia coli* cell factories for production of organic acids and alcohols. ABE, doi: 10.1007/10_2014_294
4. Dong H, Zhao C, Zhang T, Lin Z, Li Y, Zhang Y (2015) Engineering Escherichia coli cell factories for *n*-butanol production. ABE, doi: 10.1007/10_2015_306
5. Sabra W, Groeger C, Zeng A-P (2015) Microbial cell factories for diol production. ABE, doi: 10.1007/10_2015_330
6. Qin H, Xu J-W, Xiao J-H, Tang Y-J, Xiao H, Zhong J-J (2015) Cell factories of higher fungi for useful metabolite production. ABE, doi: 10.1007/10_2015_335

Contents

Adv Biochem Eng Biotechnol (2016) 155: 1–54
DOI 10.1007/10_2015_320

Published Online: 25 February 2015

Metabolic Regulation and Coordination of the Metabolism in Bacteria in Response to a Variety of Growth Conditions

Kazuyuki Shimizu

Abstract Living organisms have sophisticated but well-organized regulation system. It is important to understand the metabolic regulation mechanisms in relation to growth environment for the efficient design of cell factories for biofuels and biochemicals production. Here, an overview is given for carbon catabolite regulation, nitrogen regulation, ion, sulfur, and phosphate regulations, stringent response under nutrient starvation as well as oxidative stress regulation, redox state regulation, acid-shock, heat- and cold-shock regulations, solvent stress regulation, osmoregulation, and biofilm formation, and quorum sensing focusing on *Escherichia coli* metabolism and others. The coordinated regulation mechanisms are of particular interest in getting insight into the principle which governs the cell metabolism. The metabolism is controlled by both enzyme-level regulation and transcriptional regulation via transcription factors such as cAMP–Crp, Cra, Csr, Fis, P_{II}(GlnB), NtrBC, CysB, PhoR/B, SoxR/S, Fur, MarR, ArcA/B, Fnr, NarX/L, RpoS, and (p)ppGpp for stringent response, where the timescales for enzyme-level and gene-level regulations are different. Moreover, multiple regulations are coordinated by the intracellular metabolites, where fructose 1,6-bisphosphate (FBP), phosphoenolpyruvate (PEP), and acetyl-CoA (AcCoA) play important roles for enzyme-level regulation as well as transcriptional control, while α-ketoacids such as α-ketoglutaric acid (αKG), pyruvate (PYR), and oxaloacetate (OAA) play important roles for the coordinated regulation between carbon source uptake rate and other nutrient uptake rate such as nitrogen or sulfur uptake rate by modulation of cAMP via Cya.

Keywords Acetate overflow metabolism · Acid shock · Catabolite regulation · Heat shock · Nitrogen regulation · Osmoregulation · Oxidative stress · Oxygen limitation · Phosphate regulation · Redox regulation · Stringent response · Sulfur regulation

K. Shimizu (✉)
Kyushu Institute of Technology, Iizuka, Fukuoka 820-8502, Japan
e-mail: shimi@bio.kyutech.ac.jp; shimi@ttck.keio.ac.jp

K. Shimizu
Institute of Advanced Biosciences, Keio University, Tsuruoka, Yamagata 997-0017, Japan

Abbreviations

Metabolites

CIT	Citrate
E4P	Erythrose-4-phosphate
FBP	Fructose-1,6-bisphosphate
F1P	Fructose 1-phosphate
F6P	Fructose-6-phosphate
G6P	Glucose-6-phosphate
GAP	Glyceraldehyde-3-phosphate
GOX	Glyoxylate
ICI	Isocitrate
KDPG	2-keto-3-deoxy-6-phosphogluconate
αKG	α-ketoglutarate
MAL	Malate
OAA	Oxaloacetate
PEP	Phosphoenolpyruvate
6PG	6-phosphogluconate
PYR	Pyruvate

Protein and enzymes

Ack	Acetate kinase
Acs	Acetyl-coenzyme A synthetase
Adk	Adenylate kinase
CS	Citrate synthase
Cya	Adenylate cyclase
EI	Enzyme I
EII	Enzyme II
Fdp	Fructose bisphosphatase
FDH	Formate dehydrogenase
Fhl	Formate hydrogen lyase
GAD	Glutamate decarboxylase
G6PDH	Glucose-6-phosphate dehydrogenase
GAPDH	Glyceraldehyde-3-phosphate dehydrogenase
GOGAT	Glutamate synthase
GS	Glutamine synthetase
HPr	Histidine-phosphorylatable protein
Hyc	Hydrogenase
ICDH	Isocitrate dehydrogenase
Icl	Isocitrate lyase
KGDH	α-ketoglutaric acid dehydrogenase
LDH	Lactate dehydrogenase
Mez	Malic enzyme
MS	Malate synthase
NOX	NADH oxidase

Pck	Phosphoenolpyruvate carboxykinase
PDH	Pyruvate dehydrogenase
Pfk	Phosphofructokinase
PGDH	6-phosphogluconate dehydrogenase
Pgi	Phosphoglucose isomerase
Pox	Pyruvate oxidase
Ppc	Phosphoenolpyruvate carboxylase
Pps	Phosphoenolpyruvate synthase
Pta	Phosphotransacetylase
Pyk	Pyruvate kinase
SOD	Superoxide dismutase

Others

ED pathway	Entner–Doudoroff pathway
EMP pathway	Embden–Meyerhof–Parnas pathway
PMF	Proton motive force
PP pathway	Pentose phosphate pathway
PTS	Phosphotransferase system
ROS	Reactive oxygen species
TCA cycle	Tricarboxylic acid cycle

Contents

1 Introduction

The living organisms on earth survive by manipulating the cell system in response to the change in growth environment by sensing signals of both external and internal states of the cell. The complex signaling networks interconvert signals or stimuli for the cell to function properly. The transfer of information in signal transduction pathways and cascades is evolved to respond to the variety of growth environment. Metabolism is the core for energy generation (catabolism) and cell synthesis (anabolism). Metabolic network, defined as the set and topology of metabolic biochemical reactions within a cell, plays an essential role for the cell to survive, where it is under organized control. The set of enzymes changes dynamically in accordance with the change in growth environment and the cell's state. The enzymes that form the metabolic pathways are subject to multiple levels of regulation, and it is becoming more and more important to deeply understand the overall regulation mechanism. This may be made by integrating different levels of OMICS information with the help of systems biology approach. Although huge amount of information is embedded in genome, only a subset of the pathways among possible topological networks is active at certain point in time under certain growth condition. The keen interest is how it is managed by the cell with coordination of the metabolism in response to the change in growth condition.

Recent investigation on the metabolism is widespread ranging from bacteria to human, where much attention is focused on cancer cell metabolism [1–4]. The metabolic capabilities allow various organisms to grow in various limiting conditions and environmental niches in the ecological biosphere [5–7]. Many efforts have been focused on the emerging challenges in sustainable energy, and green society, as well as pharmaceuticals for human health by modifying the metabolic pathways [8–15].

Deep understanding on the metabolic regulation mechanism is essential for all these efforts for manipulating and redesigning the metabolism, and it is critical to understand the basic principles which govern the cell metabolism [16–20]. Such principles may be in common to various organisms, or some set of organisms, while some are the specific to the organism of concern.

Biochemical logic of metabolic pathways may be determined based on key biochemical constraints such as thermodynamic favorability, availability of enzymatic mechanisms, and physicochemical properties of pathway intermediates [16]. More specifically, there might be a connection between an organism's growth environment and thermodynamic and biochemical properties for the determination of pathways [17]. How do organisms select the pathways among available pathways? For example, there are several glycolysis pathways such as Embden–Meyerhof–Parnas (EMP) pathway and Entner–Doudoroff (ED) pathway, but how is the pathway selected among them? The glucose metabolism may reflect a trade-off between a pathway's energy (ATP) yield and the amount of enzymatic protein required to catalyze the pathway flux. From this point of view, some microorganisms such as *Zymomonas mobilis* and *Pseudomonas* sp. mainly utilize ED pathway instead of most popular energy-intensive EMP pathway due to less requirement of enzymatic protein together with thermodynamic preference [17].

In fact, the decision may be made not only by the above consideration, but it is also made by transcriptional regulation together with global regulators or transcription factors. Moreover, some specific metabolites are also involved for the coordinated regulation of the metabolism. Here, an overview is given for the metabolic regulation of microbes with special interest on the coordination of regulation systems.

2 Transport of Nutrient and Waste

The gram-negative bacteria such as *Escherichia coli* have two concentric membranes called outer and inner (or cytosolic) membranes surrounding the cytoplasm with the space called periplasm between these two membranes (Fig. 1). These membranes constitute a hydrophobic barrier against polar compounds. The outer membrane contains channel proteins, called porin proteins, where porins are substrate-specific, ion-selective, or nonspecific channels that allow the influx of small hydrophilic nutrient molecules and the efflux of waste products, antibiotics, and inhibitors [21, 22]. Among porins, OmpC and OmpF are the most abundant porins under typical growth condition. Their relative abundance changes depending on the osmolarity, temperature, and growth phase. Glucose is transported through these porins by diffusion when glucose concentration is higher above 0.2 mM [23, 24], while under glucose limitation, another porin LamB is induced, where LamB has high affinity to glucose [24].

The porin genes are under control of two-component system such as EnvZ–OmpR system, where EnvZ is an inner membrane sensor kinase, and OmpR is the cytoplasmic response regulator. In response to the environmental signals such as osmolarity, pH, temperature, nutrients, and toxins, EnvZ phosphorylates OmpR

Fig. 1 Glucose transport via PTS and other transporters

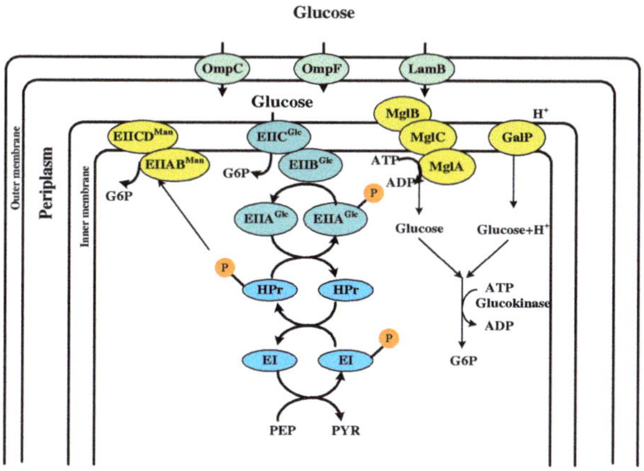

(OmpR-P), and OmpR-P activates the expression of such porin genes as *ompC* and *ompF* [21]. The outer membrane is important to protect the cell from harsh environmental condition.

The first step in the metabolism of carbohydrates is the transport of these molecules into the cytosol through inner or cytosolic membrane. Various carbohydrates can be transported by several mechanisms in bacteria [25]. The CO_2 and ammonia (NH_3)/ammonium (NH_4^+) may be transported through membrane by diffusion, where AmtB, which has the high affinity to ammonia, is induced and plays an important role for ammonia uptake under ammonia limitation. Without such a system, the important N-source, ammonia, is diffused out of the cell.

Primary transport of sugars is driven by ATP, while secondary transport is driven by electrochemical gradients of the translocated molecules across the membrane via symporters or anti-porters [26]. Sugar uptake by group translocation is unique for bacteria and is involved in the phosphotransferase system (PTS) as explained in the next section. So far, 21 PTSs have been identified in *E. coli*, while others are transported by non-PTS transporters [27]. The transporters for acids are rarely identified except formate (Foc) and acetate (ActP, YjcH), where other acids may be transported by diffusion. It is of interest to recognize [16] that charged compounds such as pyruvate, glycerate, and lactate may not be easily pass through hydrophobic lipid membrane [28–30]. On the other hand, uncharged molecules such as glucose, fructose, dihydroxyacetone, and glyceraldehydes can diffuse more freely through membrane [30–32]. Thus, a charged moiety such as phosphate group serves to reduce or prohibit their escape from the cell [33, 34]. From this point of view, it is of interest to see the fact that all the glycolysis, pentose phosphate (PP) pathway, and ED pathway metabolites are phosphorylated except pyruvate, while TCA cycle metabolites are not phosphorylated.

3 PTS and Carbon Catabolite Regulation

In the metabolic regulation, carbon catabolite regulation (CCR) plays far important roles from the points of view of energy acquisition and biomass synthesis, where PTS pays an important role [35]. In PTS, the phosphate is transferred from phosphoenolpyruvate (PEP) via successive phosphorelay reactions in turn by EI, HPr, $EIIA^{Glc}$, and $EIICB^{Glc}$ to glucose (Fig. 1) [36]. Unphosphorylated $EIIA^{Glc}$ inhibits the uptake of other carbohydrates by the so-called inducer exclusion [37, 38], while phosphorylated $EIIA^{Glc}$ ($EIIA^{Glc}$-P) activates adenylate cyclase (Cya), which generates cyclic AMP (cAMP) from ATP, and leads to an increase in the intracellular cAMP level, where cAMP combines with the global regulator Crp (cAMP receptor protein) yielding cAMP–Crp complex, and plays an essential role for catabolite regulation. If the concentration ratio between PEP and pyruvate (PYR) (PEP/PYR) is high, $EIIA^{Glc}$ is predominantly phosphorylated, while if this ratio is low, phosphorylated $EIIA^{Glc}$ ($EIIA^{Glc}$-P) is dephosphorylated [39]. Catabolite repression occurs not only by PTS-oriented regulation, but also by such catabolites as α-ketoacids as will be explained for the metabolic coordination.

Consider how the catabolite regulation affects the metabolism in response to the change in glucose consumption rate or the perturbation in the glycolysis activity. For this, it is useful to consider the continuous culture. In the continuous culture, as the dilution rate or the specific growth rate was increased, the glycolytic flux or the specific glucose consumption rate increases, and fructose 1,6-bisphosphate (FBP) concentration increases [40, 41]. The increased FBP allosterically enhances the activity of Pyk and Ppc by feed-forward control (Fig. 2a). Then, PEP concentration tends to be decreased due to the activation of Pyk and Ppc. PEP molecule allosterically inhibits Pfk activity by feedback regulation, and thus, the decrease in PEP concentration causes Pfk activity to be increased, and the glycolysis flux further increases, and in turn, FBP concentration increases more. Roughly speaking, in accordance with the increase in the glucose uptake rate, the intracellular metabolite pools of the upper glycolysis from glucose 6-phosphate (G6P) to glyceraldehydes 3-phosphate (GAP) increase, while those lower glycolysis from GAP to PEP decreases [40]. As stated above, there is a one-to-one relationship between the upper glycolysis flux and the FBP concentration. Moreover, the fluxes of lower glycolysis and the feed-forward activation of FBP on Pyk show the similar

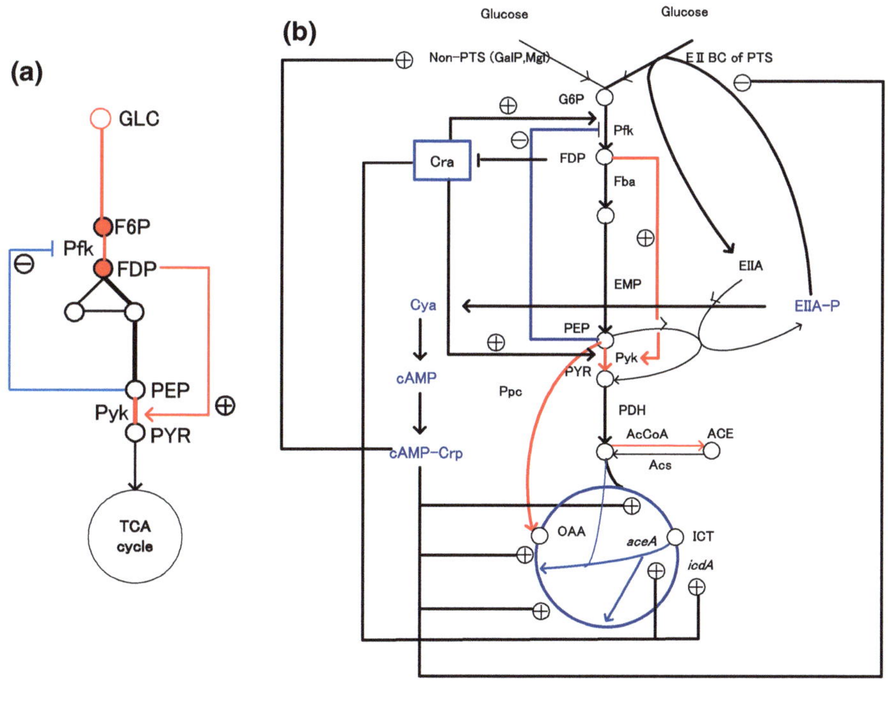

Fig. 2 Catabolite regulation of the central metabolism: **a** enzyme-level regulation, **b** overall enzyme-level and transcriptional regulations

relationship in the case of wild-type strain, where the metabolic fluxes may be considered to be sensed by molecular systems as "flux sensors" [41–43]. The meaning of this notion is that instead of preparing many nutrient-specific receptors for the cell to sense the variety of environmental signals, the flux-sensing system simply recognizes the fluxes by the specific intracellular metabolite as integral signal.

On the other hand, the decrease in PEP makes PEP/PYR ratio to be decreased. This causes $EIIA^{Glc}$-P to be dephosphorylated and in turn less activates Cya, and thus, cAMP is less formed. As a result, cAMP–Crp level decreases, which decreases the expression of *ptsG* (Table 1), and this causes the decrease in the glucose uptake rate. This forms a negative feedback loop for the initial increase in the glucose uptake rate [44, 45] (Fig. 2b). This indicates that PTS plays an essential role from the robustness point of view. Moreover, the decrease in cAMP–Crp also represses the expression of the TCA cycle genes (Table 1) (Fig. 2b).

In addition to cAMP–Crp, the catabolite repressor/activator protein (Cra) plays also an important role in the control of carbon flow in *E. coli* [46], where the carbon uptake and glycolysis genes are repressed, while gluconeogenic pathway genes are

Table 1 Effect of global regulators on the metabolic pathway gene expressions

Global regulator	Metabolic pathway genes
Cra	+: aceBAK, cydB, fbp, icdA, pckA, pgk, ppsA −: acnB, adhE, eda, edd, pfkA, pykF, zwf
Crp (cAMP–Crp)	+: aceEF, acnAB, acs, focA, fumA, gltA, lpdA, malT, manXYZ, mdh, mlc, pckA, pdhR, pflB, pgk, ptsG, sdhCDAB, sucABCD, ugpABCEQ, −: cyaA, lpdA, rpoS
ArcA/B	+: cydAB, focA, pflB −: aceBAK, aceEF, acnAB, cyoABCDE, fumAC, gltA, icdA, lpdA, mdh, nuoABCDEFGHIJKLMN, pdhR, sdhCDAB, sodA, sucABCD
Mlc	−: crr, manXYZ, malT, ptsG, ptsHI
PdhR	−: aceEF, lpdA
CsrA	+: eno, pfkA, pgi, pykF, tpiA −: fbp, glgC, glgA, glgB, pgm, ppsA, pckA,
Fur	−: entABCDEF, talB, sodA
RpoS	+: acnA, acs, ada, appAR, appB, argH, aroM, dps, bolA, fbaB, fumC, gabP, gadA, gadB, katE, katG, ldcC, narY, nuv, pfkB, osmE, osmY, poxB, sodC, talA, tktB, ugpE, C, xthA, yhgY, −: ompF
SoxR/S	+: acnA, cat, fumC, fur, sodA, sox, zwf
OxyR	+: ahpC, ahpF, katG:
PhoR/B	+: phoBR, phoA-psiF, asr, pstSCAB-phoU −: phoH, phnCHN, ugpA, argP
Fnr	+: acs, focA, frdABCD, pflB, yfiD −: acnA, cyoABCDE, cydAB, fumA, fnr, icdA, ndh, nuoABCEFGHIJKLMN, sdhCDAB, sucABCD

activated by Cra (Table 1). As mentioned above, the increase in the glucose uptake rate causes an increase in FBP concentration, where FBP inhibits Cra activity [47]. This causes the increase of the expression of the glycolysis genes such as *pfkA* and *pykF*, while it represses the expression of gluconeogenetic pathway genes such as *fbp*, *ppsA*, and *pckA* (Table 1), which implies acceleration of the glycolytic fluxes (Fig. 2b). The decrease in Cra activity also affects TCA cycle genes such that *icdA* and *aceA* gene expression is repressed, and thus, TCA cycle is further repressed by this mechanism (Fig. 2b). The increase in the glycolysis activity and the decrease in the TCA cycle activity cause more acetate production (Fig. 2b). This is the mechanism of acetate overflow metabolism in *E. coli*. The decrease in cAMP–Crp level also represses *acs* expression (Table 1), and this also causes acetate accumulation.

The question may arise as to why TCA cycle activity must be repressed in accordance with the increased activity of glycolysis. The cell growth rate might be more enhanced if the TCA cycle is not repressed, gaining more ATP from the increased NADH production at the TCA cycle. In fact, although the TCA cycle activity is repressed as mentioned above with the increase in glucose uptake rate [48–50], the absolute TCA cycle fluxes on the mmol basis tend to increase because the specific glucose consumption rate is increased [50]. This means that the respiration is activated with the increased production of NADH, and more reactive oxygen species (ROS) are generated in the respiratory chain and cause damage to the cell. This is the reason why the TCA cycle has to be repressed at higher glucose consumption rate. Of course, the cell furnishes a detoxifying system for ROS, where the transcription factors such as SoxR/S and OxyR play important roles for such oxidative stress regulation as will be mentioned later. The reducing equivalent, NADPH, plays an important role for detoxification of ROS as implied by the fact that SoxS activates *zwf* gene expression (Table 1). In the case of yeast and higher organisms, glutathione plays the similar role, where its production is enhanced by NADPH. The 6PGDH and G6PDH activities have been known to be the growth rate dependent [51], where the activation of the oxidative PP pathway is considered to be due to NADPH requirement for biosynthesis. However, the oxidative stress may also affect the activation of the oxidative PP pathway at higher cell growth rate. Moreover, some microbes such as *E. coli* produce NADPH at ICDH as well (instead of NADH) in the TCA cycle, where this is also related to the oxidative stress regulation as will be mentioned later.

4 Acetate Overflow Metabolism and the Reduction of Acetate Formation in *E. coli*

A major obstacle for the commercial scale production of useful recombinant proteins is the undesirable by-products formation. Among them, acetate formation is by far crucial in the case of *E. coli*, because it retards the cell growth, and it inhibits protein formation. Moreover, such by-product formation causes a diversion of carbon that might otherwise have generated biomass or protein product [52].

The living organisms utilize energy in a highly efficient manner. However, some portion of energy is utilized for the cell maintenance, thus affecting the biomass yield. Recent studies on exact carbon balance together with thermodynamic arguments indicate that cells might have another avenue for energy utilization, where such phenomenon is called "energy spilling" [53, 54]. Under aerobic condition, *E. coli* mainly converts carbon source into biomass and CO_2, where the production of CO_2 is the loss of carbon, which determines the cell yield. A notable amount of carbon is also lost as acetate in particular at higher cell growth rate [49, 50, 55]. Another carbon wasting occurs toward pyrimidine pathway intermediates such as dihydroorotate, carbamoylaspartate, and orotate [54].

In *E. coli*, acetate is formed from acetyl-coenzyme A (AcCoA) by phosphotransacetylase (Pta) and acetate kinase (Ack) and from pyruvate by pyruvate oxidase (Pox) [56]. Acetate can be metabolized to AcCoA either by the reversed reactions of Pta–Ack or by acetyl-CoA synthetase (ACS) (Fig. 3a). Acetate formation has been known to be due to metabolic imbalance, also known as overflow metabolism as mentioned before, where the rate of acetyl-coenzyme A (AcCoA) formation via glycolysis surpasses the capacity of the TCA cycle in *E. coli* [57]. The mechanism of reducing the accumulated AcCoA, thus reducing acetate formation, is embedded in the cell metabolism. Namely, the anaplerotic pathway enzyme Ppc is allosterically activated by FBP and AcCoA [58], where the accumulated AcCoA activates Ppc, thus reducing PEP concentration, and in turn, the incoming flux through Pyk–PDH reduces, while the outgoing flux through citrate synthase (CS) increases caused by the increased OAA due to the activated Ppc

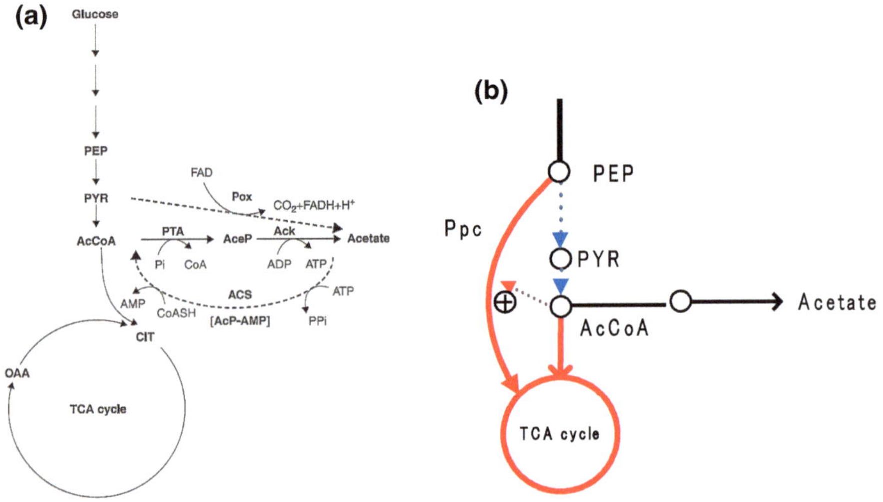

Fig. 3 Acetate metabolism: **a** acetate producing and consuming pathways, **b** homeostasis of AcCoA

(Fig. 3b). This means that AcCoA pool is in principle kept constant by such homeostatic regulation.

Since recombinant protein production is related to biomass formation, much attention has been paid on high cell density culture with reduced acetate formation [59]. The conventional approach to avoid or reduce acetate formation and attain high cell density culture is to reduce the glycolysis flux by keeping the substrate concentration at low level by fed-batch operation. The similar approach has been considered for baker's yeast cultivation to avoid ethanol production caused by Crabtree effect. The typical fed-batch culture is performed by employing feed-forward-type exponential and constant feeding profile [60], or feedback type of DO-stat [61, 62] and pH-stat [63]. In any case, feed-forward strategy must be compensated by feedback control. However, DO-stat and pH-stat strategies are based on the signals of substrate limitation, and the substrate-limited growth causes excessive energy consumption for the cell maintenance [64]. Moreover, in large-scale fed-batch culture, the substrate is distributed in space in the culture broth, and the gradients of substrate concentrations affect the cell growth, product formation, and cell viability [60, 65–67].

Another approach to reduce the glucose uptake rate is to modulate the substrate uptake pathway such as PTS, where the PTS mutation together with activation of other transporters can contribute for recombinant protein production with reduced acetate formation [52]. However, the cell growth rate inherently decreases in such mutants. Moreover, the regulation system as illustrated in Fig. 2b is not guaranteed for the robustness.

Since the main reason of acetate overflow metabolism is the repression of the TCA cycle caused by the increased substrate uptake rate as explained above, one idea is to activate TCA cycle by the knockout of *arcA* gene which codes for ArcA, and activate Acs by *pka* gene knockout [68]. However, the cell yield decreases and the cell growth rate decreases as will be also mentioned later for the roles of redox regulators such as ArcA/B.

5 Catabolite Regulation for the Uptake of Various Carbon Sources

The metabolite regulation differs depending on the carbon sources used, and this also affects acetate metabolism. Here, consider this for the typical carbon sources often used for biofuels and biochemicals production such as glycerol, xylose, fructose, and arabinose.

Glycerol, a rather "energy-poor" carbon source, has been paid recent attention for the production of biofuels and biochemicals such as 1,3-propanediol, 2,3-butanediol, ethanol, n-butanol, organic acids, and polyol, since it is a by-product of the biodiesel production [69–73]. In *E. coli*, glycerol is transported and phosphorylated to produce dihydroxy acetone phosphate (DHAP) of the central metabolism via the pathways encoded by *glpF*, *glpK*, and *glpD*, where ATP (or in certain cases PEP) is used for the phosphorylation at GlpK reaction, while NADH is produced at GlpD reaction (Fig. 4a). These genes are under catabolite regulation by cAMP–Crp, so that glycerol is assimilated after glucose was depleted if glucose coexists. In the case of using glycerol as a single carbon source, cAMP–Crp increases by the mechanism as mentioned before due to the increase in the phosphorylated $EIIA^{Glc}$, where cAMP–Crp induces *glpF*, *glpK*, and *glpD* genes via *glpR*. Proteomic and enzymatic assay studies for the case of using rich media, fructose bisphosphatase (Fdp), the gluconeogenesis enzyme in the upper glycolysis is activated for the fluxes toward PP pathway, while glyoxylate shunt is repressed [74]. Since FBP concentration decreases in the case of using glycerol as a carbon source, Cra is activated, and this together with up-regulation of cAMP–Crp causes *pckA* gene expression as well as TCA cycle gene expression to be up-regulated [75]. Moreover, *pykA* gene expression is activated instead of *pykF* gene in the case of using glycerol [75], where *pykF* is repressed by Cra, while *pykA* is activated by AMP.

The metabolism slightly changes depending on the strain. For example, in *E. coli* JM101 cultivated in minimal medium using glycerol as a carbon source, co-consumption of acetate and glycerol occurs, where the acetate produced via Pox is utilized via Acs and glyoxylate shunt, and little acetate is produced [73]. This may be caused by the higher levels of cAMP–Crp and Cra.

In the case of using glycerol as a carbon source, the glycerol uptake rate is low, and thus, the cell growth rate is also low. This is the main drawback of using glycerol as a carbon source in practice. The slow uptake rate of glycerol is due to allosteric inhibition of GlpK by FBP, where this may be considered to avoid the toxic methylglyoxal production caused by the accumulation of DHAP (Fig. 4a) [76]. The glycerol uptake rate can be increased by modulating GlpK by evolutional mutation with relaxing of feedback inhibition of GlpK by FBP [76]. However, as the glycerol uptake rate is increased, and the cell growth rate is increased, the phosphate of PEP or EI-P may be used for the phosphorylation at GlpK reaction, and thus, the phosphorylation level of $EIIA^{Glc}$ decreases, and in turn, cAMP level decreases and represses TCA cycle, causing acetate overflow metabolism [77].

In the case of using fructose, it is transported by fructose PTS, which has its own HPr-like protein domain called FPr. Namely, the phosphate of PEP is first transferred to EI (as EI-P), but then, this phosphate is transferred to FPr instead of HPr, and in turn, the phosphate is transferred via fructose-specific $EIIA^{Frc}$ and $EIIBC^{Frc}$ to fructose, where phosphorylated fructose becomes fructose 1-phosphate (F1P), where F1P inhibits Cra activity [47]. The *fruBKA* operon is under the control of cAMP–Crp, and thus, glucose is preferentially consumed by glucose PTS when glucose coexists, while this operon is repressed by Cra (Fig. 4b) [78]. Because of this, *Cra* gene knockout enables co-consumption of glucose and fructose with fructose to be consumed faster as compared to glucose (Fig. 5) [79]. Why was the fructose consumed faster than glucose, although the glycolysis activity increases upon *Cra* gene knockout [80, 81]? The possible scenario might be as follows: Namely, the activated FPr in *Cra* mutant competes with HPr (for glucose phosphorylation) for the phosphate of EI-P. Since phosphorylation of $EIIA^{Glc}$ via HPr

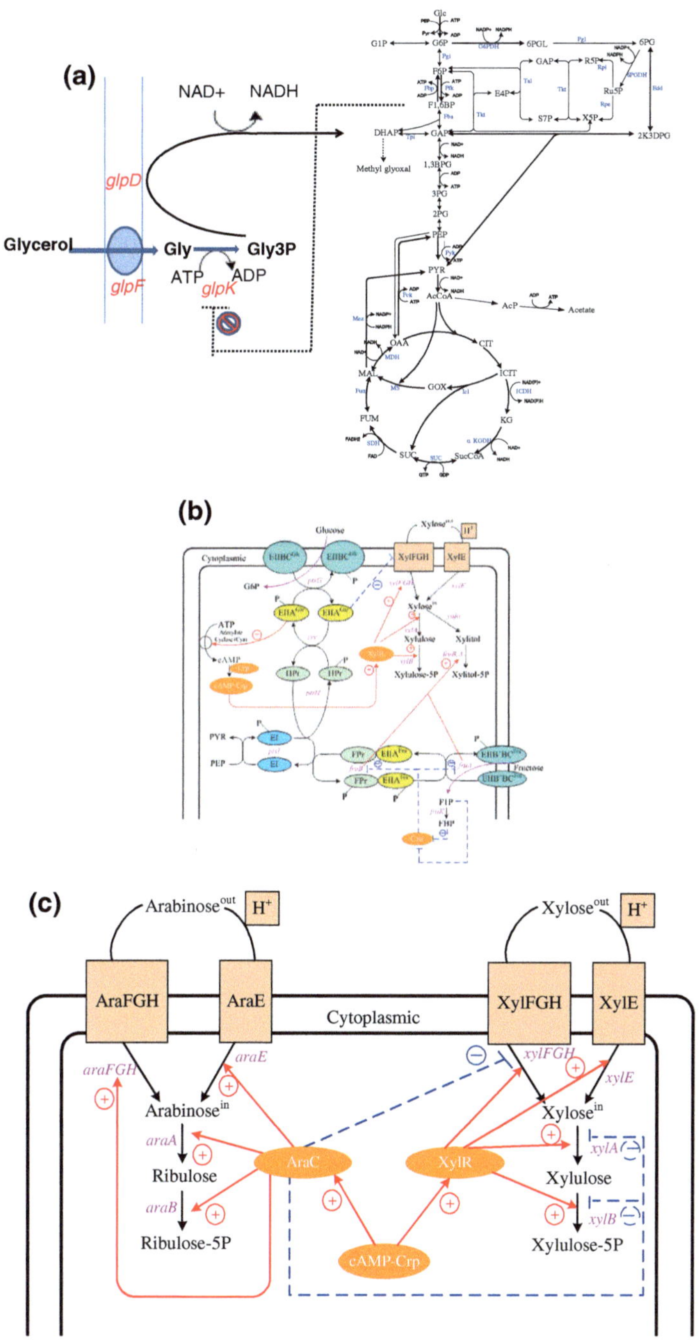

Fig. 4 Carbon sources assimilation and the hierarchical regulation for multiple carbon sources: **a** glycerol assimilating pathways, **b** assimilation of glucose, fructose, and xylose and their regulations [304], **c** assimilation of arabinose and xylose and their regulations [304]

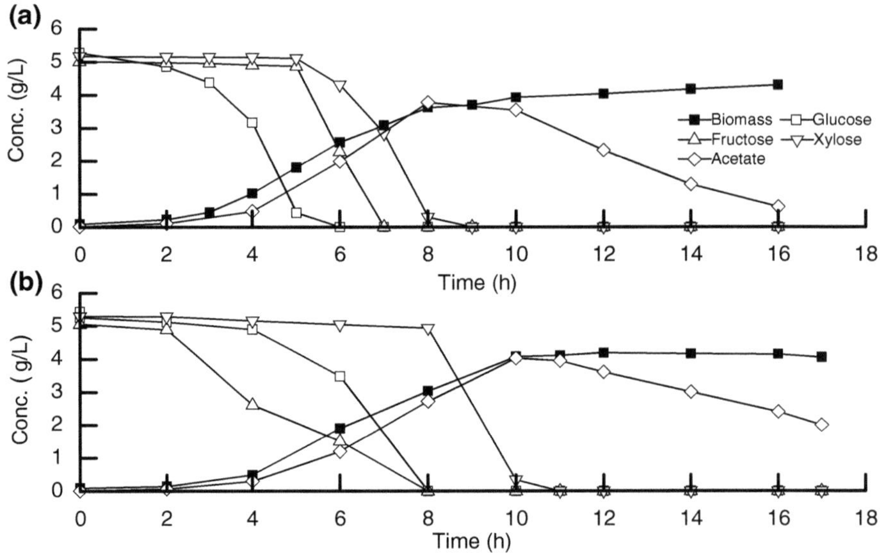

Fig. 5 Batch aerobic culture of wilt-type *E. coli* **a** and its *cra* mutant **b** for multiple carbon sources of glucose, fructose, and xylose [79]

becomes lower [82], the glucose uptake rate decreases as compared to the wild-type strain [79]. However, it is not still clear about the detailed mechanism on why catabolite repression is relaxed by *Cra* gene knockout.

In the case of using xylose as a carbon source, it is transported either by an ATP-dependent high-affinity ABC transporter encoded by *xylFGH* or by ATP-independent low-affinity proton symporter encoded by *xylE* (Fig. 4b) [83, 84]. In the case of xylose utilization, the transcription factor XylR regulates *xylAB*/*xylFGH* [85], where *xylR* is under the control of cAMP–Crp, and thus, catabolite repression occurs when glucose coexists, where glucose is preferentially consumed first. In the case when fructose coexists, phosphorylation of fructose is made via FPr, reducing the phosphorylation of HPr, and cAMP–Crp level becomes lower due to less phosphorylation of $EIIA^{Glc}$, and in turn, fructose is consumed faster as compared to xylose consumption (Fig. 5) [79]. In the case of anaerobic fermentation, NADH reoxidation and substrate-level phosphorylation for ATP generation are important, and ATP generation by Ack pathway is critical for survival in the case of using only xylose as a carbon source [86].

As for the assimilation of arabinose, the transporters are encoded by *araE*, *araFGH*, and *araJ* [87–89], while *araBAD* encodes arabinose catabolic enzymes. The arabinose system is under control of cAMP–Crp and AraC, where *araC* gene expression is activated by L-arabinose [90, 91].

As mentioned previously, the selective carbon source assimilation among available carbon sources forms a hierarchy, where glucose utilization is the highest priority in *E. coli.* Among pentose sugars such as xylose and arabinose, arabinose is the preferred carbon source as compared to xylose [92, 93], where the transcription factor AraC represses *xylR*, while *araC* is under control of cAMP–Crp (Fig. 4c) [93]. This hierarchy changes by the specific gene knockout such as *Cra* gene knockout as mentioned above.

6 Transition of the Metabolism During Batch Culture

Since most of the industrial fermentations are conducted in batch mode, it is important to understand how the metabolism changes with respect to time, where various nutrient limitation and environmental stresses change and affect the metabolism [18, 19]. The ^{13}C-metabolic flux analysis is useful to track the change in the metabolic flux distribution during batch cultivation [94]. The typical growth condition changes from glucose-rich to acetate-rich condition and changes further to carbon-starved condition in the batch culture (Fig. 6). This requires a significant reorganization of the central metabolism. Since it invokes network-wide metabolic adaptation, it is one of the current targets of systems biology [42, 95].

Although the molecular mechanism underlying the metabolic transition from glucose to acetate has been extensively investigated in *E. coli* [56], its dynamics with respect to the sequence and timing of the molecular events have been poorly understood. Recently, this has been clarified to some extent, where the timescales for gene expression and enzyme regulation with metabolites are different [96]. Since it is critical for the cell to efficiently and quickly reprogram the metabolism for efficiently assimilating the nutrients under ever-changing environmental condition, the cell must have the elaborate managing system to cope with the environmental changes. In particular, enzyme-level regulation plays an important role for this, where Pyk and Ppc are allosterically activated by FBP as mentioned before. After glucose depletion, FBP concentration decreases accordingly, and Ppc and Pyk activities decrease by allosteric regulation, and PEP consumption is almost completely turned off. These make PEP concentration to be increased, and this buildup of PEP is kept nearly constant during certain period, and this may serve to quickly uptake the glucose by PTS if it becomes available again (Fig. 6b) [94, 97]. This mechanism is important for the fed-batch culture compensated by DO-stat or pH-stat, where carbon limitation often occurs periodically, and the uptake of carbon source can be made quickly and efficiently without delay giving little damage to the cell, while this may not be the case for the cells without having such regulation mechanism.

During batch culture, several transcription factors such as cAMP–Crp, Cra, Fis, Csr, RpoS, RpoD together with RelA and SpoT play important roles for catabolite and nutrient regulation in coordination among them depending on the growth phases. Considering the roles of such transcription factors, where cAMP–Crp and Cra have already been explained, carbon storage regulator (Csr) is explained in the next section.

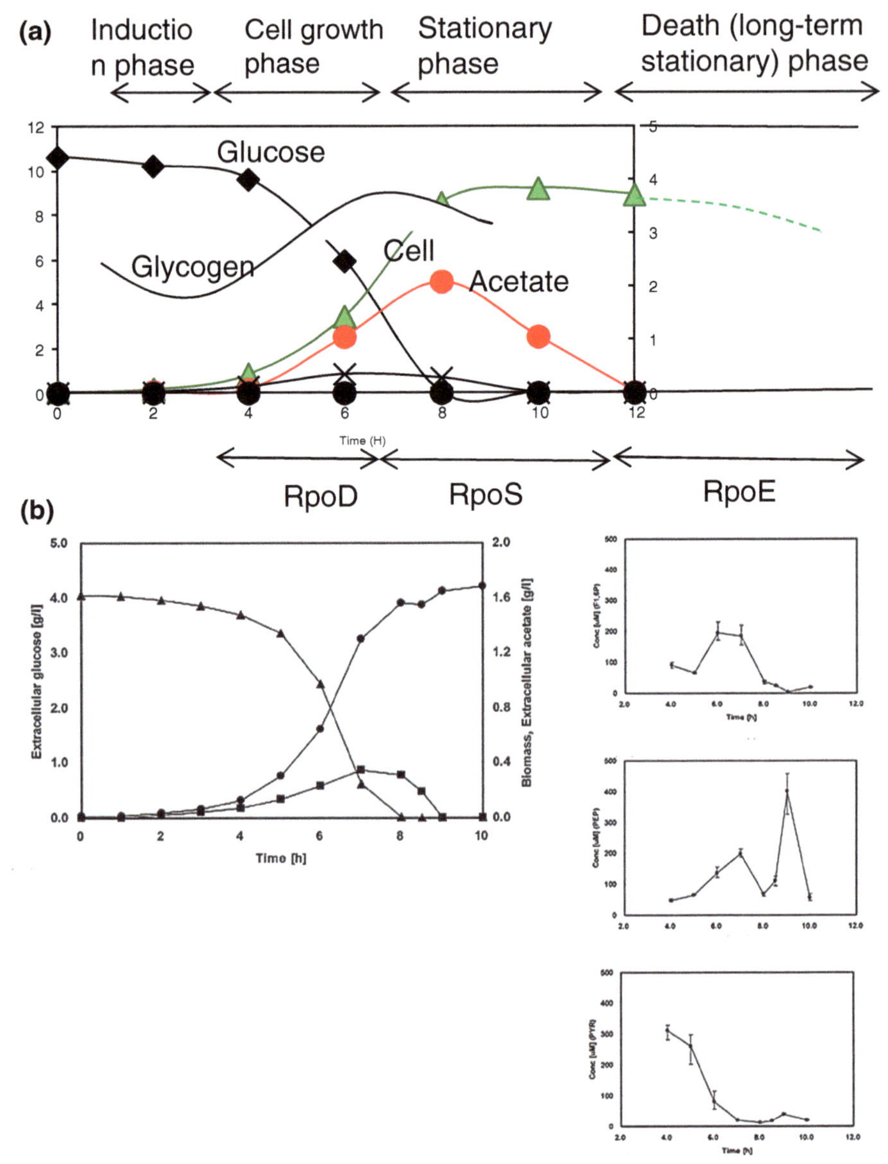

Fig. 6 Typical time profile for the batch culture of *E. coli*: **a** change in growth phase and σ factors, **b** change in intracellular metabolite concentrations for FDP, PEP, and PYR [94]

In *E. coli*, Fis (factor for inversion stimulation) is the most abundant during exponential growth phase [98]. Fis levels peak during early growth phase and thereafter decrease until they become very low during stationary phase [99], where Fis transcription is repressed by the stringent response [100], and Fis is subject to growth rate control [101]. The stringent control and the growth control all require the stringent response regulator DksA [101], where (p)ppGpp (guanosine 3′,5′-bisphosphate including penta-phosphate pppGpp) and DksA interact with RNAP [102]. Fis plays a widespread role in signaling conditions of high nutritional control and outfitting the cells for efficient nutrient uptake and rapid growth [99]. Fis also plays a role in signaling poor nutritional condition, where in response to amino acid starvation, Fis is subject to severe and rapid negative control by the stringent response [101].

Bacteria generally have distinct strategies for the starvation in different nutrient sources. The individual hunger responses may be superimposed on a common protective starvation response [103]. Carbon limitation occurs at the onset of the stationary phase and leads to amino acid limitation, which requires the signaling pathways via RelA and SpoT during carbon and amino acid limitation [60]. During stringent response, nutrient limitation leads to the accumulation of ppGpp [104], which may bind to RNA polymerase [105], where ribosomal RNA and proteins are negatively regulated by ppGpp, which implies that protein biosynthesis declines, and in turn, the cell growth rate decreases. During amino acid limitation, (p)ppGpp is mediated by RelA. The accumulation of (p)ppGpp depends on the dual activity of SpoT as (p)ppGpp-hydrolase or ppGpp synthetase (Fig. 7). SpoT is activated in response to fatty acid starvation, carbon source starvation, diauxic shifts, phosphate limitation, ion limitation, hyper-osmotic shock, and oxidative stress [106].

The alamone ppGpp is involved in the regulation of σ^S on the transcriptional and posttranscriptional level [107], where ppGpp concentration increases with lower growth rates and affects RpoS, and ppGpp accumulates immediately after onset of nutrient starvation. The elevation of σ^S negatively regulates σ^D-dependent house-keeping genes [108]. Moreover, ppGpp influences the competition between different stress-related sigma factors in the binding of RNA polymerase core enzyme at the expense of σ^D [109] and RNA polymerase availability [60]. RpoS plays an important role at the stationary phase or carbon-starved conditions as well as other

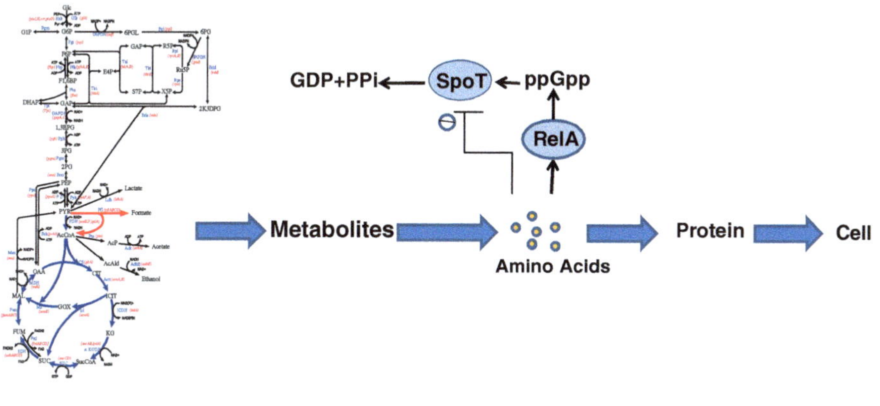

Fig. 7 Stringent response under amino acid limitation and regulation

stress conditions in *E. coli* [107, 110, 111]. Under normal situation with rich media, RpoS is rapidly degraded by ClpXP proteases, and the proteolytic activity of this enzyme is considerably reduced [107, 110, 111]. RpoS tends to increase as the cultivation proceeds from late growth phase to the stationary phase of the batch culture [112, 113].

After the stationary phase in the batch culture, the death phase and long-term stationary phase follow [114]. During the stationary phase, nutrient becomes exhausted, and waste products gradually accumulate, which may become a stress to the cell, and this eventually leads to the death phase in which the number of viable and culturable cells declines. Since majority of the cells in the death phase are viable but nonculturable or dying, nutrients from a portion of such cells are released into the medium. The released nutrients support the survival of the remaining culturable cells, and viable and culturable cells can survive for months or years in the long-term stationary phase [115, 116]. The σ^E-dependent cell lysis is to eliminate damaged cells in the stationary phase in *E. coli* [116], where the cell lysis proceeds in the cascade of $\sigma^E \rightarrow$ expression of *micA* and *rybB* $\rightarrow$ reduction in Omp proteins in the outer membrane $\rightarrow$ disintegration of outer membrane [117]. The cell lysis cascade appears to be related to the oxidative stress in the early stationary phase [118].

7 Carbon Storage Regulation

In the batch culture, the glycogen decreases during early growth phase or induction phase and increases at the late growth phase (Fig. 6) [119]. In the typical batch culture, glucose consumption rate is low during early growth phase or the induction phase, which may be due to glycogen utilization during unbalanced nutrient condition. Moreover, glycogen is accumulated at the late growth phase when the carbon source is going to be limited, which may be due to the preparation of carbon source as glycogen to be used under carbon source starvation. Csr plays an important role for such phenomenon.

The Csr system influences a variety of physiological processes such as central carbon metabolism, biofilm formation, motility, peptide uptake, virulence and pathogenesis, quorum sensing, and oxidative stress response [120–123]. Csr is controlled by the RNA binding protein CsrA, a posttranscriptional global regulator that regulates mRNA stability and translation [123], where CsrA is regulated by two sRNAs such as CsrB and CsrC [124–126]. CsrA regulates the central carbon metabolism and glycogenesis such that glycogen synthesis pathway gene expression, as well as gluconeogenic pathway gene expression, is repressed, while glycolysis gene expression is activated [120, 127] (Table 1).

Two sigma factors such as σ^{70} and σ^S affect *csrA* gene expression [122, 128]. In fact, the strong positive effects of ppGpp and DksA on *csrB/C* transcription and negative effects of CsrA on RelA expression and (p)ppGpp accumulation during stringent response [129]. This suggests that CsrB through CsrA directly regulates

DksA, thereby forming a positive feedback loop, and also, DksA and ppGpp activate the expression of *csr* genes [129], indicating the links between CsrA/B and the stringent response [129, 130].

The *csr* gene knockout affects the central metabolism such that the glycolysis activity is repressed (and the oxidative PP pathway is activated) together with acetate formation in the case of using glucose as a carbon source [131]. In the case of using gluconate as a carbon source, ED pathway is exclusively used for *csrA* mutant [131]. In the above cases, FBP concentration decreases [131], and thus, Cra may be activated and affects the metabolism by repressing glycolysis genes.

The *csrA* gene disruption also causes a significant increase in PEP concentration, since CsrA activates *pykF* gene expression, while it represses *pckA* and *ppsA* genes. The precursors of shikimate pathway for aromatic amino acids formation are a single E4P and two PEP molecules, and thus, over-expression of *tktA* with *csrA* gene disruption enhances phenylalanine biosynthesis [132].

Moreover, biofuels production can be improved by the over-expression of CsrB by activating native fatty acid and heterologous n-butanol and isoprenoid pathways [130]. In particular, CsrB-mediated degradation of CsrA drives over-expression of *glgCAP* operon, which results in the accretion of the storage polysaccharide glycogen.

8 Nitrogen Regulation

Next to carbon (C) source metabolism, nitrogen (N) metabolism is important to understand the cell metabolism. The N-regulation is controlled by σ^{54} encoded by *rpoN*. The main players in the hierarchical network for nitrogen metabolism and regulation are the ammonia transporter AmtB and a glutamine transporter GlnHPQ, metabolic pathways such as glutamate dehydrogenase (GDH) encoded by *gdhA*, glutamine synthetase (GS) encoded by *glnA*, and glutamate synthase (GOGAT) encoded by *gltBD*, the two bifunctional enzymes such as adenylyl transferase/adenylyl-removing enzyme (ATase) and uridylyl transferase/uridylyl removing enzyme (UTase), the two-component regulatory system composed of the histidine protein kinase, nitrogen regulator II (NR_{II}) encoded by *glnL* and the response regulator I (NR_I) encoded by *glnG*, three global transcriptional regulators such as nitrogen assimilation control (Nac) protein, leucine-responsive regulatory protein (Lrp), and Crp, the glutaminases, and the nitrogen phosphotransferase system [133].

N-source such as ammonia (NH_3)/ammonium (NH_4^+) is predominantly assimilated at glutamate dehydrogenase (GDH) reaction, where α-ketoglutarate (αKG) is converted to glutamic acid (Glu), where NADPH is required for this reaction (Fig. 8). Then, glutamate is converted to glutamine (Gln) at glutamate synthetase (GS) reaction, where NH_3/NH_4^+ and ATP are required for this reaction. Thus, the flux goes from αKG via Glu to Gln, and thus, Gln accumulates under excess ammonia condition. Under N-limitation, the expression of *gdhA*, which encodes GDH, is repressed by Nac, and thus, Gln concentration decreases, and αKG

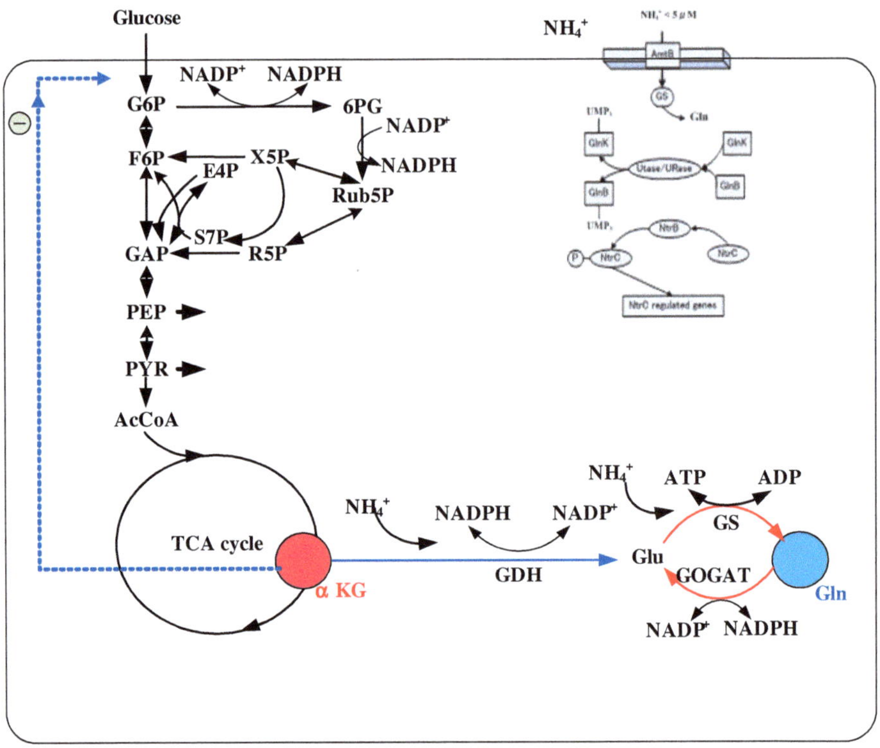

Fig. 8 Ammonia transport via AmtB, ammonia assimilation, and the effect of αKG on the glucose PTS under N-limitation

accumulates, where glutamate is formed from Gln by glutamate synthase (GOGAT) reaction (Fig. 8). Namely, under N-limitation, GS/GOGAT cycle plays an important role.

Intracellular ammonium is assimilated into biomass in two steps: Namely, it is first captured in the form of glutamic acid using carbon skeleton of αKG via GS/GOGAT cycle. Then, N-group in glutamate is transferred by aminotransferase reactions to synthesize other amino acids thus incorporating into biomass, while recycling the carbon skeleton back to αKG [134]. The αKG pool, which integrates imbalance between the ammonium assimilation flux and the biomass incorporation flux, activates AmtB [135–137] via GlnK. If ammonia level drops, then the rate of ammonia assimilation will drop immediately, which results in αKG accumulation [138]. When extracellular ammonia concentration is low around 5 μM or less, ammonia is captured and transported into the cell via AmtB and is converted to glutamine by GS, and UTase uridylylates both GlnK and GlnB [139] (Fig. 9). When extracellular NH_4^+ concentration is more than 50 μM, glutamine pool rises, and UTase deuridylylates GlnK and GlnB. Then, GlnK complexes with AmtB, thereby inhibiting the transport via AmtB, and ammonia may enter by diffusion. P_{II}

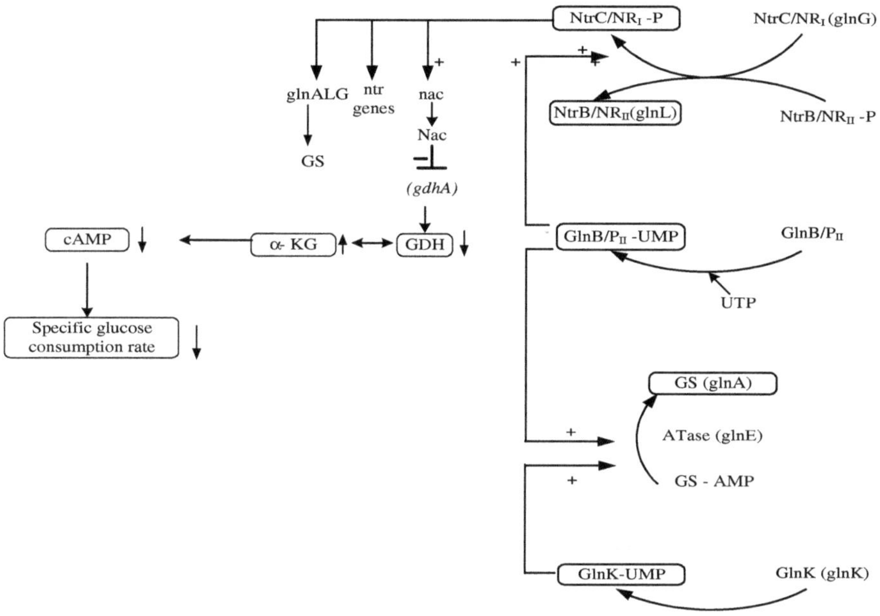

Fig. 9 Overall nitrogen regulation mechanism under N-limitation

(GlnB) interacts with NtrB (NR_{II}) and activates its phosphatase activity leading to dephosphorylation of NtrC (NR_I), and NtrC-dependent gene expression ceases (Fig. 9) [139].

Lrp regulates the expression of such genes as involved in catabolism and anabolism of amino acids (AAs). In particular, leucine indicates AA sufficiency, and it is affected by Lrp, where Lrp does not restrict to leucine but the other AAs such as isoleucine, histidine, and threonine. Lrp may activate *gltBD* and pyridine nucleotide transhydrogenase [133].

When arginine is abundant, the transcription factor ArgR binds to arginine to repress arginine biosynthesis enzymes [140] and activates arginine degradation enzymes [141]. This regulation is also subject to the NtrC regulation.

The GlnHPQ enables active transport of glutamine into the cell with higher specificity, where *glnH* is the structure gene for the periplasmic binding protein, *glnP* gene codes for the membrane-bound glutamine permease, and *glnQ* codes for the ATP hydrolyzing component of ABC transporter system [133].

Two of the major signal transduction systems of N and C metabolisms are identified as P_{II} (GlnB) and PTS. Because of the important roles in the regulatory functions, P_{II} and PTS can be regarded as the central processing units of N and C metabolisms, respectively. The P_{II} protein senses αKG and ATP and thus links the state of central carbon and energy metabolism for the control of N assimilation [142]. N assimilation is regulated by P_{II}-Ntr system together with Crp, providing a regulatory network between C and N assimilation in *E. coli* [143, 144]. The C and

N metabolisms may be linked by energy metabolism, where P_{II} controls N assimilation by acting as a sensor of adenylate energy charge. Moreover, αKG serves as a cellular signal of C and N status and strongly regulates P_{II} functions [145]. Gln and αKG are the signal metabolites for nitrogen and carbon status, respectively, and these signals regulate GS adenylylation state and nitrogen regulator I (NR_I or NtrC) phosphorylation state [146]. Nitrogen shortage is reflected by the reduced Gln levels and increased αKG level [138, 147]. This ratio is substantially constant under C-limitation, where this constant ratio is the result of tight regulation of ammonia assimilation to match exactly the carbon uptake rate. This ratio is insensitive to variations in protein levels of the core circuit and to the N-utilization rate, and this robustness depends on bifunctional enzyme adenylyl transferase [148].

During N-limitation, a sudden increase in nitrogen availability results in immediate increase in glucose uptake, and αKG plays an important role for this, where αKG directly reduces the glucose uptake under N-limitation by inhibiting EI of PTS (Fig. 8) [149]. This implies the followings: (1) αKG inhibition of sugar uptake is for all PTS sugars by inhibiting EI but not carbohydrate-specific EII; (2) this is performed without perturbing the concentrations of the glycolytic intermediates such as G6P, PEP, and PYR; (3) inhibition of EI by αKG leads to reduced amount of phosphorylated $EIIA^{Glc}$ and decreases cAMP level, where the effect of αKG on cAMP production is caused by the difference in $EIIA^{Glc}$ phosphorylation rather than a difference in substrate availability [149]. Not only αKG but also other α-ketoacids such as OAA and PYR play also the similar roles and affect not only PTS but also the cAMP level by Cya [150]. Moreover, αKG is a promiscuous enzymatic regulator that competitively inhibits citrate synthase (CS) of the TCA cycle and 3PG dehydrogenase for serine biosynthesis and further controls aspartate production by product inhibition of transaminase under N-limitation [149]. αKG noncompetitively inhibits EI and Pps, while PtsP (EI homolog in the nitrogen PTS) is insensitive to αKG.

In addition to carbohydrate PTS, most proteobacteria possess a paralogous system such as nitrogen phosphotransferase system PTS^{Ntr}, where it consists of EI^{Ntr} encoded by *ptsP*, NPr encoded by *ptsO*, and $EIIA^{Ntr}$ encoded by *ptsN*, which are paralogues to the carbohydrate PTS components such as EI, HPr, and EIIA, respectively [151–153]. *E. coli* PTS^{Ntr} plays a role in relation to K^+ uptake, where dephosphorylated $EIIA^{Ntr}$ binds to and regulates the low-affinity K^+ transporter TrkA [154] and the K^+-dependent sensor kinase KdpD [153, 155]. K^+ regulates global gene expression involving both σ^{70}- and σ^S-dependent promoters [156]. Moreover, dephosphorylated $EIIA^{Ntr}$ modulates the phosphate starvation response through interaction with sensor kinase PhoR [157]. Dephosphorylated form of PTS^{Ntr} interacts with and inhibits LpxD, which catalyzes biosynthesis of lipidA of the lipopolysaccharide (LPS) layer [158].

Although the physiological role of PTS^{Ntr} has not been well known, glutamine and αKG reciprocally regulate the phosphorylation state of the PTS^{Ntr} by direct effects on EI^{Ntr} autophosphorylation. This implies that PTS^{Ntr} senses nitrogen availability [159].

9 Sulfur Regulation

Under sulfur (S) limitation, at least three metabolites such as sulfide, the reduction product of sulfate used for cysteine biosynthesis; N-acetylserine, the only precursor of cysteine; and adenosine 5'-phosphosulphate (APS), the first intermediate in sulfate assimilation, are involved for the metabolic regulation [160, 161]. Under S-limitation, the concentrations of sulfide and APS decrease, while N-acetylserine pool increases. The two regulators CysB and Cbl mediate homeostatic responses to S-limitation, where these responses help *E. coli* to scavenge trace amounts of cysteine and sulfate, preferred S sources, or the alternative S sources such as glutathione and various alkaline sulfonate including taurine. S-limitation affects methionine metabolism, synthesis of FeS clusters, and oxidative stress.

Like NtrC for N-regulation, CysB is the primary regulator for homeostatic responses to S, and it is required for the synthesis of Cbl [162]. CysB is positively controlled by N-acetylserine and negatively controlled by sulfide or thiosulfate [161], and Cbl is negatively controlled by APS [160]. It is of interest that *cbl* gene is transcribed from *nac* promoter under N-limitation [163]. The *ddp* operon is activated by NtrC, and there might be a cross-regulation between S-limitation and N-limitation [164].

10 Phosphate Regulation

The phosphate (P) metabolism is also quite important from the energy generation and phosphorelay regulation points of view. The phosphorous compounds serve as major building blocks of many biomolecules and have important roles in signal transduction [165]. Depending on the concentration of environmental phosphate, *E. coli* controls phosphate metabolism through Pho regulon, which forms a global regulatory circuit involved in a bacterial phosphate management [165, 166]. The PhoR/PhoB two-component system plays important roles in detecting and responding to the changes of the environmental phosphate concentration [167]. Namely, under phosphate limitation, the phosphate is transferred by an ABC transporter composed of PstSCAB for the high-affinity capture of P_i, and the phosphate is then transferred to PhoR (PhoR-P), and in turn, PhoB is phosphorylated by PhoR. The phosphorylated PhoB acts as the response regulator and regulates Pho Box genes such as *eda*, *phnCDEFGHIJKLMNOP*, *phoA*, *phoR/B*, *phoE*, *phoH*, *psiE*, *pstSCAB*, *phoU*, and *ugpBAECQ* [168]. When P_i is rich or in excess, P_i is taken up by the low-affinity transporter Pit, and PhoR, Pst, and PhoU together turn off the Pho regulon by dephosphorylating PhoB. The sensor protein CreC (PhoM) can phosphorylate PhoB, while acetyl phosphate can also directly phosphorylate PhoB [166]. The overall regulation mechanism is complex, and it is not so clear how the phosphate limitation affects the metabolism [169].

The promoters of the Pho genes are recognized by σ^D-associated RNA polymerase. A mutation in *rpoS* significantly increases the level of AP (alkaline phosphatase) activity, and the over-expression of σ^S inhibits it [170]. The Pho regulon is thus evolved to maintain a trade-off between cell nutrition and cell survival during P_i starvation [170].

11 Metal Ion Regulation and Oxidative Stress Regulation

Iron is ubiquitous and the fourth most abundant element on earth and supports the metabolism of living organisms on the planet [171]. Iron is involved in the formation and destruction of ROS such as superoxide ($O_2^{\cdot -}$), peroxidase (H_2O_2 and ROOH), and free radicals ($^{\cdot}OH$ and $^{\cdot}OR$) usually generated as toxic by-products of aerobic metabolism in a cascade of monovalent reductions from molecular oxygen. Although certain amounts of iron and ROS are required for the cell to survive, the excess amounts cause stress to the cell leading to the cell death [172].

The metal ion levels are often sensed by metal-sensing regulatory RNA, which encodes metal-sensing proteins involved in the transport and storage of intracellular metals [173, 174]. In the native environment, the cell continuously faces iron deficiency, where metal ion functions as cofactor in many of the cellular constituents such as flavoproteins, and therefore, the cell furnishes the mechanism for iron uptake and storage system [175, 176]. However, excess iron causes toxicity by catalyzing the formation of reactive free radicals through Fenton/Haber–Weiss reaction [177]. In combination with inability to convert NADH to NAD^+ in the respiration, a decrease in endogenous O_2^- causes reductive stress and in turn activates Fur (ferric uptake regulator) [178]. Fur generally represses ion transport and ion siderophore biosynthetic genes when complexed with ferrous ion. Under ion limitation, ion dissociates from Fur, where Fur requires binding to Fe^{2+} to become active. O_2^- deactivates Fur after its conversion to H_2O_2 by superoxide dismutase (SOD) through Fenton reaction ($H_2O_2 + Fe^{2+} \rightarrow HO^{\cdot} + OH^- + Fe^{3+}$) [179]. Therefore, a decrease in endogenous O_2^- increases the availability of Fe^{2+}, through a decrease in H_2O_2 level, and in effect activates Fur [180]. Namely, Fur senses the reductive stress and protects Fe–S clusters to be safe from damage by ROS. It is essential for the cell to use iron economically, and this is attained by siderophore synthesis and iron transport regulation [181]. Iron transport and siderophore (e.g. enterobactin) pathway genes such as *talB* and *entF* are repressed by Fur [182–184], and enterobactin may be produced in *fur* mutant *E. coli* [185]. There are functional interactions between carbon and ion utilization via Crp and Fur, where many ion transport genes and several catabolic genes are subject to dual control [186].

The widely conserved multiple antibiotic resistant regulator (MarR) family of transcription factors modulates bacterial detoxification in response to antibiotics such as fluoroquinolones and β-lactams such as ampicillin, tetracycline, and chloramphenicol, as well as toxic chemicals and synthesis of virulence determinants. [187]. MarR senses copper (II) as a signal to cope with stress caused by antibiotics, etc., where copper (II) oxidizes a cysteine residue on MarR to generate disulfide bonds between two MarR dimers, thereby inducing tetramer formation and dissociation of MarR from its cognate promoter DNA [188].

The microbial cell responds to oxidative stress by inducing antioxidant proteins such as SOD and catalase, where those are regulated by OxyR and SoxR/S [189]. SoxR is a member of the MarR family of metal-binding transcription factors, and it exists in solution as a homodimer with each subunit containing a [2Fe–2S] cluster. These clusters are in the reduced state in inactivated SoxR, and their oxidation activates SoxR as a powerful transcription factor [190]. The active form of SoxR activates transcription of *soxS* gene, where SoxS belongs to the AraC/XylS family.

Although the respiration is universal among all aerobic organisms, inefficient electron transfer via the respiratory complexes results in one electron reduction of diatomic oxygen, a phenomenon known to generate toxic ROS [191]. Since NADPH plays an important role for detoxification of ROS, some prokaryotic microorganisms such as *E. coli* produce NADPH at ICDH in the TCA cycle together with the reactions at G6PDH, 6PGDH, and possibly at Mez.

The αKG is a key participant in the detoxification of ROS with concomitant formation of succinate, where succinate is a biomarker for oxidative stress [191]. Moreover, NADPH-producing ICDH is activated, while NADH-producing KGDH is deactivated in *Pseudomonas fluorescens* [191]. This indicates that for both prokaryotic and eukaryotic cells, the TCA cycle acts both as a scavenger and generator of ROS, and its modulation is important for regulating ROS [191]. The TCA cycle can both regulate their formation and decomposition, where the concomitant accumulation of succinate may act as a potent signal for this [191].

The proper understanding on the regulation of ROS homeostasis gives a way for medical applications [172]. Namely, iron- and ROS-dependent cell death may be considered for cancer treatment. As mentioned above for bacteria, high NADPH production with low ROS levels is essential for tumor cell proliferation and survival [192–194]. NADPH is required for glutathione homeostasis, which indicates that tumor cells require a highly reduced environment for survival. Therefore, one idea for pushing cancer cells to sentence or death is the decrease of the glutathione levels and/or the increase of the oxidative stress levels [195].

12 Redox State Regulation

Global regulators such as Fnr (fumarate nitrate reduction), Arc (anoxic respiration control) system, and Nar (nitrate reduction) are mainly responsible for the regulation under oxygen limitation and other electron acceptors in the culture environment, where Fnr directly senses molecular oxygen and plays a role under anaerobic condition [196], in coordination with ArcA/B system, where Fnr activates *arcA* gene expression. Under oxygen limitation, Fnr binds a $[4Fe–4S]^{2+}$ cluster and becomes a transcriptionally active dimeric form. Molecular oxygen can oxidize the ion–sulfur cluster of the corresponding region, resulting in monomerization of the protein and subsequent loss of its ability to bind DNA [197]. The ArcA/B system plays a role under both anaerobic and micro-aerobic conditions [198, 199], where it is composed of ArcA, the cytosolic response regulator, and ArcB, the membrane-bound sensor kinase. The ArcA/B two-component system responds to the redox state of the membrane-associated redox carriers such as quinones in the respiratory chain [200, 201]. The quinone pool decreases under oxygen limitation and causes ArcB to be self-phosphorylated (ArcB-P), and then, ArcB-P transphosphorylates ArcA (Fig. 10) [202]. The ArcA-P then represses the expression of the TCA cycle and the glyoxylate shunt genes (Table 1). Moreover, the genes that encode the primary dehydrogenases such as *glpD, lctPRD, aceE,F,* and *lpdA* are also repressed by ArcA (Table 1). The *cyoABCDE* operon is repressed by both ArcA and Fnr, while *cydAB* operon is activated by ArcA and repressed by Fnr (Fig. 10) [203].

The expression of *pfl* genes which encode pyruvate formate lyase, Pfl, is activated by ArcA and Fnr, whereas *aceE,F* and *lpdA* which encode PDHc are repressed by ArcA under oxygen limitation (Fig. 10). The formate can be excreted

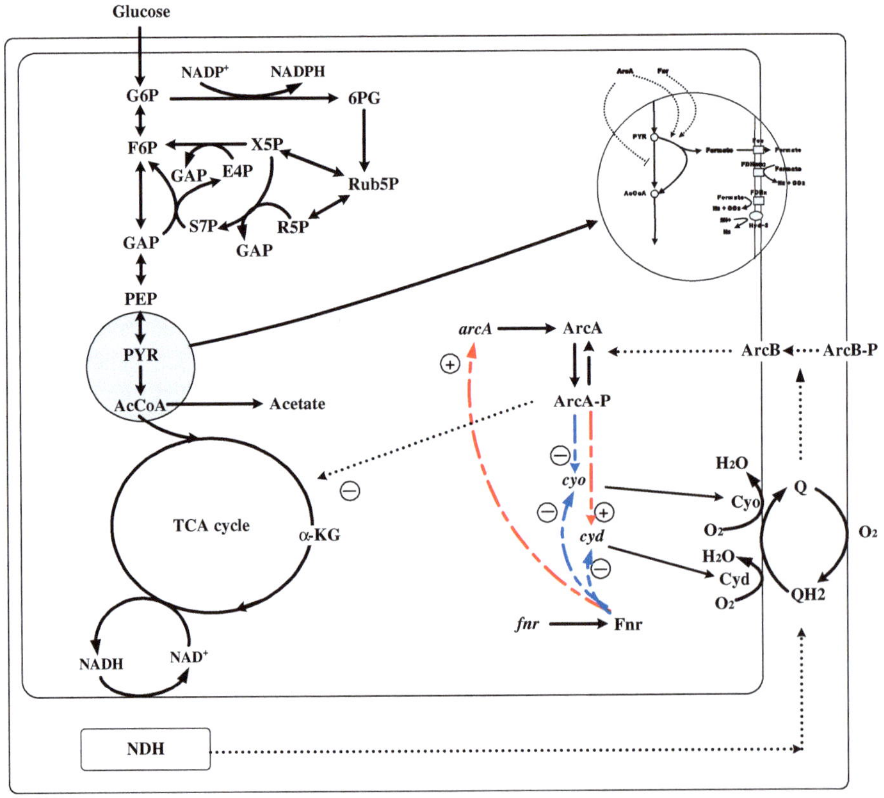

Fig. 10 Respiratory and metabolic pathway regulation by ArcA/B and Fnr

via Foc or converted to hydrogen via formate dehydrogenase, FDH_H, and formate hydrogen lyase, Fhl, and deletion of FocAB, FDH_N, and hydrogenase Hyd (Fig. 10) [204, 205]. Moreover, the flux from PYR to AcCoA is blocked in *pfl* mutants (Δ*pflA* or Δ*pflB*), and pyruvate exclusively goes to lactate formation via LDH reaction [206, 207]. Moreover, Fnr activates *frd* gene expression, while repressing *sdh* gene expression, resulting in branched pathways for TCA cycle under anaerobic condition.

As mentioned before, the TCA cycle activity is repressed as the glucose consumption rate was increased due to lower level of cAMP–Crp, which in turn causes acetate overflow metabolism. This also occurs by the higher redox ratio [208]. This phenomenon can be relaxed by activating TCA cycle by *arcA/B* genes knockout [198, 204, 209]. The activated TCA cycle produces more NADH and allosterically inhibits CS and ICDH activities [210]. Thus, the NADH oxidation by the expression of *nox* gene coding for NADH oxidase, NOX, in the *arc*A mutant further reduces the acetate formation, resulting in the increased recombinant protein production [211], while nicotinic acid and Na nitrate may also activate TCA cycle [212]. The activation of the TCA cycle causes the decrease in the cell yield due to higher production of CO_2 in the TCA cycle.

Many bacteria utilize oxygen as the terminal electron acceptor, but they can switch to other acceptors such as nitrate under oxygen limitation. The reducing equivalents such as NADH are reoxidized in the respiratory chain, where oxygen, nitrate, fumarate, and dimethyl sulfoxide can be the electron acceptors. Nar plays a role when nitrate is present under oxygen limitation. Nar belongs to the two-component redox regulation systems, where it comprises a membrane sensor (NarX) that acts as a kinase causing phosphorylation of the regulator (NarL) under certain conditions [202]. The Nar system activates such genes as nitrate reduction encoding nitrate and nitrite reductases and represses such genes as *frd* genes for fumarate reductase.

13 Acid-shock Response

The cell such as *E. coli* has the regulation systems in response to acidic condition [213–216]. Some of these depend on the available extracellular amino acids such as glutamate, arginine, and lysine, where the intracellular proton is consumed by the reductive decarboxylation of the amino acid followed by the excretion of γ-amino butyric acid (GABA) from cytoplasm to the periplasm by the anti-porter that also imports the original amino acid [213]. *E. coli* is acid resistant by glutamate decarboxylase system, where *gadA* and *gadB* encode glutamate decarboxylase isozymes and *gadC* encodes glutamate/GABA anti-porter (Fig. 11). Glutamate decarboxylase is activated in response to acid, osmotic, and stationary phase signals. The GADAB forms a glutamate-dependent acid response system, where the process of decarboxylation consumes an intracellular proton and helps maintain pH homeostasis. There are other similar acid-resistant systems for the case of using

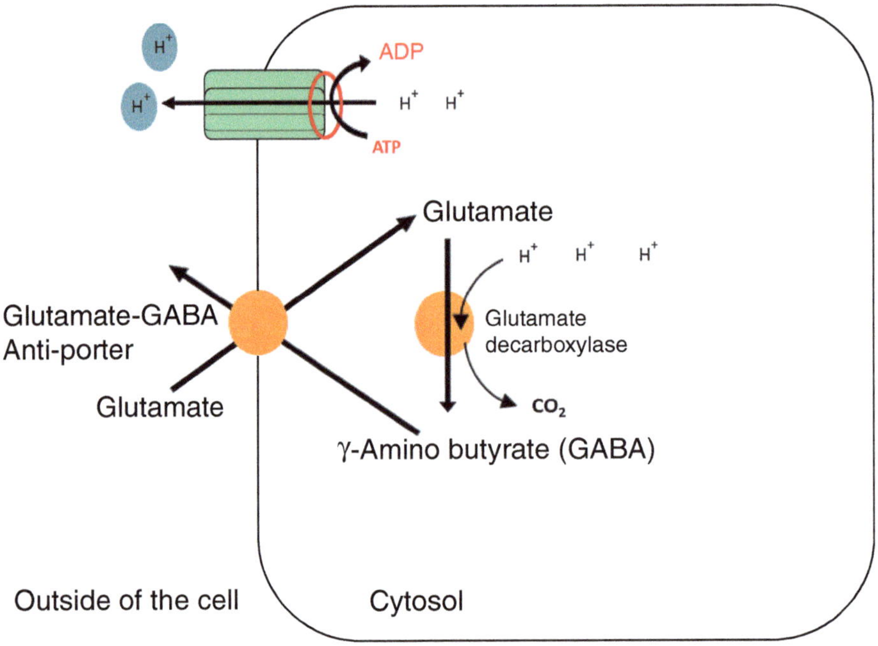

Fig. 11 Acid-shock regulation by amino acid decarboxylase and reversed PMF

arginine instead of glutamate by arginine decarboxylase, where the anti-porter is AdiC in this case [217, 218], and for the case of using lysine by lysine decarboxylase [218]. The cells grown in media rich with amino acids such as LB are acid resistant because of the above mechanism [213].

Moreover, ATPase is involved in acid regulation system [216], where ATPase is usually utilized for the protons in the periplasm move into the cytosol across the cell membrane producing ATP from ADP and P_i by the negative proton motive force (PMF). Since the basic problem of acid stress is the accumulated proton in the cytosol, this proton may be pumped out through ATPase by hydrolyzing ATP with reversed proton move due to positive PMF at low pH such as pH 2 or 3 [216]. Without amino acid in the media, this acid response system is activated by utilizing ATPase [215, 219], where the positive PMF pumps out extra protons (H^+) from the cytoplasm using ATP (Fig. 11) [215]. This proton homeostasis by PMF is conserved in large class of organisms.

RpoS that increases at the late growth phase and the stationary phase and Crp are involved in acid resistance [213, 220]. As implied by the involvement of Crp, the acid-resistant system is repressed when glucose is present. The acidic pH lowers cAMP levels in exponentially growing cells in the minimal glucose medium. Since cAMP–Crp represses RpoS, this may elevate RpoS and increases the expression of *gadX*. The overall regulation system seems to be quite complex involving EvgS/A, B1500, PhoQ/PhoP, GadX, GadW, etc. [221].

OmpR may be a key regulator for acid adaptation, and thus, *ompR* mutant is sensitive to acid exposure [222]. The acid-inducible *asr* gene is regulated by PhoR/B, and thus, *phoR/phoB* deletion mutant fails to induce *asr* gene expression [223].

In order to keep pH constant, alkali such as NaOH is supplied during the cell growth in practice, which results in the increase in sodium ion (Na^+), where *nhaA* gene encoding the Na^+/H^+ anti-porter membrane protein and *nhaR* gene encoding the NhaA regulatory protein can be over-expressed in *pflB* mutant, showing performance improvement for lactate fermentation [224].

14 Heat-shock Stress Response

The organisms respond to a sudden temperature upshift by increasing the synthesis of a set of proteins. This phenomenon is called as heat-shock response, where this does not restrict to the temperature upshift, but also other stresses such as solvent stress. The heat-shock proteins play important roles in the assembly and disassembly of macromolecular complex such as GroE, the intracellular transport such as Hsp70, transcription such as σ^{70}, proteolysis such as Lon, and translation such as lysyl tRNA synthetase. The heat-shock response in *E. coli* is mediated by σ^{32} encoded by *rpoH*. Among them, *groEL, dnaK*, and *htpG* encode major chaperones such as Hsp 60, Hsp 70, and Hsp 90. ClpP, Lon, and HtrC are involved in the proteolysis. DnaK, DnaJ, GrpE, and RpoH are involved in the autoregulation of heat-shock response. DnaK prevents the formation of inclusion bodies by reducing aggregation and promotion of proteolysis of misfolded proteins. A bichaperone system involving DnaK and ClpB mediates the solubilization or disaggregation of proteins. GroEL operates protein transit between soluble and insoluble protein fractions and participates positively in disaggregation and inclusion body formation. Small heat-shock proteins such as IbpA and IbpB protect heat-denatured proteins from irreversible changes in association with inclusion bodies [225, 226].

Hoffmann et al. [227] investigated the metabolic adaptation of *E. coli* during temperature-induced recombinant protein production and showed that cAMP–Crp-controlled LpdA of pyruvate dehydrogenase complex (PDHc) and SdhA in the TCA cycle are highly induced. Namely, the TCA cycle is activated due to increased level of cAMP–Crp at higher temperature. In *E. coli*, heat-shock protein synthesis rates peak at about 5–10 min after the temperature upshift and then decline to a new steady-state level [228]. The σ^{70} is itself a heat-shock protein, and the increase in its concentration after heat shock may contribute to its decline in heat-shock protein synthesis. DnaK contributes to the shutoff of the high-level synthesis of heat-shock proteins [229]. The heat shock activates *crp* gene expression, and in turn, Crp activates *mlc* gene which codes for Mlc [230], and thus, the glucose consumption rate decreases (Fig. 12) [231]. This also causes cAMP level to be increased (Fig. 12).

Acetate production is affected by higher temperature. Transcription of *acs* gene occurs from two σ^{70}-dependent promoters such as distal promoter *acs* P_1 and proximal promoter *acs* P_2 [232, 233]. The cAMP–Crp binds two sites within the *acs*

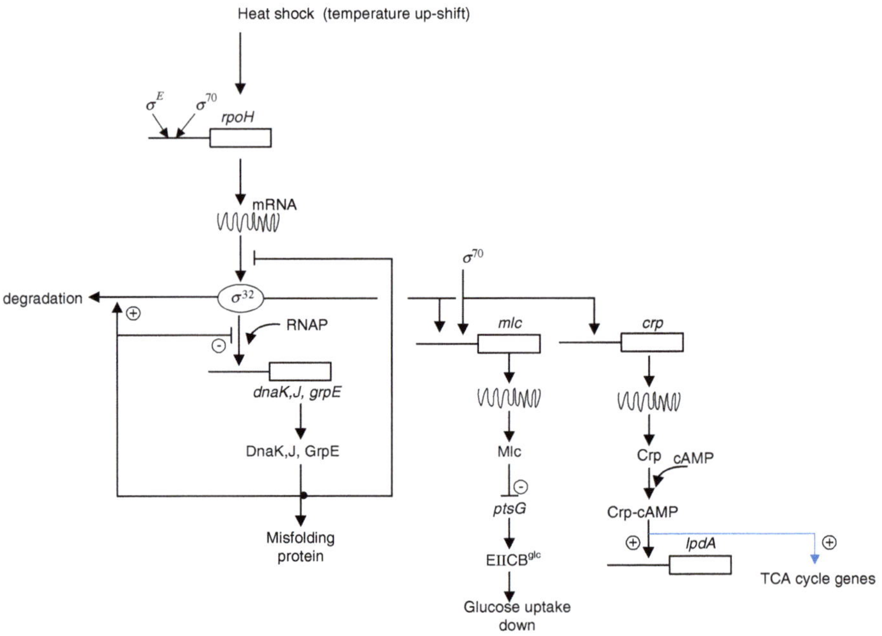

Fig. 12 Overall heat-shock regulation mechanism

regulatory region. However, Fis and Ihf independently modulate Crp-dependent activation of *acs* P_2 transcription [234, 235].

The respiration is activated during the temperature upshift [227], and *sod* is induced in response to the oxidative stress imposed by dioxygen or by the redox-active compounds such as viologens or quinones caused by the temperature upshift [236]. This phenomenon may be also caused by the activated TCA cycle.

15 Cold-shock Response

Upon temperature downshift from 37 to 15 °C, the major cold-shock proteins such as CspA, CspB, and CspG are induced, where cold-shock proteins are able to bypass the inhibitory effect of the antibiotics such as kanamycin and chloramphenicol [237]. Although thermoregulatory mechanism is not well understood, the adaptation of the cell to low temperature such as 20–23 °C requires coordinated and multifunctional response, where RpoS and the small regulatory RNA DsrA are involved in both cold-shock and biofilm formation genes [238] as well as flagella biosynthesis and motility genes [239].

16 Solvent Stress Regulation

The biofuels production by microorganisms has been paid recent attention. However, many biofuels are toxic to microorganisms and reduce the cell viability through damage to the cell membrane and interference with essential physiological processes. Several attempts have been made to improve the tolerance to biofuels, where biofuel export systems, heat-shock proteins, and membrane modifications have been considered [240]. The effect of biofuels on the cell is through hydrophobicity of the cytoplasmic membrane, where the accumulation of solvent in the cytoplasmic membrane increases permeability of membrane, diminishes energy transduction, interferes with membrane protein function, and increases fluidity [240–243]. This may cause the release of ATP, ions, phospholipids, RNA, and proteins, and thus, the cell growth is depressed due to disturbances on ATP production by diminished PMF. Moreover, the increase in fluidity affects the nutrient transport as well as energy transduction.

Toxicity levels vary depending on the microbes and the types of biofuels and biochemicals. In general, longer chain alcohols are more toxic than short-chain alcohols. Efflux pumps are membrane transporters that recognize and export toxic compounds from the cell by PMF, where this is important for the cell to survive by exporting bile salts, antimicrobial drugs, and solvents. The *acrAB–tolC* pump in *E. coli* provides tolerance to hexane, heptanes, octane, and nonane [244]. Efflux pumps are effective for increasing tolerance and production of biofuels, in particular, for long-chain alcohols, but those are not effective for exporting short-chain alcohols such as 1-propanol and isobutanol [245].

The heat-shock proteins are up-regulated in response to short-chain alcohols [180, 246], and heat-shock protein refolding genes such as *rpoH, dnaJ, htpG,* and *ibpAB* are up-regulated [247], while *groESL, dnaKJ, hsp18, and hsp90* are up-regulated in *Chrostridium acetobutylicum* [248]. Over-expression of heat-shock proteins may increase tolerance against biofuels [249, 250].

In general, solvents disrupt the cell membrane structure and have a strong impact on physiological function and eventually leading to the cell death [251]. To overcome this problem, solvent-tolerant microbes change the composition of the fatty acids from *cis* to *trans* unsaturated fatty acids catalyzed by *cis–trans* isomerase (*cti*), thus decreasing membrane fluidity, preventing the entry of solvents into the cell [252, 253]. In addition, modifications to phospholipid headgroups or phospholipid chain length increase solvent tolerance [246].

In relation to solvent stresses caused by the accumulation of biofuels in the culture broth, the primary role to protect the cell from such stress is made by outer membrane porin proteins. Since cytosolic membrane is also under stress condition, respiration and membrane proteins as well as general stress response mechanism are affected [243]. ROS highly increase in response to the stress caused by n-butanol in *E. coli* [247].

17 Osmoregulation

The bacterial cell is surrounded by the cell envelope, where the plasma membrane is responsible for the transport of ions such as H^+, Na^+, and K^+, and various substrates or nutrients, and metabolites to maintain homeostasis. The bacterial cells exchange such components together with energy and information with their surroundings by the appropriate sensing and responding mechanisms [254]. Under osmotic stress condition, a number of transport systems for ions such as K^+, and compatible solutes such as proline betaine and the precursor choline are activated [255]. The typical two-component histidine kinase/response regulator system such as KdpD/KdpE is ubiquitous in various bacteria [256], where it regulates *kdpFABCDE* operon including the Kdp ATPase and active K^+ uptake system. Namely, KdpD/KdpE system responds to K^+ limitation and salt stress [257–259]. As also mentioned before, EnvZ/OmpR two-component system regulates the expression of the porin genes such as *ompC* and *ompF* encoding outer membrane porins in relation to osmolarity.

The cytoplasmic or inner membrane is impermeable to most large and poler solutes, while these are compensated for by freely diffusing water molecules, and thus, the transmembrane concentration gradients are developed for such compounds. The resulting changes in cellular volume and turgor pressure exert strong mechanical force on the cytoplasmic membrane and associated proteins and preclude the cell growth [260]. To cope with osmotic stress, bacteria adapt their intracellular osmolarity [254] or increase the cell wall stability [261]. The salt stress tolerance is mediated by flux control of water across the cell membrane, adjustments of intracellular potassium levels, synthesis of disaccharide trehalose, and/or transport of small molecule osmoprotectants [262].

In principle, bacterial cells respond to environmental or growth conditions by immediate protein or enzyme-level regulation, and by slow gene transcriptional regulation via transcription factors. In summary [260], in response to sudden changes in osmotic pressure, *E. coli* controls in- and outflux of water and other small molecules by activating aquaporins as an immediate response [263]. It regulates intracellular potassium concentrations by adjusting the potassium transporters such as Kup, KdpFABC, or TrkA for transient adaptation to short-term osmotic stress [264]. In the case of prolonged osmotic stress, *E. coli* takes up the osmotolerants such as glycine betaine and proline from the environment via ABC transporter encoded by *proVWX* or synthesizes glycine betaine from the extracellular precursor choline [265–267]. If no extracellular compatible solutes are available, *E. coli* induces expression of trehalose 6-phosphate synthase (OtsA) and phosphatase (OtsB) to produce high intracellular concentrations of the nonreducing disaccharide trehalose from the precursors such as UDP-glucose and G6P in response to long-term resistance to sustained osmotic stress [268–270].

There is indeed an interaction between trehalose and membrane lipid head groups, but this effect is insufficient to fully account for the resistance of membrane against strong osmotic stress. Upon osmotic stress, bacteria adjust their intracellular

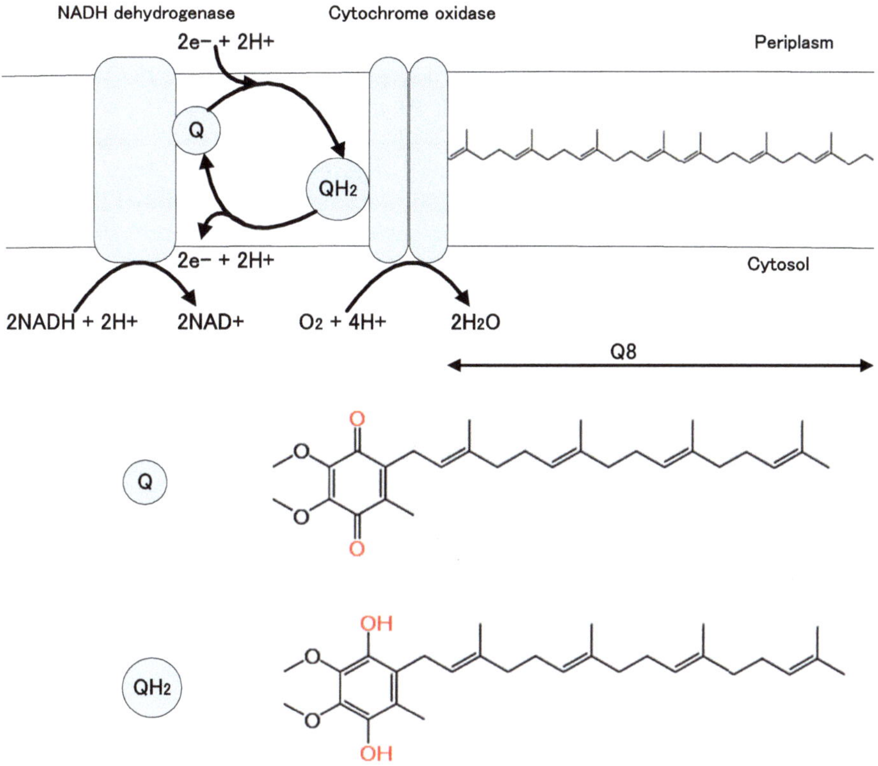

Fig. 13 Electron transport carrier quinone and quinol in the respiratory chain and the role of osmoprotection

osmolarity and modify their cell wall structure [260]. For this, polyisoprene lipids may also contribute to osmoprotection by increasing resistance to high-salt conditions in the cytoplasmic membrane and in the membrane bilayers of liposomes in *E. coli* [260]. Coenzyme Q functions as an electron and proton carrier in aerobic respiration and has an additional crucial role as a chain-breaking antioxidant [271]. The long polyisoprenyl tail of CoQ_n functions to anchor this lipid in the membranes of cells (Fig. 13), where n designates the number of five carbon isoprene units such as CoQ_6 in *S. cerevisiae*, CoQ_8 in *E. coli*, and CoQ_{10} in human [262]. In *E. coli*, CoQ_8 level becomes significantly high in response to high-salt condition (Fig. 13) [260].

18 Biofilm, Motility by Flagella, and Quorum Sensing

Biofilm formation is one of the important microbial survival strategies, where biofilm development involves attachment of bacteria to surfaces and cell–cell adhesion to form microcolonies. This is useful for the cell to protect against predators and antibiotics [272]. The attachment of bacteria to abiotic and biotic surfaces is made by motility, proteinaceous adhesion, and a cell-bound polysaccharide such as PGA (poly-β-1,6-N-acetyl-D-glucosamine), where PGA is a cell-bound exopolysaccharide adhesion [272]. As mentioned before, Csr plays important roles for biofilm formation, where *pga* operon involved in PGA formation and excretion is negatively regulated by CsrA. CsrA also negatively regulates c-di-GMP, a second messenger involved in biofilm formation and motility [273]. Curli are extracellular proteinaceous structures extending from the cell surface for attachment during biofilm development [274]. Curli filaments are activated by CsgD, where it is inversely correlated with flagella synthesis. The master regulator of flagella synthesis is $FlhD_2C_2$, which activates the genes involved in motility and chemotaxis [275]. McaS (multicellular adhesion sRNA) represses CsgD expression, while activates FlhD and PgaA [275], and thus regulates the synthesis of curli flagella and polysaccharide. Moreover, biofilm formation is under catabolite repression by cAMP and Crp [276].

Quorum sensing is a cell-to-cell communication [277], where the signal molecules are homoserine lactones (AHL) synthesized by LuxI-type enzyme. At high cell density cultivations, LuxR-type regulator plays a role for the positive feedback in association with AHL when its concentration exceeds a threshold level [120]. The quorum sensing is the sensing of cell density, where in *E. coli*, CyaR represses *luxS* gene which encodes autoinducer-2 synthase [278].

19 Systems Biology Approach

In order to deepen our knowledge on metabolic regulation and for the efficient metabolic engineering, it is quite useful to develop the appropriate metabolic models which describe the dynamic behavior of the intracellular metabolite concentration [279, 280]. Although kinetic models for the glycolysis and the PP pathways have been developed for *E. coli* [281], it is better to include TCA cycle, thus covering the whole main metabolism, which enables the simulation of the aerobic batch and continuous cultivations. Since the fluxes of the main metabolic pathways can be computed with respect to time by such a model, the cell growth rate may be reasonably predicted by taking into account the experimental observation that the cell growth rate is correlated with the specific ATP production rate [282]. More importantly, it is highly desirable to incorporate the effects of transcription factors on the enzymatic reactions to simulate the transition of the metabolism during the batch culture [42, 283, 284]. This type of model can be used to simulate acetate overflow metabolism and co-consumption of multiple sugars in relation to catabolite regulation [285]. As shown in Fig. 14, the trend of the transcription factors such as cAMP–Crp and Cra, and the pathway activities can be well predicted with such a model for the case of continuous culture as mentioned in Sect. 3, and it is quite useful to understand the complicated metabolic regulation mechanism [285].

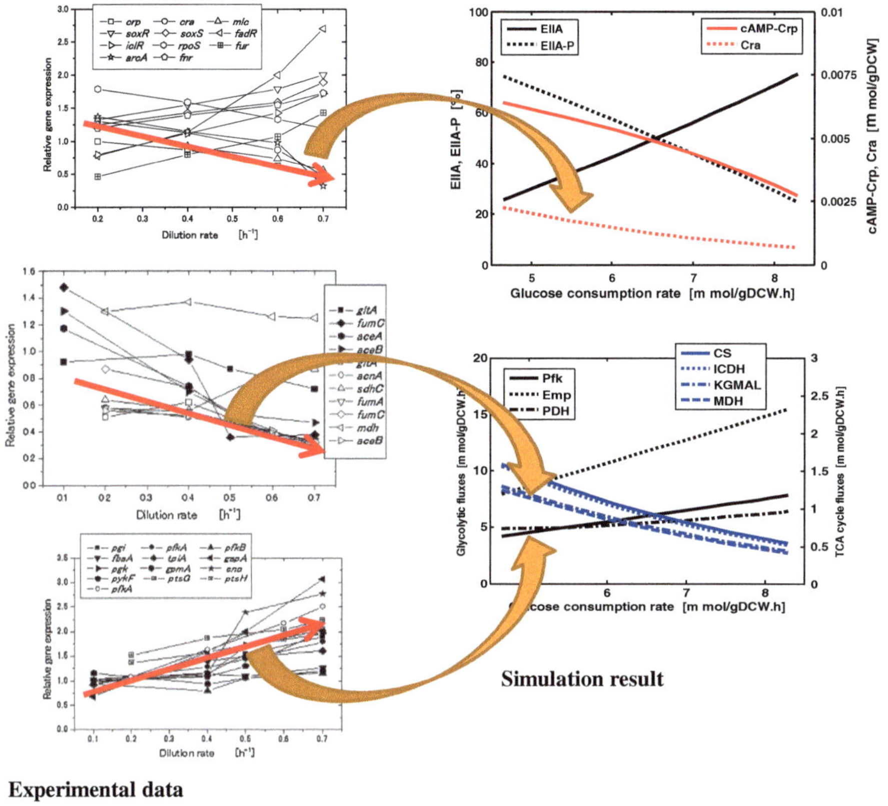

Fig. 14 Comparison of simulation result [45] with experimental data [48] for the continuous culture

20 Concluding Remarks

As seen above, the global regulators are responsive to the specific stimuli. Examples of such pleiotropic TFs in *E. coli* are Crp, a primary sensor for C-availability; NtrBC, a sensor for N-availability; PstSCAB and PhoR, the sensor for P-availability; CysB, the sensor for S-availability; and Fur, the sensor for ion availability. Functional interactions among such regulators must coordinate the activities of the metabolon so that the supply of one type of nutrient matches the supply of other nutrients [286]. Thus, multiple links between C and N metabolism have been identified [287]. Other functional links between C and S metabolism [288], and between C and ion metabolism [289, 290] have been identified. Moreover, the links between S and N limitations have been also identified [291].

In general, bacteria in nature live far away from the optimal growth condition, where multiple stresses are imposed on the cell. Therefore, the cell must have the ability to sense, integrate, and respond to the variety of stresses for survival.

Although little is known about "cross-stress" protection, cross-stress dependencies are ubiquitous, highly interconnected and may emerge within short time frames [292]. In fact, high degree of overlap was observed in the transcriptional profiling for different stresses such as starvation, osmotic, and acidic stresses [293], as well as starvation and heat-shock or oxidative stress [294, 295], where high osmolarity and high temperature induce the oxidative stress regulons such as SoxRS and OxyR [296, 297]. The responses to n-butanol share the same high overlap with those in heat-shock, oxidative, and acidic stresses [298].

As mentioned in this article, the specific metabolites such as FBP, PEP, PYR, OAA, AcCoA, and αKG in the main metabolic pathways play important roles for metabolic regulation. This implies that these metabolites play roles for the coordinated and integrated metabolic regulation. The regulation system ranges from relatively rapid interactions such as enzyme-level regulation by allosteric binding of the specific molecules or posttranslational modification to slow interactions such as transcriptional regulation via transcription factors. It is important to get deep insight into the whole cellular metabolic systems not only by molecular biology and biochemistry, but also by systems biology approach, and apply this for the efficient metabolic engineering.

Among intracellular metabolites, α-ketoacids such as αKG, OAA, and PYR turn to be mater regulators for catabolite regulation and coordination of different regulations [299]. Namely, when favoured carbon sources are depleted, α-ketoacid levels fall, and cAMP increases to stimulate other carbon catabolite machinery. When preferred nutrients are abundant, the cell growth rate becomes higher with lower cAMP level, while if they are scarce, the cell growth rate declines with higher cAMP level. This change in growth rate is accompanied by a change in cellular composition, where ribosomes are needed for rapid protein production at higher growth rate, while more metabolic enzymes for nutrient assimilation (catabolism) are needed at lower cell growth rate [300, 301]. There is a linear relationship between the total protein composition of a cell and its growth rate, where this can be extended beyond ribosomes to metabolic enzymes [150]. Under N- or S-limitation or other nutrient limitation, α-ketoacids such as αKG accumulate and inhibit carbon assimilation, where there is less need for carbon catabolic enzymes and more demand for those involved in such nutrient assimilation. When anabolic nutrients are in excess, αKG concentration decreases, cAMP level increases, and carbon catabolic enzymes increase to accelerate carbon assimilation. In the end, the physiological function of cAMP signaling goes beyond simply enabling hierarchical utilization of carbon sources, but also controls the function of the proteome [150, 299]. The energy level also affects carbon uptake rate [20, 302].

As mentioned before, cAMP–Crp and Cra play important roles for carbon catabolite regulation in *E. coli*, where either PTS or non-PTS sugars are ranked for assimilation in a hierarchy [303–305]. This may be caused by the cAMP–Crp level and the promoter activities of the corresponding promoters of the transporters [303–305].

The roles of cAMP–Crp are not only limited to carbon catabolite regulation. Among many transcription factors, Crp plays significantly important roles in the wide range of regulations such as osmoregulation [306], osmotolerance [307], oxidative stress [308], acid tolerance [309], acetate tolerance [310], ethanol tolerance [311], and butanol tolerance [312] as well as catabolite regulation. Therefore, it may be of interest to modulate such transcription factors [313, 314] for the development of next-generation cell factories.

It is of surmount interest to understand how the cell growth rate is regulated, since such information gives us a hint for improving the cell growth rate, and thus increasing the protein or metabolic production. For this, it is important to recognize at which regulation levels affect the cell growth rate, where posttranscriptional control of protein abundances and posttranslational control of flux rates are dominated [315].

Moreover, it is also important to understand the effect of the specific pathway gene mutation on the metabolic regulation in addition to the effect of growth condition. In fact, ^{13}C-metabolic flux analysis has been extensively employed for the metabolic flux distributions of pathogens such as *Mycobacterium tuberculosis* [316] and the specific gene knockout mutant *E. coli* [48, 317, 318], where the flux information is located on top of different levels of information, manifested as the result of metabolic regulation, and central to understanding the metabolism. It is, therefore, of interest to investigate the effect of the specific gene knockout on the metabolism as well in view of metabolic regulation for the design of next-generation cell factories.

References

Introduction

1. Jone RG, Thompson CB (2009) Tumor suppressors and cell metabolism: a recipe for cancer growth. Genes Dev 23:537–548
2. Hsu PP, Sabatini DM (2008) Cancer cell metabolism: Warburg and beyond. Cell 134:703–707
3. Kroemer G, Pouyssegur J (2008) Tumor cell metabolism: cancer's Achilles' heel. Cancer Cell 13:472–482
4. Folger O et al (2011) Predicting selective drug targets in cancer through metabolic networks. Mol Syst Biol 7:501
5. McInerney MJ, Sieber JR, Gunsalus RP (2009) Syntrophy in anaerobic global carbon cycles. Curr Opin Biotechnol 20:623–632
6. Dolfing J, Jiang B, Henstra AM, Stams AJ, Plugge CM (2008) Syntrophic growth on formate: a new microbial niche in anoxic environments. Appl Environ Microbiol 74:6126–6131
7. Lin LH et al (2006) Long-term sustainability of a high-energy, low-diversity crustal biome. Science 314:479–482
8. Peralta-Yahya PP, Zhang F, del Cardayre SB, Keasling JD (2012) Microbial engineering for the production of advanced biofuels. Nature 488:320–328

9. Atsumi S, Hanai T, Liao JC (2008) Non-fermentative pathways for synthesis of branched-chain higher alcohols as biofuels. Nature 451:86–90
10. Yim H et al (2011) Metabolic engineering of *Escherichia coli* for direct production of 1,4-butanediol. Nat Chem Biol 7:445–452
11. Dellomonaco C, Clomburg JM, Miller EN, Gonzalez R (2011) Engineered reversal of the β-oxidation cycle for the synthesis of fuels and chemicals. Nature 476:355–359
12. Bond-Watts BB, Bellerose RJ, Chang MC (2011) Enzyme mechanism as a kinetic control element for designing synthetic biofuel pathways. Nat Chem Biol 7:222–227
13. Shen CR et al (2011) Driving forces enable high-titer anaerobic 1-butanol synthesis in *Escherichia coli*. Appl Environ Microbiol 77:2905–2915
14. Keasling JD (2010) Manufacturing molecules through metabolic engineering. Science 330:1355–1358
15. Shimizu K (2014) Biofuels and biochemicals production by microbes. NOVA Publishing Co., USA
16. Bar-Even A, Flamholz A, Noor E, Mil R (2012) Rethinking of glycolysis: on the biochemical logic of metabolic pathways. Nat Chem Biol 8:509–517
17. Flamholz A, Noor E, Bar-Even A, Liebermeister W, Milo R (2013) Glycolytic strategy as a tradeoff between energy yield and protein cost. PNAS USA 110(24):10039–10044
18. Shimizu K (2013) Metabolic regulation of a bacterial cell system with emphasis on *Escherichia coli* metabolism. *ISRN Biochemistry,* doi:10.1155/2013/645983 (Article ID 645983)
19. Shimizu K (2014) Metabolic regulation of *Escherichia coli* in response to nutrient limitation and environmental stress conditions. Metabolites 4:1–35
20. Chubukov V, Gerosa L, Kochanowski K, Sauer U (2014) Coordination of microbial metabolism. Nat Rev 12:327–340

Transport of Nutrient and Waste

21. de la Cruz MA, Fernandez-Mora M, Guadarrama C et al (2007) LeuO antagonizes H-NS and StpA-dependent repression in *Salmonella enterica* ompS1. Mol Microbiol 66(3):727–743
22. Nikaido H (2003) Molecular basis of bacterial outer membrane permeability revisited. Microbiol Mol Biol Rev 67(4):593–656
23. Nikaido H, Vaara M (1985) Molecular basis of bacterial outer membrane permeability. Microbiol Rev 49(1):1–32
24. Death A, Ferenci T (1994) Between feast and famine: endogenous inducer synthesis in the adaptation of *Escherichia coli* to growth with limiting carbohydrates. J Bacteriol 176 (16):5101–5107
25. Gunnewijk MGW, van den Bogaard PTC, Veenhoff LM et al (2001) Hierarchical control versus autoregulation of carbohydrate utilization in bacteria. J Mol Microbiol Biotechnol 3 (3):401–413
26. Poolman B, Konings WN (1993) Secondary solute transport in bacteria. Biochim Biophy Acta 1183(1):5–39
27. Deutscher J, Francke C, Postoma PW (2006) How phosphotransferase system-related protein phosphorylation regulates carbohydrate metabolism in bacteria. Microbiol Mol Biol Rev 70 (4):939–1031
28. Palm K, Luthman K, Ros J, Grasjo J, Artursson P (1999) Effect of molecular charge on intestinal epithelial drug transport: pH-dependent transport of cationic drugs. J Pharmacol Exp Ther 291:435–443
29. Chakrabarti AC, Deamer DW (1992) Permeability of lipid bilayers to amino acids and phosphate. Biochim Biophys Acta 1111:171–177
30. Chakrabarti AC, Deamer DW (1994) Permeation of membranes by the neutral form of amino acids and peptides: relevance to the origin of peptide translocation. J Mol Evol 39:1–5

31. Finkelstein A (1976) Water and nonelectrolyte permeability of lipid bilayer membranes. J Gen Physiol 68:127–135
32. Winiwarter S et al (1998) Correlation of human jejunal permeability (in vivo) of drugs with experimentally and theoretically derived parameters. A multivariate data analysis approach. J Med Chem 41:4939–4949
33. Davis BD (1958) On the importance of being ionized. Arch Biochem Biophys 78:497–509
34. Westheimer FH (1987) Why nature chose phosphates. Science 235:1173–1178

PTS and Carbon Catabolite Regulation

35. Gorke B, Stulke J (2008) Carbon catabolite repression in bacteria: many ways to make the most out of nutrients. Nat Rev Microbiol 6:613–624
36. Inada T, Kimata K, Aiba H (1996) Genes Cells 1:293–301
37. Gosset G (2005) Improvement of *Escherichia coli* production strains by modification of the phosphoenolpyruvate: sugar phosphotransferase system. Microb Cell Fact 4:14
38. Hogema BM, Arents JC, Bader R, Eijkemans K et al (1998) Inducer exclusion in *Escherichia coli* by non-PTS substrates: the role of the PEP to pyruvate ratio in deter-mining the phosphorylation state of enzyme IIA^{Glc}. Mol Microbiol 30:487–498
39. Bettenbrock K, Sauter T, Jahreis K, Klemling A, Lengeler JW, Gilles ED (2007) Correlation be-tween growth rates, EIIACrr phosphorylation, and intra-cellular cyclic AMP levels in *Escherichia coli* K-12. J Bacteriol 189:6891–6900
40. Schaub J, Reuss M (2008) In vivo dynamics of glycolysis in *E. coli* shows need for gowth-rate dependent metabolome analysis. Biotechnol Prog 24:1402–1407
41. Kochanowski K, Volkmer B, Gerosa L, van Rijsewijk HBR, Schmidt A, Heinemann M (2013) Functioning of a metabolic flux sensor in *Escherichia coli.* PNAS USA 110:1130–1135
42. Kotte O, Zaugg JB, Heinemann M (2010) Bacterial adaptation through distributed sensing of metabolic fluxes. Mol Syst Biol 6:355
43. Huberts DH, Niebel B, Heinemann M (2012) A flux-sensing mechanism could regulate the switch between respiration and fermentation. FEMS Yeast Res 12(2):118–128
44. Kremling A, Bettenbrock K, Gilles ED (2008) A feed-forward loop guarantees robust behavior in *Escherichia coli* carbohydrate uptake. Bioinformatics 24:704–710
45. Matsuoka Y, Shimizu K (2013) Catabolite regulation analysis of *Escherichia coli* for acetate overflow mechanism and co-consumption of multiple sugars based on systems biology approach using computer simulation. J Biotechnol 168:155–173
46. Shimada T, Yamamoto K, Ishihama A (2011) Novel members of the Cra regulon involved in carbon metabolism in *Escherichia coli.* J Bacteriol 193(3):649–659
47. Saier MH Jr, Ramseier TM (1996) The catabolite repressor/activator (Cra) protein of enteric bacteria. J Bacteriol 178:3411–3417
48. Ishii N, Nakahigashi K, Baba T, Robert M, Soga T, Kanai A, Hirasawa T, Naba M, Hirai K, Hoque A, Ho PY, Kakazu Y, Sugawara K, Igarashi S, Harada S, Masuda T, Sugiyama N, Togashi T, Hasegawa M, Takai Y, Yugi K, Arakawa K, Iwata N, Toya Y, Nakayama Y, Nishioka T, Shimizu K, Mori H, Tomita M (2007) Multiple high-throughput analyses monitor the response of *E. coli* to perturbations. Science 316:593–597
49. Valgepea K, Adamberg K, Nahku R, Lahtvee PJ, Arike L, Vilu R (2010) Systems biology approach reveals that overflow metabolism of acetate in *Escherichia coli* is triggered by carbon catabolite repression of acetyl-CoA synthetase. BMC Syst Biol 4:166
50. Yao R, Hirose Y, Sarkar D, Nakahigashi K, Ye Q, Shimizu K (2011) Catabolic regulation analysis of *Escherichia coli* and its *crp*, *mlc*, *mgsA*, *pgi* and *ptsG* mutants. Microb Cell Fact 10:67

51. Wolf RE, Prather DM, Shea FM (1979) Growth-rate dependent alteration of 6-phosphogluconate dehydrogenase and glucose 6-phosphate dehydrogenase levels in *Escherichia coli* K-12. J Bacteriol 139:1093–1096

Acetate Overflow Metabolism and Reduction of Acetate Formation in *E. coli*

52. Wong MS, Wu S, Causey TB, Bennet GN, San KY (2008) Reduction of acetate accumulation in *Escherichia coli* cultures for increased recombinant protein production. Metab Eng 10:97–108
53. Russel JB et al (2007) The energy spilling reactions of bacteria and other organisms. J Mol Microbiol Biotechnol 13:1–11
54. Valgepea K, Adamberg K, Vilu R (2011) Decrease of energy spilling in *Escherichia coli* continuous cultures with rising specific growth rate and carbon wasting. BMC Syst Biol 5:106
55. Kayser A, Weber J, Hecht V, Rinas U (2005) Metabolic flux analysis of *Escherichia coli* in glucose-limited continuous culture. I. Growth-rate dependent metabolic efficiency at steady state. Microbiology 151:693–706
56. Wolfe AJ (2005) The acetate switch. Microbiol Mol Biol Rev 69:12–50
57. Majewski RA, Domach MM (1990) Simple constrained-optimization view of acetate overflow in *Escherichia coli*. Biotech Bioeng 35:732–738
58. Yang C, Hua Q, Baba T, Mori H et al (2003) Analysis of *E. coli* anaplerotic metabolism and its regulation mechanism from the metabolic responses to alter dilution rates and *pck* knockout. Biotechnol Bioeng 84:129–144
59. Lee SY (1996) High cell-density cultivation of *Escherichia coli*. Trends Biotechnol 14:98–105
60. Hardiman T, Lemuth K, Keller MA, Reuss M, Siemann-Herzberg M (2007) Topology of the global regulatory network of carbon limitation in *Escherichia coli*. J Biotechnol 132:359–374
61. Konstantinov K, Kishimoto M, Seki T, Yoshida T (1990) A balanced DO-stat and its application to the control of acetic acid excretion by recombinant *Eshcherichia coli*. Biotechnol Bioeng 36(7):750–758
62. Ferreira AR, Ataide F, von Stosch M, Jias JML et al (2012) Application of adaptive DO-stat feeding control to *Pichia pastoris* X33 cultures expressing a single chain antibody fragment (scFv). Bioprocess Biosyst Eng 35(9):1603–1614
63. Li K-T, Liu D-H, Chu J, Wang Y-H et al (2008) An efficient and simplified pH-stat control strategy for the industrial fermentation of vitamin B12 by *Pseudomonas denitrificans*. Bioprocess Byosyst Eng 31:605–610
64. Pirt SJ (1982) Maintenance energy: a general model for energy-limited and energy-sufficient growth. Arch Microbiol 133:300–302
65. Hewitt, CJ, Caron NV, Nienow AW, McFarlane CM (1999) Use of multi-staining flow cytometry to characterise the physiological state of *Escherichia coli* W3110 in high cell density fed-batch cultures. *Biotechnol Bioeng* 63:705–711
66. Hewitt CJ, Caron NV, Axelsson B, McFarlane CM, Nienow AW (2000) Studies related to the scale-up of high-cell-density *E. coli* fed-batch fermentations using multiparameter flow cytometry: effect of a changing microenvironment with respect to glucose and dissolved oxygen concentration. *Biotechnol Bioeng* 70(4):381–390
67. Hewitt CJ, Nebe-Von-Caron G (2001) An industrial application of multiparameter flow cytometry: Assessment of cell physiological state and its application to the study of microbial fermentations. Cytometry 44:179–187

68. Peebo K, Valgepea K, Nahku R, Riis G, Õun M, Adamberg K, Vilu R (2014) Coordinated activation of PTA-ACS and TCA cycles strongly reduces overflow metabolism of acetate in *Escherichia coli*. Appl Microbiol Biotechnol, doi: 10.1007/s00253-014-5613-y

Catabolite Regulation for the Uptake of Various Carbon Sources

69. Vasudevan P, Briggs M (2008) Biodiesel production–current state of the art and challenges. J Ind Microbiol Biotechnol 35:421–430
70. Dharmadi Y, Murarka A, Gonzalez R (2006) Anaerobic fermentation of glycerol by *Escherichia coli*: a new platform for metabolic engineering. Biotechnol Bioeng 94:821–829
71. Clomburg JM, Gonzalez Ramon (2013) Anaerobic fermentation of glycerol: a platform for renewable fuels and chemicals. Trends Biotechnol 31(1):20–28
72. Almeida JRM, Fávaro LCL, Betania F, Quirino BF (2012) Biodiesel biorefinery: opportunities and challenges for microbial production of fuels and chemicals from glycerol waste. Biotechnol Biofuels 5:48
73. Martínez-Gómez K, Flores N, Castañeda HM, Martínez-Batallar G, Hernández-Chávez G, Ramírez OT, Gosset G, Encarnación S, Bolivar F (2012) New insights into *Escherichia coli* metabolism: carbon scavenging, acetate metabolism and carbon recycling responses during growth on glycerol. Micob Cell Fact 11:46
74. Peng L, Shimizu K (2003) Global metabolic regulation analysis for *Escherichia coli* K12 based on protein expression by 2-dimensional electrophoresis and enzyme activity measurement. Appl Microbiol Biotechnol 61:163–178
75. Oh MK, Liao JC (2000) Gene expression profiling by DNA microarrays and metabolic fluxes in *Escherichia coli*. Biotechnol Prog 16:278–286
76. Applebee MK, Joyce AR, Conrad TM, Pettigrew DW, Palsson BO (2011) Functional and metabolic effects of adaptive glycerol kinase (GLPK) mutants in *Escherichia coli*. J Biol Chem 286(26):23150–23159
77. Cheng K-K, Lee B-S, Masuda T, Ito T, Ikeda K, Hirayama A, Deng L, Dong J, Shimizu K, Soga T, Tomita M, Palsson BO, Robert M (2014) Global metabolic network reorganization by adaptive mutations allows fast growth of *Escherichia coli* on glycerol. Nat Commun 5:3233
78. Kornberg HL (2001) Routes for fructose utilization by *Escherichia coli*. J Mol Microbiol Biotechnol 3:355–359
79. Yao R, Kurata H, Shimizu K (2013) Effect of *cra* gene mutation on the metabolism of *Escherichia coli* for a mixture of multiple carbon sources. Adv Biosci Biotechnol 4:477−486
80. Sarkar D, Shimizu K (2008) Effect of *cra* gene knockout together with other genes knockouts on the improvement of substrate consumption rate in *Escherichia coli* under microaerobic conditions. Biochem Eng J 42:224–228
81. Sarkar D, Siddiquee KAZ, Arauzo-Bravo MJ, Oba T, Shimizu K (2008) Effect of *cra* gene knockout together with *edd* and *iclR* genes knockout on the metabolism in *Escherichia coli*. Arch Microbiol 190:559–571
82. Crasnier-Mednansky M, Park MC, Studley WK, Saier MH Jr (1997) Cra-mediated regulations of *Escherichia coli* adenylate cyclase. Microbiology 143:785–792
83. Griffith JK, Baker ME, Rouch DA, Page MG, Skurray RA, Paulsen IT, Chater KF, Baldwin SA, Henderson PJ (1992) Membrane transport proteins: implications of sequence comparisons. Curr Opin Cell Biol 4:684–695
84. Sumiya M, Davis EO, Packman LC, McDonald TP, Henderson PJ (1995) Molecular genetics of a receptor protein for D-xylose, encoded by the gene *xylF*, in *Escherichia coli*. Receptors Channels 3:117–128
85. Song S, Park C (1997) Organization and regulation of the D-xylose operons in *Escherichia coli* K-12: XylR acts as a transcriptional activator. J Bacteriol 179:7025–7032

86. Hasona A, Kim Y, Healy FG, Ingram LO, Shanmugam KT (2004) Pyruvate formate lyase and acetate kinase are essential for anaerobic growth of *Escherichia coli* on xylose. J Bacteriol 186:7593–7600
87. Novotny CP, Englesberg E (1966) The L-arabinose permease system in *Escherichia coli* B/r. Biochim Biophys Acta 117:217–230
88. Schleif R (1969) An L-arabinose binding protein and arabinose permeation in *Escherichia coli*. J Mol Biol 46:185–196
89. Doyle ME, Brown C, Hogg RW, Helling RB (1972) Induction of the *ara* operon of *Escherichia coli* B-r. J Bacteriol 110:56–65
90. Davidson CJ, Narang A, Surette MG (2010) Integration of transcriptional inputs at promoters of the arabinose catabolic pathway. BMC Syst Biol 4:75
91. Madar D, Dekel E, Bren A, Alon U (2011) Negative auto-regulation increases the input dynamical-range of the arabinose system of *Escherichia coli*. BMC Syst Biol 5:111
92. Desai TA, Rao CV (2010) Regulation of arabinose and xylose metabolism in *Escherichia coli*. Appl Environ Microbiol 76:1524–1532
93. Groff D, Benke PI, Batth TS, Bokinsky G, Petzold CJ, Adams PD, Keasling JD (2012) Supplementation of intracellular XylR leads to co-utilization of hemicellulose sugars. Appl Environ Microbiol 78:2221–2229

Transition of the Metabolism during Batch Culture

94. Toya Y, Ishii N, Nakahigashi K, Tomita M, Shimizu K (2010) ^{13}C-metabolic flux analysis for batch culture of *Escherichia coli* and its Pyk and Pgi gene knockout mutants based on mass isotopomer distribution of intracellular metabolites. Biotechnol Prog 26:975–992
95. Schuetz R, Zamboni N, Zampieri M, Heinemann M et al (2012) Multidimensional optimality of microbial metabolism. Science 336:601–604
96. Enjalbert B, Letisse F, Portais J-C (2013) Physiological and molecular timing of the glucose to acetate transition in *Escherichia coli*. Metabolites 3:820–837
97. Xu Y-F, Amador-Noguez D, Reaves ML, Feng XJ et al (2012) Ultrasensitive regulation of anapleurosis via allosteric activation of PEP carboxylase. Nat Chem Biol 8:562–568
98. Ali Azam T, Iwata A, Nishimura A, Ueda S, Ishihama A (1999) Growth phase-dependent variation in protein composition of the *Escherichia coli* nucleoid. J Bacteriol 181:6361–6370
99. Bradley MD, Beach MB, Jason de Koning AP, Pratt TS, Osuna R (2007) Effects of Fis on *Escherichia coli* gene expression during different growth stages. Microbiology 153:2922–2940
100. Mallik P, Pratt TS, Beach MB, Bradley MD, Undamatla J, Osuna R (2004) Growth phase-dependent regulation and stringent control of fis are conserved processes in enteric bacteria and involve a single promoter (fis P) in *Escherichia coli*. J Bacteriol 186:122–135
101. Mallik P, Paul BJ, Rutherford ST, Gourse RL, Osuna R (2006) DksA is required for growth phase-dependent regulation, growth rate-dependent control, and stringent control of fis expression in *Escherichia coli*. J Bacteriol 188:5775–5782
102. Paul BJ, Berkmen MB, Gourse RL (2005) DksA potentiates direct activation of amino acid promoters by ppGpp. PNAS USA 102:7823–7828
103. Ferenci T (2001) Hungry bacteria—definition and properties of a nutritional state. Environ Microbiol 3:605–611
104. Braeken K, Moris M, Daniels R, Vanderleyden J, Michiels J (2006) New horizons for (p) ppGpp in bacterial and plant physiology. Trends Microbiol 14(1):45–54
105. Artsimovitch I, Patlan V, Sekine S, Vassylyeva MN, Hosaka T, Ochi K, Yokoyama S (2004) Structural basis for transcription regulation by alarmone ppGpp. Cell 117:299–310
106. Kanjee U, Ogata K, Houry WA (2012) Direct binding targets of the stringent response alarmone (p)ppGpp. Mol Micobiol 85(6):1029–1043

107. Hengge-Aronis R (2002) Signal transduction and regulatory mechanisms involved in control of the σ^S (RpoS) subunit of RNA polymerase. Microbiol Mol Biol Rev 66:373–395
108. Patten CL, Kirchhof MG, Schertzberg MR, Morton RA, Schellhorn HE (2004) Microarray analysis of RpoS-mediated gene expression in *Escherichia coli* K-12. Mol Genet Genom 272 (5):580–591
109. Jishage M, Kvint K, Shingler V, Nystrom T (2002) Regulation of σ factor competition by the alarmone ppGpp. Genes Dev 16:1260–1270
110. Lacour S, Landini P (2004) σ^S-dependent gene expression at the onset of stationary phase in *Escherichia coli*: function of σ^s-dependent genes and identification of their promoter sequences. J of Bacteriol 186(21):7186–7195
111. Vijayakumar SRV, Kirchhof MG, Patten CL, Schellhorn HE (2004) RpoS-regulated genes of *Escherichia coli* identified by random *lac*Z fusion mutagenesis. J of Bacteriol 186(24):8499–8507
112. Rahman M, Shimizu K (2008) Altered acetate metabolism and biomass production in several *Escherichia coli* mutants lacking *rpoS*-dependent metabolic pathway genes. Mol BioSyst 4 (2):160–169
113. Rahman M, Hasan MM, Shimizu K (2008) Growth phase-dependent changes in the expression of global regulatory genes and associated metabolic pathways in *Escherichia coli*. Biotechnol Lett 30:853–860
114. Llorens NJM, Tormo A, Martínez-García E (2010) Stationary phase in gram-negative bacteria. FEMS Microbiol Rev 34:476–495
115. Desnues B, Cuny C, Gregori G, Dukan S, Aguilaniu H, Nyström T (2003) Differential oxidative damage and expression of stress defence regulons in culturable and non-culturable *Escherichia coli* cells. EMBO Rep 4:400–404
116. Nagamitsu H, Murata M, Kosaka T, Kawaguchi J, Mori H, Yamada M (2013) Crucial roles of MicA and RybB as vital factors for σ^E-Dependent cell lysis in *Escherichia coli* long-term stationary phase. J Mol Microbiol Biotechnol 23:227–232
117. Murata M, Noor R, Nagamitsu H, Tanaka S, Yamada M (2012) Novel pathway directed by sigma E to cause cell lysis in *Escherichia coli*. Genes Cells 17:234–247
118. Noor R, Murata M, Yamada M (2009) Oxidative stress as a trigger for growth phase-specific sigma E-dependent cell lysis in *Escherichia coli*. J Mol Microbiol Biotechnol 17:177–187

Carbon Storage Regulation

119. Yamamotoya T, Dose H, Tian Z, Faure A, Toya Y, Honma M, Igarashi K, Nakahigashi K, Soga T, Mori H, Matsuno H (2012) Glycogen is the primary source of glucose during the lag phase of *E. coli* proliferation. Biochim Biophys Acta 1824:1442–1448
120. Timmermans J, van Melderen L (2010) Post-transcriptional global regulation by CsrA in bacteria. Cell Mol Life Sci 67(17):2897–2908
121. Jonas K, Edwards AN, Ahmad I, Romeo T, Romling U, Melefors O (2010) Complex regulatory network encompassing the Csr, c-di-GMP and motility systems of *Salmonella typhimurium*. Environ Microbiol 12(2):524–540
122. Yakhnin H, Pandit P, Petty TJ, Baker CS, Romeo T, Babitzke P (2007) CsrA of *Bacillus subtilis* regulates translation initiation of the gene encoding the flagellin protein (hag) by blocking ribosome binding. Mol Microbiol 64(6):1605–1620
123. Romeo T, Vakulskas CA, Babitzke P (2013) Post-transcriptional regulation on a global scale: from and function of Csr/Rsm systems. Environ Microbiol 15(2):313–324
124. Dubey AK, Baker CS, Romeo T, Babitzke P (2005) RNA sequence and secondary structure participate in high-affinity CsrA-RNA interaction. RNA 11(10):1579–1587
125. Suzuki K, Wang X, Weilbacher T et al (2002) Regulatory circuitry of the CsrA/CsrB and BarA/UvrY systems of *Escherichia coli*. J of Bacteriol 184(18):5130–5140

126. Weilbacher T, Suzuki K, Dubey AK et al (2003) A novel sRNA component of the carbon storage regulatory system of *Escherichia coli*. Mol Microbiol 48(3):657–670
127. Baker CS, Morozov I, Suzuki K, Romeo T, Babitzke P (2002) CsrA regulates glycogen biosynthesis by preventing translation of *glgC* in *Escherichia coli*. Mol Microbiol 44 (6):1599–1610
128. Dong T, Schellhorn HE (2009) Control of RpoS in global gene expression of *Escherichia coli* in minimal media. Mol Genet Genomics 281:19–33
129. Edwards AN, Patterson-Fortin LM, Vakulskas CA, Mercante JW, Potrykus K, Vinella D, Camacho MI, Fields JA, Thompson SA, Georgellis D et al (2011) Circuitry linking the Csr and stringent response global regulatory systems. Mol Microbiol 80:1561–1580
130. McKee AE, Rutherford BJ, Chivian DC, Baidoo EK, Juminaga D, Kuo D, Benke PI, Dietrich JA, Ma SM, Arkin AP, Petzold CJ, Adams PD, Keasling JD, Chhabra SR (2012) Manipulation of the carbon storage regulator system for metabolite remodeling and biofuel production in *Escherichia coli*. Microb Cell Fact 11:79
131. Revelles O, Millard P, Nougayrede J-P, Dpbrindt U, Osward E, Letisse F, Portais, J-C (2013) The carbon storage regulator (Csr) system exerts a nutrient-specific control over central metabolism in *Escherichia coli* strain Nissle 1917. Plos One 8(6):e66386 (1–12)
132. Tatarko M, Romeo T (2001) Disruption of a global regulatory gene to enhance central carbon flux into phenylalanine biosynthesis in *Escherichia coli*. Current Microbiol 43(1):26–32

Nitrogen Regulation

133. Van Heeswijk WC, Westerhoff HV, Boogerd FC (2013) Nitrogen assimilation in *Escherichia coli*: putting molecular data into a systems perspective. Microbiol Mol Biol Rev 77(4):628–695
134. Kim M, Zhang Z, Okano H, Yan D, Groisman A, Hwa T (2012) Need-based activation of ammonium uptake in *Escherichia coli*. Mol Syst Biol 8:616
135. Gruswitz F, O'Connell J, Stroud RM (2007) Inhibitory complex of the transmembrane ammonia channel, AmtB, and the cytosolic regulatory protein, GlnK, at 1.96 A. PNAS USA 104:42–47
136. Radchenko MV, Thornton J, Merrick M (2010) Control of AmtB-GlnK complex formation by intracellular levels of ATP, ADP, and 2-oxoglutarate. J Biol Chem 285:31037–31045
137. Truan D, Huergo LF, Chubatsu LS, Merrick M, Li XD, Winkler FK (2010) A new P(II) protein structure identifies the 2-oxoglutarate binding site. J Mol Biol 400:531–539
138. Yuan J, Doucette CD, Fowler WU, Feng X-J, Piazza M, Rabitz HA, Wingreen NS, Rabinowitz JD (2009) Metabolomics-driven quantitative analysis of ammonia assimilation in *E. coli*. Mol Syst Biol 5:302
139. Ninfa AJ, Jiang P, Atkinson MR, Peliska JA (2000) Integration of antagonistic signals in the regulation of nitrogen assimilation in *Escherichia coli*. Curr Topics Cell Regul 36:31–75
140. Gerosa L, Kochanowski K, Heinemann M, Sauer U (2013) Dissecting specific and global transcriptional regulation of bacterial gene expression. Mol Syst Biol 9:658
141. Kiupakis AK, Reitzer L (2002) ArgR-independent induction and ArgR-dependent superinduction of the *astCADBE* operon in *Escherichia coli*. J Bacteriol 184(11):2940–2950
142. Commichau FM, Forchhammer K, Stulke J (2006) Regulatory links between carbon and nitrogen metabolism. Curr Opin Microbiol 9(2):167–172
143. Mao XJ, Huo YX, Buck M, Kolb A, Wang YP (2007) Interplay between CRP-cAMP and PII-Ntr systems forms novel regulatory network between carbon metabolism and nitrogen assimilation in *Escherichia coli*. Nucleic Acids Res 35(5):1432–1440
144. Kumar R, Shimizu K (2010) Metabolic regulation of *Escherichia coli* and its *gdhA*, *glnL*, *gltB*, *D* mutants under different carbon and nitrogen limitations in the continuous culture. Microb Cell Fact 9:8

145. Jiang P, Ninfa AJ (2007) *Escherichia coli* PII signal transduction protein controlling nitrogen assimilation acts as a sensor of adenylate energy charge in vitro. Biochemistry 46(45): 12979–12996
146. Ninfa J, Jiang P (2005) PII signal transduction proteins: sensors of -ketoglutarate that regulate nitrogen metabolism. Curr Opin Microbiol 8(2):168–173
147. Brauer MJ, Yuan J, Bennet BD, Lu W, Kimball E, Botstein D, Rabinowitz JD (2006) Conservation of the metabolomic response to starvation across two divergent microbes. PNAS USA 103:19302–19307
148. Hart Y, Madar D, Yuan J, Bren A, Mayo AE, Rabinowitz JD, Alon U (2011) Robust control of nitrogen assimilation by a bifunctional enzyme in *E. coli*. Moleculae Cell 41:117–127
149. Doucette CD, Schwab DJ, Wingreen NS, Rabinowitz JD (2011) Alpha-ketoglutarate coordinates carbon and nitrogen utilization via enzyme i inhibition. Nat Chem Biol 7: 894–901
150. You C, Okano H, Hui S, Zhang Z, Kim M, Gunderson CW, Wang Y-P, Lenz P, Yan D, Hwa T (2013) Coordination of bacterial proteome with metabolism by cyclic AMP signaling. Nature 500:301–306
151. Powell BS, Court DL, Inada T, Nakamura Y, Michotey V, Cui X et al (1995) Novel proteins of the phosphotransferase system encoded within the *rpoN* operon of *Escherichia coli*. Enzyme IIANtr affects growth on organic nitrogen and the conditional lethality of an erats mutant. J Biol Chem 270:4822–4839
152. Peterkofski A, Wang G, Seok Y-J (2006) Parallel PTS systems. Arch Biochem Biophys 453 (1):101–107
153. Pflüger-Grau K, Görke B (2010) Regulatory roles of the bacterial nitrogen-related phosphotransferase system. Trend Microbiol 18(5):205–214
154. Lee CR, Cho SH, Yoon MJ, Peterkofsky A, Seok YJ (2007) *Escherichia coli* enzyme IIA^{Ntr} regulates the K^+ transporter TrkA. PNAS USA 104:4124–4129
155. Lüttmann D, Heermann R, Zimmer B, Hillmann A, Rampp IS, Jung K, Görke B (2009) Stimulation of the potassium sensor KdpD kinase activity by interaction with the phosphotransferase protein IIA^{Ntr} in *Escherichia coli*. Mol Micobiol 72(4):978–994
156. Lee C-R, Cho S-H, Kim H-J, Kim M, Peterkofsky A, Seok Y-J (2010) Potassium mediates *Escherichia coli* enzyme IIA^{Ntr}-dependent regulation of sigma factor selectivity. Mol Microbiol 78(6):1468–1483
157. Lüttmann D, Göpel Y, Görke B (2012) The phosphotransferase protein $EIIA^{Ntr}$ modulates the phosphate starvation response through interaction with histidine kinase PhoR in *Escherichia coli*. Mol Microbiol 86(1):96–110
158. Kim H-J, Lee C-R, Kim M, Peterkofsky A, Seok Y-J (2011) Dephosphorylated NPr of the nitrogen PTS regulates lipid a biosynthesis by direct interaction with LpxD. Biochem Biophys Res Commun 409(3):556–561
159. Lee C-R, Park Y-H, Kim M, Kim Y-R, Park S, Peterkofsky A, Seok Y-J (2013) Reciprocal regulation of the autophosphorylation of enzyme I^{Ntr} by glutamine and α-ketoglutarate in *Escherichia coli*. Mol Microbiol 88(3):473–485

Sulfur Regulation

160. Bykowski T, van der Ploeg JR, Iwanicka-Nowicka R, Hryniewicz MM (2002) The switch from inorganic to organic sulphur assimilation in *Escherichia coli*: adenosine 5′-phosphosulphate (APS) as a signalling molecule for sulphate excess. Mol Microbiol 43:1347–1358
161. Kredich NM (1996) Biosynthesis of cysteine. In: Neidhardt FC, Curtiss R III, Ingraham JL, Lin ECC, Low KB, Magasanik B, Reznikoff WS, Riley M, Schaechter M, Umbarger HH (eds) *Escherichia coli* and *Salmonella*: cellular and molecular biology, vol 1, 2nd edn. ASM Press, Washington DC, pp 514–527

162. Iwanicka-Nowicka R, Hryniewicz MM (1995) A new gene, *cbl*, encoding a member of the LysR family of transcriptional regulators belongs to *Escherichia coli cys* regulon. Gene 166:11–17
163. Zimmer DP, Soupene E, Lee HL, Wendisch VF, Khodursky BJ, Peter R, Bender A, Kustu S (2000) Nitrogen regulatory protein controlled genes of *Escherichia coli*: scavenging as a defense against nitrogen limitation. PNAS USA 97:14674–14679
164. Gyaneshwar P, Paliy O, McAuliffe J, Popham DL, Jordan MI, Kustu S (2005) Sulfur and Nitrogen Limitation in *Escherichia coli* K-12: specific homeostatic responses. J Bacteriol 187 (3):1074–1090

Phosphate Regulation

165. Wanner BL (1996) Phosphorus assimilation and control of the phosphate regulon. In: Neidhardt FC, Curtiss IIIR, Ingraham JL et al (eds) *Escherichia coli* and salmonella: cellular and molecular biology, pp 1357–1381. ASM Press, Washington DC
166. Wanner BL (1993) Gene regulation by phosphate in enteric bacteria. J Cellr Biochem 51 (1):47–54
167. Baek JH, Lee SY (2007) Transcriptome analysis of phosphate starvation response in *Escherichia coli*. J Microbiol Biotechnol 17(2):244–252
168. Van Dien SJ, Keasling JD (1998) A dynamic model of the *Escherichia coli* phosphate-starvation response. J Theoret Biol 190(1):37–49
169. Marzan LW, Shimizu K (2011) Metabolic regulation of *Escherichia coli* and its phoB and phoR gene knockout mutants under phosphate and nitrogen limitations as well as acidic condition. Microb Cell Fact 10:39
170. Spira B, Silberstein N, Yagil E (1995) Guanosine 3′,5′-bispyrophosphate (ppGpp) synthesis in cells of *Escherichia coli* starved for Pi. J Bacteriol 177(14):4053–4058

Metal Ion Regulation and Oxidative Stress Regulation

171. Frey PA, Reed GH (2012) The ubiquity of iron. ACS Chem Biol 7:1477–1481
172. Dixon SJ, Stockwell BR (2014) The role of iron and reactive oxygen species in cell death. Nat Chem Biol 10:9–17
173. Dann CEIII, Wakeman CA, Sieling CL, Baker SC, Irnov I, Winker WC (2007) Structure and mechanism of a metal-sensing regulatory RNA. Cell 130:878–892
174. Helmann JD (2007) Measuring metals with RNA. Mol Cell 27:859–860
175. Zheng M, Doan B, Schneider TD, Storz G (1999) OxyR and SoxRS regulation of fur. J Biotechnol 181:4639–4643
176. McHugh JP, Rodríguez-Quiñones F, Abdul-Tehrani H, Svistunenko DA, Poole RK, Cooper CE, Andrews SC (2003) Global iron-dependent gene regulation in *Escherichia coli*. J Biol Chem 278:29478–29486
177. Storz G, Imlay JA (1999) Oxidative stress. Curr Opin Microbiol 2:188–194
178. Jovanovic G, Lloyde LJ, Stumpf MP, Mayhew AJ, Buck M (2006) Induction and function of the phase shock protein extracytoplasmic stress response in *Escherichia coli*. J Biol Chem 281:21147–21161
179. Blanchard JR, Wholey WY, Conlon EM, Pomposiello PJ (2007) Rapid changes in gene expression dynamics in response to superoxide reveal SoxRS dependent and independent transcriptional network. PLoS One 2:e1186
180. Brynildsen MP, Liao JC (2009) An integrated network approach indentifies the isobutanol response network of *Escherichia coli*. Mol Syst Biol 5:34

181. Braun V, Hantke K, Koster W (1998) Bacterial ion transport: mechanisms, genetics, and regulation. In: Sigel A, Sigel H (eds) Metal ions in biological systems. Marcel Dekker, New York, pp 67–145
182. Azpiroz MF, Laviña M (2004) Involvement of enterobactin synthesis pathway in production of Microcin H47. Antimicrob Agents Chemother 48:1235–1241
183. Semsey S, Andersson AM, Krishna S, Jensen MH, Masse E, Sneppen K (2006) Genetic regulation of fluxes: iron homeostasis of *Escherichia coli*. Nucleic Acids Res 34:4960–4967
184. Hantash FM, Ammerlaan M, Earhart CF (1997) Enterobactin synthase polypeptides of *Escherichia coli* are present in an osmotic-shock-sensitive cytoplasmic locality. Microbiol 143:147–156
185. Kumar R, Shimizu K (2011) Transcriptional regulation of main metabolic pathways of *cyoA*, *cydB*, *fnr*, and *fur* gene knockout *Escherichia coli* in C-limited and N-limited aerobic continuous cultures. Microb Cell Fact 10:3
186. Zhang Z, Gosset G, Barabote R, Gonzalez CS, Cuevas WA, Saier MH Jr (2005) Functional interactions between the carbon and iron utilization regulators, Crp and Fur, in *Escherichia coli*. J Bacteriol 187:980–990
187. Perera IC, Grove A (2010) Molecular mechanisms of ligand-mediated attenuation of DNA binding by MarR family transcriptional regulators. J Mol Cell Biol 2:243–254
188. Hao Z, Lou H, Zhu R, Zhu J, Zhang D, Zhao BS, Zeng S, Chen X, Chan J, He C, Chen PR (2014) The multiple antibiotic resistance regulator MarR is a copper sensor in *Escherichia coli*. Nat Chem Biol 10:21–28
189. Pomposiello PJ, Demple B (2001) Redox-operated genetic switches: the SoxR and OxyR transcription factors. Trends Biotechnol 19(3):109–114
190. Gaudu P, Weiss B (1996) SoxR, a [2Fe–2S] transcription factor, is active only in its oxidized form. PNAS USA 93(19):10094–10098
191. Mailloux RJ, Bériault R, Lemire J, Singh R, Chénier RR, Hamel RD, Appanna VD (2007) The tricarboxylic acid cycle, an ancient metabolic network with a novel twist. Plos One 2(8): e690
192. DeNicola GM et al (2011) Oncogene-induced Nrf2 transcription promotes ROS detoxification and tumorigenesis. Nature 475:106–109
193. Son J et al (2013) Glutamine supports pancreatic cancer growth through a KRAS-regulated metabolic pathway. Nature 496:101–105
194. Maddocks OD et al (2013) Serine starvation induces stress and p53-dependent metabolic remodeling in cancer cells. Nature 493:542–546
195. Trachootham D, Alexandre J, Huang P (2009) Targeting cancer cells by ROS-mediated mechanisms: a radical therapeutic approach ? Nat Rev Drug Discov 8:579–591

Redox State Regulation

196. Gunsalus RP (1992) Control of electron flow in *Escherichia coli*: coordinated transcription of respiratory pathway genes. J Bacteriol 174(22):7069–7074
197. Salmon K, Hung SP, Mekjian K, Baldi P, Hatfield GW, Gunsalus RP (2003) Global gene expression profiling in *Escherichia coli* K12: the effects of oxygen availability and FNR. J Biol Chem 278(32):29837–29855
198. Alexeeva S, Hellingwerf KJ, de Mattos MJT (2003) Requirement of ArcA for redox regulation in *Escherichia coli* under microaerobic but not anaerobic or aerobic conditions. J Bacteriol 185(1):204–209
199. Zhu J, Shalel-Levanon S, Bennett G, San KY (2006) Effect of the global redox sensing/regulation networks on *Escherichia coli* and metabolic flux distribution based on C-13 labeling experiments. Metab Eng 8(6):619–627
200. Georgellis D, Kwon O, Lin ECC (2001) Quinones as the redox signal for the arc two-component system of bacteria. Science 292(5525):2314–2316

201. Malpica R, Franco B, Rodriguez C, Kwon O, Georgellis D (2004) Identification of a quinone-sensitive redox switch in the ArcB sensor kinase. PNAS USA 101(36):13318–13323
202. Constantinidou C, Hobman JL, Grifiths L, Patel MD, Penn CW, Cole JA, Overton TW (2006) A reassessment of the FNR regulon and transcriptomic analysis of the effects of nitrate, nitrite, NarXL, and NarQP as *Escherichia coli* K12 adapts from aerobic to anaerobic growth. J Biol Chem 281(8):4802–4815
203. Kessler D, Knappe J (1996) Anaerobic dissimilation of pyruvate. In: Neidhardt FC, Curtiss R, Ingraham JI et al (eds) *E. Coli* and salmonella: cellular and molecular biology, 2nd edn, vol 1, pp 199–205. ASM Press, Washington, DC
204. Toya Y, Nakahigashi K, Tomita M, Shimizu K (2012) Metabolic regulation analysis of wild-type and *arcA* mutant *Escherichia coli* under nitrate conditions using different levels of omics data. Mol Biosyst 8:2593–2604
205. Maeda T, Sanchez-Torres V, Wood TK (2007) Enhanced hydrogen production from glucose by metabolically engineered *Escherichia coli*. Appl Microbiol Biotechnol 77:879–890
206. Zhu J, Shimizu K (200) The effect of *pfl* genes knockout on the metabolism for optically pure d-lactate production by *Escherichia coli*. Applied Microbiol Biotechnol 64:367–75
207. Zhu J, Shimizu K (2005) Effect of a single-gene knockout on the metabolic regulation in *E. coli* for d-lactate production under microaerobic conditions. Metabolic Eng 7:104–115
208. Vemuri GN, Altman E, Sangurdekar DP, Khodursky AB, Eiteman MA (2006) Overflow metabolism in *Escherichia coli* during steady-state growth: transcriptional regulation and effect of the redox ratio. Appl Environ Microbiol 72(5):3653–3661
209. Valgepea K, Adamberg K, Seiman A, Vilu R (2014) Escherichia coli achieves faster growth by increasing catalytic and translation rates of proteins. Mol BioSyst 9:2344–2358
210. Nizam SA, Zhu J, Ho PY, Shimizu K (2009) Effects of *arcA* and *arcB* genes knockout on the metabolism in *Escherichia coli* under aerobic condition. Biochem Eng J 44(2–3):240–250
211. Vemuri GN, Eiteman MA, Altman E (2006) Increased recombinant protein production in *Escherichia coli* strains with overexpressed water-forming NADH oxidase and a deleted ArcA regulatory protein. Biotechnol Bioeng 94(3):538–542
212. Nizam SA, Shimizu K (2008) Effects of *arc* A and *arc* B genes knockout on the metabolism in *Escherichia coli* under anaerobic and microaerobic conditions. Biocheml Eng J 42:229–236

Acid Shock

213. Foster JW (2004) Escherichia coli acid resistance: tales of an amateur acidophile. Nat Rev Microbiol 2(11):898–907
214. Stincone A, Daudi N, Rahman AS et al (2011) A systems biology approach sheds new light on *Escherichia coli* acid resistance. Nucleic Acids Res 39(17):7512–7528
215. Richard HT, Foster JW (2003) Acid resistance in *Escherichia coli*. Adv Appl Microbiol 52:167–186
216. Richard HT, Foster JW (2004) *Escherichia coli* glutamate-and arginine-dependent acid resistance systems increase internal pH and reverse transmembrane potential. J Bacteriol 86 (18):6032–6041
217. Gong S, Richard H, Foster JW (2003) YjdE (AdiC) is the arginine: agmatine antiporter essential for arginine-dependent acid resistance in *Escherichia coli*. J Bacteriol 185 (15):4402–4409
218. Iyer R, Williams C, Miller C (2003) Arginine-agmatine antiporter in extreme acid resistance in *Escherichia coli*. J Bacteriol 185:6556–6561
219. Martin-Galiano AJ, Ferrandiz MJ, de La Campa AG (2001) The promoter of the operon encoding the F0F1 ATPase of *Streptococcus pneumonia* is inducible by pH. Molr Microbiol 41:327–338

220. Castanie-Cornet MP, Foster JW (2001) *Escherichia coli* acid resistance: cAMP receptor protein and a 20 bp *cis*-acting sequence control pH and stationary phase expression of the *gadA* and *gadBC* glutamate decarboxylase genes. Microbiol 147:709–715
221. Marzan LW, Shimizu K (2011) Metabolic regulation of *Escherichia coli* and its *phoB* and *phoR* genes knockout mutants under phosphate and nitrogen limitations as well as at acidic condition. Microb Cell Fact 10:39
222. Greenberg JT, Monach P, Chou JH, Josephy PD, Demple B (1990) Positive control of a global antioxidant defense regulon activated by superoxide-generating agents in *Escherichia coli*. PNAS USA 87(16):6181–6185
223. Suziedeliene E, Suziedelis K, Garbenciute V, Normark S (1999) The acid-inducible asr gene in *Escherichia coli*: transcriptional control by the phoBR operon. J Bacteriol 181(7):2084–2093
224. Wu X, Altman R, Eiteman MA, Altman E (2013) Effect of overexpressing *nhaA* and *nhaR* on sodium tolerance and lactate production in *Escherichia coli*. J Biol Eng 7:3

Heat Shock

225. Kitagawa M, Miyakawa M, Matsumura Y, Tsuchido T (2002) *Escherichia coli* small heat shock proteins, IbpA and IbpB, protect enzymes from inactivation by heat and oxidants. Eur J Biochem 269(12):2907–2917
226. Sørensen HP, Mortensen KK (2005) Soluble expression of recombinant proteins in the cytoplasm of *Escherichia coli*. Microb Cell Fact 4:1
227. Hoffmann F, Weber J, Rinas U (2002 Metabolic adaptation of *Escherichia coli* during temperature-induced recombinant protein production: 1. Readjustment of metabolic enzyme synthesis. Biotechnol Bioeng 80(3):313–319
228. Tilly K, Erickson J, Sharma S, Georgopoulos C (1986) Heat shock regulatory gene *rpoH* mRNA level increases after heat shock in *Escherichia coli*. J Bacteriol 168(3):1155–1158
229. Tilly K, Spence J, Georgopoulos C (1989) Modulation of stability of the *Escherichia coli* heat shock regulatory factor ^{32}. J Bacteriol 171(3):1585–1589
230. Shin D, Lim S, Seok YJ, Ryu S (2001) Heat shock RNA polymerase (E ^{32}) is involved in the transcription of *mlc* and Crucial for Induction of the Mlc regulon by glucose in *Escherichia coli*. J Biol Chem 276(28):25871–25875
231. Hasan CM, Shimizu K (2008) Effect of temperature up-shift on fermentation and metabolic characteristics in view of gene expressions in *Escherichia coli*. Microb Cell Fact 7:35
232. Kumari S, Beatty CM, Browning DF et al (2000) Regulation of acetyl coenzyme A synthetase in *Escherichia coli*. J Bacteriol 182(15):4173–4179
233. Browning DF, Beatty CM, Wolfe AJ, Cole JA, Busby SJW (2002) Independent regulation of the divergent *Escherichia coli nrfA* and *acs*P1 promoters by a nucleoprotein assembly at a shared regulatory region. Mol Microbiol 43(3):687–701
234. Beatty CM, Browning DF, Busby SJW, Wolfe AJ (2003) Cyclic AMP receptor protein-dependent activation of the *Escherichia coliacs* P2 promoter by a synergistic class III mechanism. J Bacteriol 185(17):5148–5157
235. Browning DF, Beatty CM, Sanstad EA, Gunn KE, Busby SJW, Wolfe AJ (2004) Modulation of CRP-dependent transcription at the *Escherichia coli* acsP2 promoter by nucleoprotein complexes: anti-activation by the nucleoid proteins FIS and IHF. Mol Microbiol 51(1):241–254
236. Privalle CT, Fridovich I (1987) Induction of superoxide dismutase in *Escherichia coli* by heat shock. PNAS USA 84(9):2723–2726

Cold Shock

237. Etchegaray J-P, Inoue M (1999) CspA, CspB, and CspG, major cold shock proteins of *Escherichia coli*, are induced at low temperature under conditions that completely block protein synthesis. J Bacteriol 181(6):1827–1830
238. White-Ziegler CA, Um S, Perez NM, Berns AL, Malhowski AJ, Young S (2008) Low temperature (23 °C) increases expression of biofilm-, cold shock-, and RpoS-dependent genes in *Escherichia coli* K-12. Microbiol 154:148–166
239. Kim Y-H, Han KY, Lee K, Lee J (2005) Proteome response of *Escherichia coli* fed-batch culture to temperature downshift. Appl Microbiol Biotechnol 68:786–793

Solvent Stress Regulation

240. Dunlop MJ (2011) Engineering microbes for tolerance to next-generation biofuels. Biotechnol Biofuels 4:32
241. Isken S, de Bont JAM (1998) Bacteria tolerant to organic solvents. Extremophiles 2:229–238
242. Ramos JL, Duque E, Gallegos M-T, Godoy P, Ramos-Gonzalez MI, Rojas A, Teran W, Segura A (2002) Mechanisms of solvent tolerance in Gram-negative bacteria. Annu Rev Microbiol 56:743–768
243. Nicolaou SA, Gaida SM, Papoutsakis ET (2010) A comparative view of metabolite and substrate stress and tolerance in microbial bioprocessing: from biofuels and chemicals, to biocatalysis and bioremediation. Metab Eng 12:307–331
244. Takatsuka Y, Chen C, Nikaido H (2010) Mechanism of recognition of compounds of diverse structures by the multidrug efflux pump AcrB of *Escherichia coli*. PNAS USA 107:6559–6565
245. Ankarloo J, Wikman S, Nicholls IA (2010) *Escherichia coli mar* and *acrAB* mutants display no tolerance to simple alcohols. Int J Mol Sci 11:1403–1412
246. Piper P (1995) The heat-shock and ethanol stress responses of yeast exhibit extensive similarity and functional overlap. FEMS Microbiol Lett 134:121–127
247. Rutherford BJ, Dahl RH, Price RE, Szmidt HL, Benke PI, Mukhopadhyay A, Keasling JD (2010) Functional genomic study of exogenous n-butanol stress in *Escherichia coli*. Appl Environ Microbiol 76:1935–1945
248. Tomas C, Beamish J, Papoutsakis E (2004) Transcriptional analysis of butanol stress and tolerance in *Clostridium acetobutylicum*. J Bacteriol 186:2006–2018
249. Fiocco D, Capozzi V, Goffin P, Hols P, Spano G (2007) Improved adaptation to heat, cold, and solvent tolerance in *Lactobacillus plantarum*. Appl Microbiol Biotechnol 77:909–915
250. Reyes LH, Almario MP, Kao KC (2011) Genomic library screens for genes involved in n-butanol tolerance in *Escherichia coli*. PLoS One 6:e17678
251. Sikkema J, de Bont JAM, Poolman B (1995) Mechanisms of membrane toxicity of hydrocarbons. Microbiol Rev 59:201–222
252. Holtwick R, Meinhardt F, Keweloh H (1997) cis-trans isomerization of unsaturated fatty acids: cloning and sequencing of the cti gene from Pseudomonas putida P8. Appl Environ Microb 63:4292–4297
253. Kiran M, Prakash J, Annapoorni S, Dube S, Kusano T, Okuyama H, Murata N, Shivaji S (2004) Psychrophilic Pseudomonas syringae requires transmonounsaturated fatty acid for growth at higher temperature. Extremophiles 8:401–410

Osmoregulation

254. Kramer R (2010) Bacterial stimulus perception and signal transduction: response to osmotic stress. Chem Res 10:217–229
255. Wood JM (2011) Bacterial osmoregulation: a paradigm for the study of cellular homeostasis. Ann Rev Microbiol 65:215–238
256. Walderhaug MO, Polarek JW, Voelkner P, Daniel JM, Hesse JE, Altendorf K, Epstein W (1992) KdpD and KdpE, proteins that control expression of the *kdpABC* operon, are members of the two-component sensor-effector class of regulators. J Bacteriol 174:2152–2159
257. Sugiura A, Hirokawa K, Nakashima K, Mizuno T (1994) Signal-sensing mechanisms of the putative osmosensor KdpD in *Escherichia coli*. Mol Microbiol 14:929–938
258. Hamann K, Zimmann P, Altendorf K (2008) Reduction of turgor is not the stimulus for the sensor kinase KdpD of *Escherichia coli*. J Bacteriol 190:2360–2367
259. Heermann R, Jung K (2010) The complexity of the 'simple' two-component system KdpD/KdpE in *Escherichia coli*. Microbiol Lett 304:97–106
260. Sevin DC, Sauer U (2014) Ubiquinone accumulation improves osmotic-stress tolerance in *Escherichia coli*. Nat Chem Biol 10:266–272
261. Piuri M, Sanchez-Rivas C, Ruzal SM (2005) Cell wall modifications during osmotic stress in *Lactobacullus casei*. J Appl Microbiol 98:84–95
262. Clarke CF, Rowat AC, Gober JW (2014) Is CoQ a membrane stabilizer? Nat Chem Biol 10:242–243
263. Calamita G, Bishai WR, Preston GM, Guggino WB, Agre P (1995) Molecular cloning and characterization of AqpZ, a water channel from *Escherichia coli*. J Biol Chem 270:29063–29066
264. Lamins LA, Rhoads DB, Epstein W (1981) Osmotic control of kdp operon expression in *Escherichia coli*. PNAS USA 78:464–468
265. Record MT Jr, Courtenary ES, Cayley DS, Guttman HJ (1998) Responses of *E. coli* to osmotic stress: large changes in amounts of cytoplasmic solutes and water. Trends Biochem Sci 23:143–148
266. Grothe S, Krogsrud RL, McClellan DJ, Milner JL, Wood JM (1986) Proline transport and osmotic response in *Escherichia coli* K-12. J Bacteriol 166:253–259
267. Von Weyman N, Nyyssola A, Reinikainen T, Leisola M, Ojamo H (2001) Improved osmotolerance of recombinant *Escherichia coli* by de novo glycine betaine biosynthesis. Appl Microbiol Biotechnol 55:214–218
268. Giaever HM, Styrvoid OB, Kaasen I, Strom AR (1998) Biochemical and genetic characterization of osmoregulatory trehalose synthesis in *Escherichia coli*. J Bacteriol 170:2841–2849
269. Elbein AD, Pan YT, Pastuszak I, Carroll D (2003) New insights on trehalose: a multifunctional molecule. Glycobiology 13:17R–27R
270. Klein W, Ehmann U, Boos W (2991) The repression of trehalose transport and metabolism in *Escherichia coli* by high osmolarity is mediated by trehalose-6-phosphate phosphatase. Res Microbiol 142:359–371
271. Turunen M, Olsson J, Dallner G (2004) Metabolism and function of coenzyme Q. Biochim Biophys ACTA Biomembr 1660:171–199

Biofilm, Motility and Quarum Sensing

272. Wang X, Dubey AK, Suzuki K, Baker CS, Babitzke P, Romeo T (2005) CsrA post-transcriptionally represses pgaABCD, responsible for synthesis of a biofilm polysaccharide adhesin of *Escherichia coli*. Mol Microbiol 56:1648–1663

273. Hengge R (2009) Principles of c-di-GMP signalling in bacteria. Nat Rev Microbiol 7:263–273
274. Barnhart MM, Chapman MR (2006) Curli biogenesis and function. Annu Rev Microbiol 60:131–147
275. Thomason MK, Fontaine F, De Lay N, Storzl G (2012) A small RNA that regulates motility and biofilm formation in response to changes in nutrient availability in *Escherichia coli*. Mol Microbiol 84(1):17–35
276. Jackson DW, Simecka JW, Romeo T (2002) Catabolite repression of *Escherichia coli* biofilm formation. J Bacteriol 184:3406–3410
277. Waters CM, Bassler BL (2005) Quorum sensing: cell-to-cell communication in bacteria. Annu Rev Cell Dev Biol 21:319–346
278. De Lay N, Gottesman S (2009) The Crp-activated small noncoding regulatory RNA CyaR (RyeE) links nutritional status to group behavior. J Bacteriol 191:461–476

Systems Biology Approach

279. Yu Matsuoka, Shimizu K (2015) Current status and future perspectives of kinetic modeling for the cell metabolism with incorporation of the metabolic regulation mechanism. Bioreses Bioprocess 2:4
280. Almquist J, Cvijovic M, Hatzimanikatis V, Nielsen J, Jirstrand M (2014) Kinetic models in industrial biotechnology-improving cell factory performance. Metab Eng 24:38–60
281. Chassagnole C, Noisommitt-Rizzi N, Schmid JW, Mauch K, Reuss M (2002) Dynamic modeling of the central carbon metabolism of *Escherichia coli*. Biotechnol Bioeng 79:53–73
282. Kadir TA, Mannan AA, Kierzek AM, McFadden J, Shimizu K (2010) Modeling and simulation of the main metabolism in *Escherichia coli* and its several single-gene knockout mutants with experimental verification. Microb Cell Fact 9:88
283. Nishio Y, Usuda Y, Matsui K, Kurata H (2008) Computer-aided rational design of the phosphotransferase system for enhanced glucose uptake in *Escherichia coli*. Mol Syst Biol 4:160
284. Usuda Y, Nishio Y, Iwatani S, Van Dien SJ, Imaizumi A, Shimbo K, Kageyama N, Iwahata D, Miyano H, Matsui K (2010) Dynamic modeling of *Escherichia coli* metabolic and regulatory systems for amino-acid production. J Biotechnol 147:17–30
285. Matsuoka Y, Shimizu K (2013) Catabolite regulation analysis of *Escherichia coli* for acetate overflow mechanism and co-consumption of multiple sugars based on systems biology approach using computer simulation. J Biotechnol 168:155−173

Concluding Remarks

286. Gutierrez-Rios RM, Rosenblueth DA, Loza JA, Huerta AM, Glasner JD, Blattner FR, Collado-Vides J (2003) Regulatory network of *Escherichia coli*: consistency between literature knowledge and microarray profiles. Genome Res 13:2435–2443
287. Maheswaran M, Forchhammer K (2003) Carbon-source-dependent nitrogen regulation in *Escherichia coli* is mediated through glutamine-dependent GlnB signaling. Microbiology 149:2163–2172
288. Quan JA, Schneider BL, Paulsen IT, Yamada M, Kredich NM, Saier MH Jr (2002) Regulation of carbon utilization by sulfur availability in *Escherichia coli* and *Salmonella typhimurium*. Microbiology 148:123–131
289. De Lorenzo V, Herrero M, Giovannini F, Neilands JB (1998) Fur (ferric uptake regulation) protein and CAP (catabolite-activator protein) modulate transcription of *fur* gene in *Escherichia coli*. Eur J Biochem 173:537–546

290. Zhang Z, Gosset G, Barabote R, Gonzalez CS, Cuevas WA, Saier MH Jr (2005) Functional interactions between the carbon and iron utilization regulators, Crp and Fur, *Escherichia coli*. J Bacteriol 187(3):980–990
291. Gyaneshwar P, Paliy O, McAuliffe J, Popham DL, Jordan MI, Kustu S (2005) Sulfur and nitrogen limitation in *Escherichia coli* K-12: specific homeostatic responses. J Bacteriol 187 (3):1074–1090
292. Dragosits M, Mozhayskiy V, Quinones-Soto S, Park J, Tagkopoulos I (2013) Evolutionary potential, cross-stress behavior and the genetic basis of acquired stress resistance in *Escherichia coli*. Mol Syst Biol 9:643
293. Weber H, Polen T, Heuveling J, Wendisch V, Hengge R (2005) Genome-wide analysis of the general stress response network in *Escherichia coli*: sigmaS-dependent genes, promoters, and sigma factor selectivity. J Bacteriol 187:1591–1603
294. Jenkins D, Schultz J, Matin A (1988) Starvation-induced cross protection against heat or H_2O_2 challenge in *Escherichia coli*. J Bacteriol 170:3910–3914
295. Jenkins D, Auger E, Matin A (1991) Role of RpoH, a heat shock regulator protein, in *Escherichia coli* carbon starvation protein synthesis and survival. J Bacteriol 173:1992–1996
296. Gunasekera T, Csonka L, Paliy O (2008) Genome-wide transcriptional responses of *Escherichia coli* K-12 to continuous osmotic and heat stresses. J Bacteriol 190:3712–3720
297. White-Ziegler C, Um S, Pérez N, Berns A, Malhowski A, Young S (2008) Low temperature (23 degrees C) increases expression of biofilm-, cold-shock- and RpoS-dependent genes in *Escherichia coli* K-12. Microbiology 154:148–166
298. Rutherford B, Dahl R, Price R, Szmidt H, Benke P, Mukhopadhyay A, Keasling J (2012) Functional genomic study of exogenous n-butanol stress in *Escherichia coli*. Appl Environ Microbiol 76:1935–1945
299. Rabinowitz J, Silhavy TJ (2012) Metabolite turns master regulator. Nature 500:283–284
300. Klumpp S, Zhang Z, Hwa T (2009) Growth rate-dependent global effects on gene expression in bacteria. Cell 139:1366–1375
301. Scott M, Gunderson CW, Mateescu EM, Zhang Z, Hwa T (2010) Interdependence of cell growth and gene expression: origins and consequences. Science 330:1099–1102
302. Koebman BJ, Westerhoff HV, Snoep JL, Nilsson D, Jensen PR (2002) The glycolytic flux in *Escherichia coli* is controlled by the demand for ATP. J Bacteriol 184:3909
303. Luo Y, Zhang T, Wu H (2014) The transport and mediation mechanisms of the common sugars in *Escherichia coli*. Biotechnol Adv 32:905–919
304. Yao R, Shimizu K (2012) Recent progress in metabolic engineering for the production of biofuels and biochemical from renewable sources with particular emphasis on catabolite regulation and its modulation. Process Biochem 48(9):1409–1417
305. Aidelberg G, Towbin BD, Rothschild D, Dekel E, Bren A, Alon U (2014) Hierarchy of non-glucose sugars in *Escherichia coli*. BMC Syst Biol 8:133
306. Balsalobre C, Johansson J, Uhlin BE (2006) Cyclic AMP-dependent osmoregulation of crp gene expression in *Escherichia coli*. J Bacteriol 188(16):5935–5944
307. Zhang H, Chong H, Ching CB, Jiang R (2012) Random mutagenesis of global transcription factor cAMP receptor protein for improved osmotolerance. Biotechnol Bioeng 109(5):1165–1172
308. Basak S, Jiang R (2012) Enhancing *E. coli* tolerance towards oxidative stress via engineering its global regulator cAMP receptor protein (CRP). PLOS One 7(12):e51179
309. Basak S, Geng H, Jiang R (2013) Rewiring global regulator cAMP receptor protein (CRP) to improve *E. coli* tolerance towards low pH. J Biotechnol 173:68–75
310. Chong H, Yeow J, Wang I, Song H, Jiang R (2013) Improving acetate tolerance of *Escherichia coli* by rewiring its global regulator cAMP receptor protein (CRP). PLOS One 8 (10):e77422
311. Chong H, Huang L, Yeow J, Wang I, Zhang H, Song H, Jiang R (2013) Improving ethanol tolerance of Escherichia coli by rewiring its global regulator cAMP receptor protein (CRP). PLOS One 8(2):e57628

312. Zhang H, Chong H, Ching CB, Song H, Jiang R (2012) Engineering global transcription factor cyclic AMP receptor protein of *Escherichia coli* for improved 1-butanol tolerance. Appl Microbiol Biotechnol 94:1107–1117
313. Khankal R, Chin JW, Ghosh D, Cirino PC (2009) Transcriptional effects of CRP* expression in *Escherichia coli*. J Biol Eng 3:13
314. Gosset G, Zhang Z, Nayyar S, Cuevas WA, Saier MH Jr (2004) Transcriptome analysis of Crp-dependent catabolite control of gene expression in *Escherichia coli*. J Bacteriol 186 (11):3516–3524
315. Valgepea K, Adamberg K, Seiman A, Vilu R (2014) *Escherichia coli* achieves faster growth by increasing catalytic and translation rates of proteins. Mol BioSyst 9:2344–2358
316. Baste DJV, Nol K, Niedenfuhr Mendum TA, Hawkins ND, Ward JL, Beale MH, Wiechert W, McFadden J (2013) ^{13}C-Flux spectral analysis of host-pathogen metabolism reveals a mixed diet for intracellular *Mycobacterium tuberculosis*. Chem Biol 20:1–10
317. Shimizu K (2004) Metabolic flux analysis based on ^{13}C-labeling experiments and integration of the information with gene and protein expression patterns. Adv Biochem Eng Biotechnol 91:1–49
318. Shimizu K (2013) Bacterial cellular metabolic systems. Woodhead Publ. Co., Oxford

Adv Biochem Eng Biotechnol (2016) 155: 55–106
DOI 10.1007/10_2014_291

Published Online: 24 December 2014

Efficient Biocatalytic Synthesis of Chiral Chemicals

Zhi-Jun Zhang, Jiang Pan, Bao-Di Ma and Jian-He Xu

Abstract Chiral chemicals are a group of important chiral synthons for the synthesis of a series of pharmaceuticals, agrochemicals, and fine chemicals. In past decades, a number of biocatalytic approaches have been developed for the green and effective synthesis of various chiral chemicals. However, the practical application of these biocatalytic processes is still hindered by the lack of highly efficient and robust biocatalysts, which usually results in the low volumetric productivity and high cost of the bioprocesses. Further step forward of biocatalysis in industrial application strongly requires the development of versatile and highly efficient biocatalysts, aiming to increase the process efficiency and facilitate the downstream processing. Recently, the fast growth of genome sequences in the database in post-genomic era offers great opportunities for accessing numerous biocatalysts with practical application potential, and the so-called genome mining approach provides time-effective and highly specific strategy for the fast identification of target enzymes with desired properties and outperforms the traditional screening of soil samples for microbial enzyme producers of interest. A number of biocatalytic processes with industrial application potential were developed thereafter. Further development of protein engineering strategies, process optimization, and cooperative work between biologists, organic chemists, and engineers is expected to make biocatalysis technology the first choice approach for the eco-friendly, highly efficient, and cost-effective synthesis of chiral chemicals in the near future.

Keywords Biocatalysis · Data mining · Efficient biocatalytic synthesis · Chiral chemicals · High volumetric productivity

Z.-J. Zhang · J. Pan · B.-D. Ma · J.-H. Xu (✉)
Laboratory of Biocatalysis and Synthetic Biotechnology,
State Key Laboratory of Bioreactor Engineering,
East China University of Science and Technology, 130 Meilong Road,
Shanghai 200237, China
e-mail: jianhexu@ecust.edu.cn

Contents

1 Introduction

Chiral chemicals are usually required as key intermediates for the synthesis of a variety of pharmaceuticals, agrochemicals, food ingredients, flavors, and fine chemicals. Traditional chemical routes to chiral chemicals usually require harsh reaction conditions (e.g., elevated temperature, high pressure, strongly acidic, or basic condition etc.) and expensive metal-based catalysts, which have led to severe environmental problems. In the transformation of compounds bearing labile groups, tedious protection/deprotection steps are required, thereof leading to relatively low yields of the final products. Therefore, there is an urgent need for the development of green and cost-effective processes for the preparation of chiral chemicals. In past decades, there has been a significant rise in the application of biocatalysis for industrial settings since biocatalysts offer cheap, environmentally benign, excellently selective, and highly efficient alternatives to chemical routes [1–7].

However, the widespread industrial application of biocatalysis is still hindered by low volumetric productivity, unsatisfactory selectivity, and limited availability of robust biocatalyst [6]. The development of novel and robust biocatalysts still remains a great challenge in both academia and industry [8]. Great effort has been paid to develop powerful biocatalyst for biocatalytic processes with high substrate concentration in order to achieve cost-efficiency and competitiveness in practical biotransformation [6]. Traditional biocatalyst discovery is based on screening soil samples from various sources for microorganisms producing desired enzyme activity [9, 10]; however, this strategy is always time-consuming (typically 1–2 years), and the enzyme expression level in the original host strain is usually very low resulting in insufficient catalytic efficiency. Therefore, the cloning and over-expression of the target enzyme in a suitable host organism and even further protein engineering of the enzyme to meet specific application requirement is always necessary [11, 12]. Most importantly, only less than 1 % of microbes in the environment are culturable, limiting the versatility of biocatalysts explored. Due to the exponential growth of genome sequences in the database in post-genomic era, and most of them are uncharacterized for their definite biological functions, data mining offers an unprecedented opportunity for accessing novel and useful biocatalysts with industrial application potential [13, 14].

For data mining, the gene sequences of the already-known enzymes with best performance in the specific reaction are preferentially used as templates for BLAST in GenBank or SWISS-PROT database, then a series of gene sequences (preferentially uncharacterized before) with moderate sequence similarity to known enzymes are considered to be possible candidates. Multiple sequence alignment of the candidate sequences with known enzyme sequences with respect to key motifs, conserved regions, and catalytically important residues can further increase the successful rate of hits. If possible, three-dimensional structure analysis might be applied. Since the chosen candidate sequences only shows moderate similarity to known enzymes (usually lower than 80 %), they are considered to be relatively novel enzymes. The finally picked target genes are then obtained through PCR amplification from the genome DNA of the target strain or through gene synthesis, cloned into suitable expression vectors, and transformed into host strains. After protein expression and functional screening, one can get the suitable biocatalysts with desired properties.

In this chapter, efficient biocatalytic synthesis of a series of chiral pharmaceutical intermediates including carboxylic acids, epoxides, and alcohols will be discussed focusing on the newly developed biocatalytic processes using biocatalysts.

2 Bioresolution with Esterases and Lipases

Carboxylic ester-hydrolyzing enzymes, which are ubiquitous in animals, plants, and microorganisms, can catalyze the hydrolysis or formation of the ester bonds. Of them, lipases (EC 3.1.1.1, triacylglycerol hydrolases) and esterases (EC 3.1.1.3, carboxyl ester hydrolases) are two groups of biocatalysts that are widely used in industry [15–18]. Esterase and lipase can be distinguished by substrate specificity and interfacial activation [19]. The esterases hydrolyze soluble substrates such as triglycerides with fatty acids shorter than C_6, while the water-insoluble triglycerides with long-chain fatty acids are typical substrates for lipases. Furthermore, a minimum substrate concentration was needed for high activity of lipase because of the hydrophobic "lid" covering the active site of lipase.

The interests in esterase and lipase from academia and industry mainly reside in their desirable properties for practical application, such as high enantioselectivity toward a variety of substrates, robustness, and cofactor independence. They have been successfully applied in numerous industrial processes [20–23], including detergents, oils and fats, cheese making, and pharmaceutical industry. Enantioselective hydrolysis, transesterification, or synthesis of the single enantiomer ester from the racemic mixtures by esterase or lipase provides an attractive approach for the production of chiral chemicals. The intrinsic disadvantage of a maximum theoretical yield of 50 % in kinetic resolution can be overcome by the dynamic kinetic resolution (DKR) or stereoinversion of the unwanted enantiomer [24–27]. DKR combines the enzymatic resolution and racemization of the substrate in situ and has attracted great interest [28–30]. Therefore, the synthetic potential of esterase and lipase was pushed forward to a certain extent and the synthetic applications of these groups of biocatalyst are well reviewed [15, 18, 21–23]. Herein, we focus on the synthesis of important chiral chemicals by the recently exploited esterase or lipase.

2.1 *Optically Pure Alcohols*

2.1.1 *l*-Menthol

l-Menthol is one of the most important flavor components and widely used in the confectionary goods, pharmaceuticals, oral healthcare products, cosmetics, etc. Enzymatic resolution via enantioselective esterification/transesterification of racemic menthol in organic solvents [31, 32] or enantioselective hydrolysis of racemic menthol esters in aqueous medium [33, 34] is an extensively investigated approach to optically pure *l*-menthol. A prominent biocatalyst for the efficient preparation of *l*-menthol is the high substrate concentration-tolerable esterase from *Bacillus subtilis* ECU0554 (BSE), which was newly isolated from soil [35]. The low activity and poor stability of *B. subtilis* whole cells under operational conditions were overcome by overexpression of the *B. subtilis* esterase in *E. coli* BL21(DE3) [36] and immobilization by cross-linked enzyme aggregates [37], respectively. The cross-linked enzyme aggregates of BSE were recycled for the enantioselective hydrolysis of *dl*-menthol acetate with high substrate load (200 g/L, *ca.* 1.0 M) for 10 rounds, giving *l*-menthol with >94 % *ee* at conversion of >40 % (Fig. 1). In further work, the versatile BSE was also successfully employed for the production of (*S*)-1-phenyl-1,2-ethanediol [38] and *m*-substituted 1-phenylethanol acetates [39], which are valuable intermediates for pharmaceuticals and other fine chemicals.

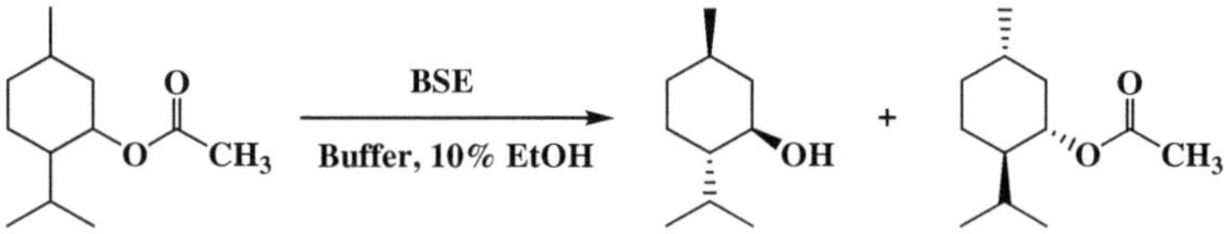

Fig. 1 Enzymatic resolution of *dl*-menthol acetate using BSE

2.1.2 (*S*)-1,2-*O*-Isopropylideneglycerol

(*S*)-1,2-*O*-Isopropylideneglycerol (IPG) is an important primary alcohol and serves as a starting material for the synthesis of β-adrenoceptor antagonists, prostaglandins, or leukotrienes. (*R*,*S*)-IPG could be selectively esterified with butyric acid in *n*-heptane by dry mycelia of *Rhizopus oryzae* and *Aspergillus oryzae*, with low enantiomeric ratio (*E*-value) of 3.4 and 8.0, respectively [40]. The yeast *Kluyveromyces marxianus* exhibited preference for the hydrolysis of (*S*)-IPG esters with moderate enantioselectivity ($E = 28$). The hydrolysis of (*R*,*S*)-IPG acetate by whole cells of the yeast was conducted in a membrane reactor, in which an ultrafiltration membrane (cut-off 10,000 Da) was used to recover the cells and released enzymes when the ee_s reached 100 %. The repeated-batch operation in the membrane reactor was run for 20 cycles, and enantiomerically pure (*R*)-IPG acetate of 19.2 g/L was recovered from 60 g/L of racemic mixture [41]. However, most of the esterases or lipases involved in the resolution of IPG ester preferred to hydrolyze (*R*)-IPG esters leaving the (*S*)-IPG esters untouched and this biocatalytic process has been extensively studied for the preparation of (*S*)-IPG (Fig. 2) [42–44]. An interesting investigation was the enantioselective hydrolysis of benzoyl-l,2-*O*-isopropylideneglycerol by *Bacillus coagulans* NCIB 9365, 1.50 g of (*S*)-IPG with 88 % *ee* was obtained from 5 g/L of benzoyl-l,2-*O*-isopropylideneglycerol at 1-L scale under the optimized reaction conditions [45]. Enzyme purification study revealed that the insufficient enantioselectivity of *B. coagulans* resting cells was attributed to the existence of two different enzymes: The partially purified enzyme A is thermostable and enantioselective toward IPG ester, while the thermolabile enzyme B is not enantioselective [46]. A simple heat treatment of the whole cells at 65 °C for 1 h remarkably increased the enantioselectivity ($E = 80-100$ for (*R*)-benzoyl-l,2-*O*-IPG, 95–96 % ee_p). Repeated-batch reaction was performed in the aforementioned membrane reactor for the preparation of enantiopure IPG by the heat-treated cells of *B. coagulans*, and no obvious activity loss was observed for 11 cycles. (*S*)-1,2-*O*-Isopropylideneglycerol benzoate of 9.55 g/L could be recovered from 24.0 g/L of

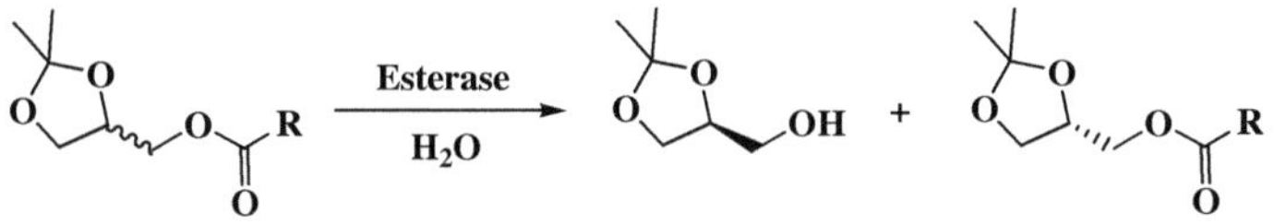

Fig. 2 Esterase-catalyzed resolutions of IPG esters

racemic substrate [47]. Another promising candidate catalyst, *E. coli* esterase YbfF, which shows 124/9.3 U/mg for (*R*)/(*S*)-IPG butyrate and 31/1.3 U/mg for (*R*)/(*S*)-IPG caprylate was reported in 2011 [48]. The moderate enantioselectivity of YbfF was improved by site-directed saturation mutagenesis and resulted in a mutant, W235I, with *E*-values of 38 and 77 for IPG butyrate and IPG caprylate, respectively [49].

2.1.3 Optically Pure Tertiary Alcohols

Optically pure tertiary alcohols (TAs) represent a group of important building blocks for the synthesis of various chiral chemicals and valuable pharmaceuticals. Enzymatic preparation of the enantiopure TAs has gained great interest due to the harsh reaction conditions of the chemical methods [50, 51]. It is still a great challenge for the resolution of sterically hindered TAs [52–54] even though the esterase- or lipase-catalyzed kinetic resolution is the most practical approach and a standard procedure for the synthesis of enantiopure secondary and primary alcohols. Furthermore, there is no corresponding ketone to be reduced to a tertiary alcohol although asymmetric reduction represents a complementary route to enzymatic resolution for the preparation of optically active secondary alcohols [6]. Thus, enzymatic kinetic resolution of TAs is of particular importance and gains great attention. The observation that the GGGX motif in the oxyanion hole is crucial for enzyme activity toward TAs [53, 55] facilitated the discovery of several GGGX-type α/β-hydrolases with hydrolytic activity toward esters of TAs [56–58]. However, the GGGX-type α/β-hydrolases exhibited low-to-moderate enantioselectivity. Modern protein engineering strategies have substantially expanded the toolbox available for the preparation of enantiopure TAs. Rational design and site-directed saturation mutagenesis have yielded several mutants of the esterase BS2 from *B. subtilis* including the G105A [55] and E188D [59] with excellent

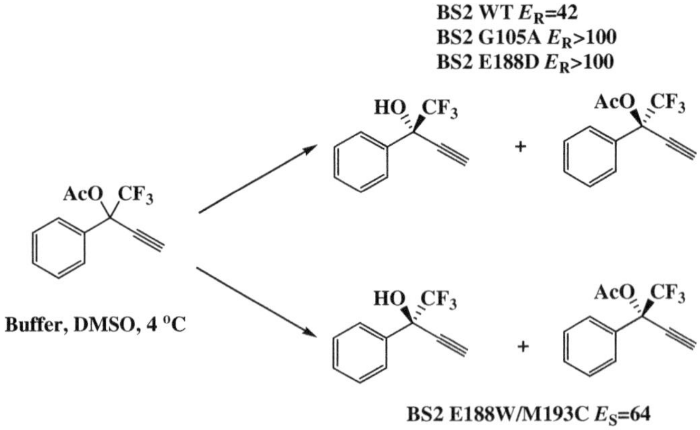

Fig. 3 Kinetic resolution of 1,1,1-trifluoro-2-phenyl-but-3-yn-1-yl acetate using the wild-type esterase and its mutants

R enantioselectivity ($E > 100$) toward 1,1,1-trifluoro-2-phenyl-but-3-yn-1-yl acetate (Fig. 3). A double-mutant, E188W/M193C, with inversed enantioselectivity ($E_S = 64$) was created by focused directed evolution approach thereafter [60]. Another synthetically useful biocatalyst (mutant EstA-AGA) with excellent enantioselectivity was provided by engineering of the *Paenibacillus barcinonensis* esterase (EstA) based on the structure-guided alignment [61]. The EstA–AGA was used to resolve the racemic 1,1,1-trifluoro-2-phenyl-but-3-yn-1-yl acetate in preparative scale with 91 % ee_s and 99 % ee_p, demonstrating the feasibility of the EstA-AGA-catalyzed kinetic resolution for the synthesis of chiral TAs.

2.2 *Optically Pure Carboxylic Acids*

2.2.1 (*S*)-(+)-2,2-Dimethylcyclopropane Carboxylic Acid

(*S*)-(+)-2,2-Dimethylcyclopropane carboxylic acid [(*S*)-(+)-DMCPA] is a key precursor for cilastatin, an excellent renal dehydropeptidase-I inhibitor. Several approaches, including the chemical asymmetric synthesis, chemical, or enzymatic resolution, have been developed to prepare (*S*)-(+)-DMCPA [62–64]. Of these methods, the esterase- or lipase-catalyzed enantioselective hydrolysis of 2,2-dimethylcarboxylate ester represents an effective and environmentally benign approach. The commercially available Novozyme 435 was utilized for the enantioselective hydrolysis of ethyl-2,2-dimethylcyclopropanecarboxylate at substrate concentration of 60 mM. In repeated-batch operations, glutaraldehyde-modified Novozyme 435 retained 76 % of its original activity after 10 repeated cycles and the ee_p was kept above 98 % throughout the process [65]. Recently, a new bacterial strain *Rhodococcus* sp. ECU1013 [66] was isolated from soil for the enantioselective hydrolysis of (*S*)-DMCPA esters from their racemic counterparts, providing an alternative useful biocatalyst for the production of (*S*)-DMCPA. By using the resting cells of *Rhodococcus* sp. ECU1013 as catalyst, up to 400 mM (±)-DMCPM was enantioselectively hydrolyzed into (*S*)-(+)-DMCPA in an organic–aqueous biphasic system (Fig. 4), with an isolated yield of 38 and 99 % ee_p. Further, heterogeneous overexpression and activity improvement of this newly discovered esterase by protein engineering are under progress in our laboratory.

2.2.2 (2*S*,3*R*)-3-Phenylglycidate Methyl Ester

(2*S*,3*R*)-3-Phenylglycidate methyl ester (PGM) is a key intermediate for the synthesis of a potent anticancer drug Taxol®. Low-to-moderate enantioselectivity and the difficulty in separation of the desired product restricted the resolution of PGM via transesterification [67, 68]. The whole cells of *Pseudomonas putida* have been reported for the enantioselective hydrolysis of (±)-PGM at substrate concentration of 50–60 mM, furnishing (2*S*,3*R*)-PGM with 99 % *ee* [69]. Recently,

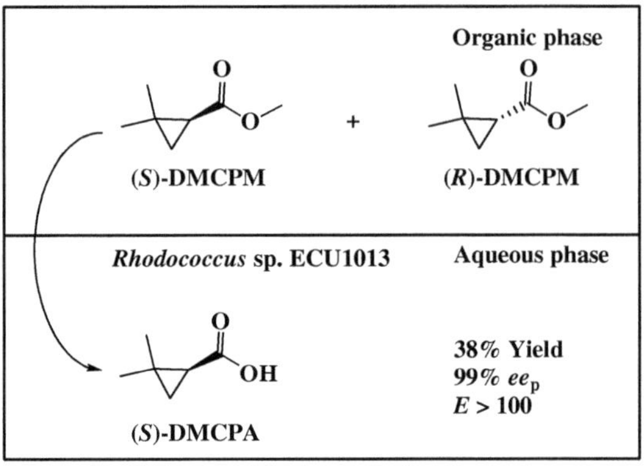

Fig. 4 Enzymatic resolution of (±)-DMCPM in organic–aqueous biphasic system

Zhou et al. reported a newly isolated bacterial strain, *Enterobacter* sp. ECU1107 for the enantioselective hydrolysis of (2*R*,3*S*)-PGM with substrate concentration of 600 mM [70]. The recovery of the desired (2*S*,3*R*)-PGM in this process could be facilitated through the decomposition of the unstable product, (2*R*,3*S*)-3-phenylglycidic acid (Fig. 5). The great potential of *Enterobacter* sp. ECU1107 for industrial production of the key precursor of pharmaceutically important Taxol was further demonstrated by the reaction on a scale of 1.0 L, yielding 11.6 g (2*S*,3*R*)-PGM with >99 % *ee*.

2.2.3 Key Intermediates to Pregabalin

Pregabalin, a marketed GABA analog, is used for the treatment of neuropathic pain and partial seizures [169] and has been launched by Pfizer as Lyrica®. The chemoenzymatic process involving biocatalytic resolution of *rac*-2-carboxyethyl-3-cyano-5-methylhexanoic acid ethyl ester (CNDE) has been considered to be the most cost-effective and greenest route for pregabalin [71]. The enzymatic resolution process adopted a commercially available lipase from *Thermomyces lanuginosus*

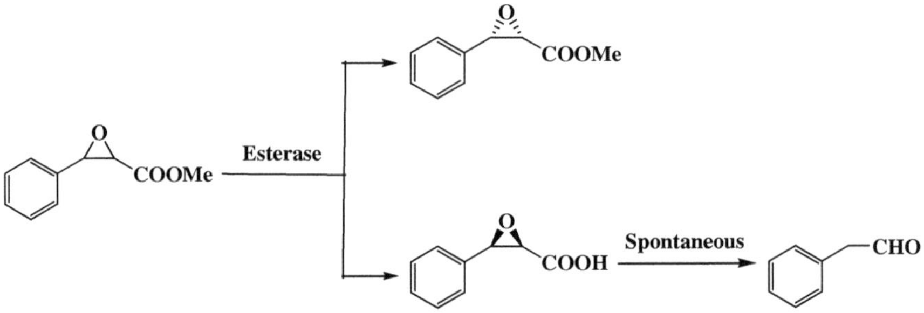

Fig. 5 Enantioselective hydrolysis of racemic methyl trans-3-phenylglycidate [(±)-PGM] using whole cells of *Enterobacter* sp. ECU1107 for the production of optically pure (2*S*,3*R*)-PGM

(Lipolase) to prepare the key enantiopure intermediate (*S*)-3-cyano-2-(ethoxycarbonyl)-5-methylhexanoic acid (*ee* > 98 %) with high yields and unprecedented high substrate load (765 g/L, 3.0 M) (Fig. 6). The undesired (*R*)-CNDE could readily be racemized using sodium ethoxide in ethanol at 80 °C. The enzymatic resolution was also tried for pilot runs at 900 kg scale and manufactured at 3.5 tons to demonstrate the scalability and consistency in performance. Besides the commercial lipase Lipolase, a newly isolated strain, *Morgarella morganii* ZJB-09203, could also be used for the resolution of CNDE [72]. Recently, an efficient route to pregabalin with higher atom economy was developed through the preparation of (*S*)-3-cyano-5-methylhexanoic acid ethyl ester [73]. The commercial lipase PS (Amano) from *Pseudomonas cepacia* was demonstrated to be the best enzyme for the hydrolytic resolution (Fig. 7). The substrate load was as high as 2.0 M (366 g/L), and (*S*)-3-cyano-5-methylhexanoic acid ethyl ester was produced in 99 % *ee* and 44.5 % yield.

2.3 Optically Pure Hydroxy Acids

Because of the dual functionality, optically pure hydroxy acids are versatile chiral synthons of particular interest in pharmaceutical industries. Optically active 2-hydroxy-phenyl acetic acid and its derivatives are the most important hydroxy acids. For example, (*R*)-2-hydroxy-2-(2′-chlorophenyl) acetic acid is the key chiral intermediate for the synthesis of (*S*)-clopidogrel, a platelet aggregation inhibitor. The enzymatic resolution catalyzed by esterase or lipase is a competitive method among the enzymatic routes to enantiomerically pure aromatic α-hydroxy acids [74].

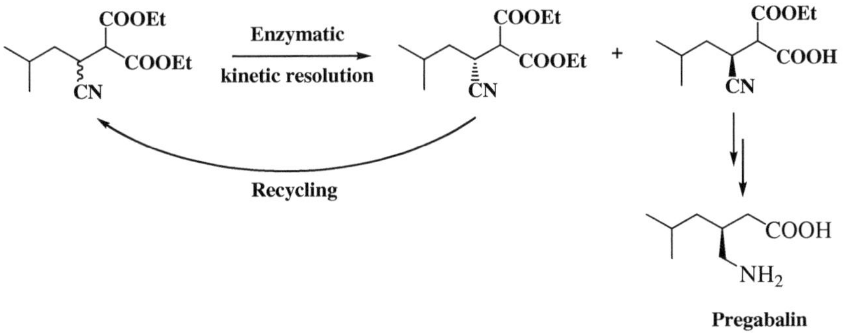

Fig. 6 Lipolase-catalyzed resolution of *rac*-2-carboxyethyl-3-cyano-5-methylhexanoic acid ethyl ester

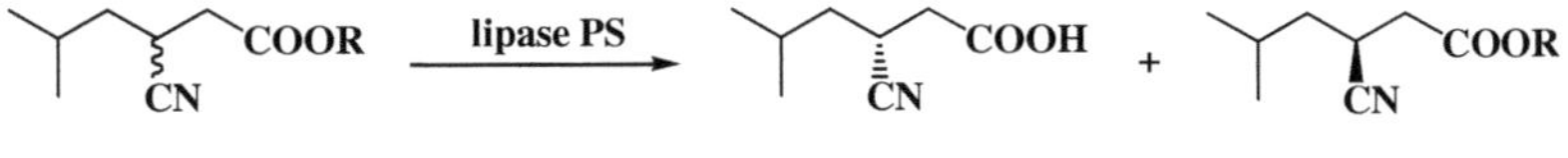

Fig. 7 Lipase-mediated resolution of *rac*-3-cyano-5-methylhexanoic acid ester

Chiral 2-hydroxy-phenyl acetic acid and its derivatives can be prepared through the enantioselective synthesis or hydrolysis of hydroxy esters or *O*-acetylated hydroxy acids (Fig. 8). The ester group can be hydrolyzed directly from racemic hydroxy esters [75, 76] or the *O*-protected hydroxy esters [77]. The enantioselective synthesis or hydrolysis of *O*-acetylated hydroxy acids receives more interest and several promising catalysts including the *P. putida* esterase [78, 79], and commercially available lipases [80, 81] have been reported for this bioprocess in last decades. A series of substituted mandelic acids were enantioselectively acetylated by lipase PS (Amano) using vinyl acetate as acyl donor [81]. A thermostable *P. putida* esterase, rPPE01, which was screened and cloned for the resolution of 2-acetoxy-phenyl acetic acid and its derivatives, exhibited excellent enantioselectivity ($E > 200$) to a series of acetylated aromatic α-hydroxy acids at a substrate concentration of 100 mM [79]. In further work, the low activity was improved by semi-rational design, giving a single-point variant (rPPE01-W187H) with remarkably increased activity and excellent enantioselectivity [82]. Meanwhile, the biocatalyst deactivation was alleviated by carefully selection of suitable substrate counterion, among the five counterions tested, K^+ showed the best stabilization effect. Finally, the resolution of 500 mM racemic potassium 2-acetoxy-2-(2′-chlorophenyl)acetic acid was successfully carried out with merely 0.5 g/L of lyophilized cells, and the conversion reached 49.9 % after 15 h with >99 % ee_p and 98.7 % ee_s (Fig. 9).

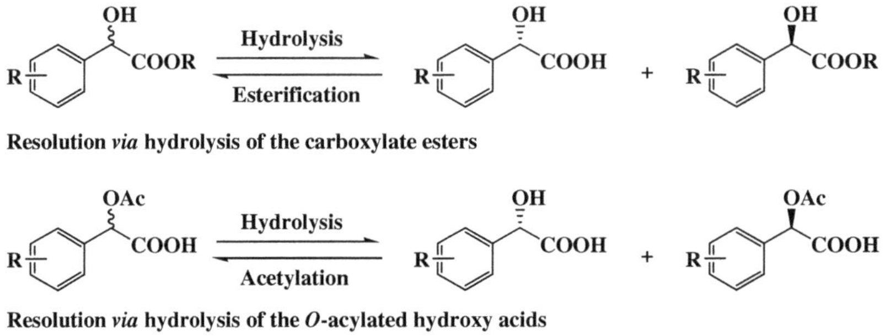

Fig. 8 Enzymatic resolution of 2-hydroxy-phenyl acetic acid and its derivatives

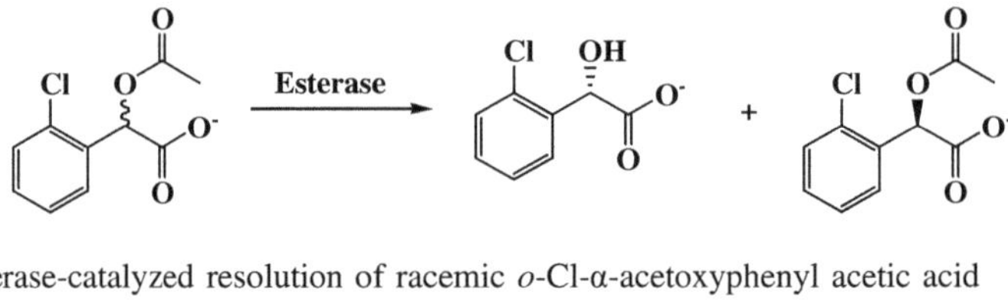

Fig. 9 Esterase-catalyzed resolution of racemic *o*-Cl-α-acetoxyphenyl acetic acid

3 Chiral Synthesis with Epoxide Hydrolases

Enantiopure epoxides, as well as their corresponding vicinal diols are important chiral building blocks in organic synthesis. Epoxide hydrolase (EH)-catalyzed stereoselective hydrolysis of racemic or *meso*-epoxides are important methods for the preparation of enantiopure epoxides or corresponding vicinal diols.

Epoxides have two adjacent oxirane carbon atoms, and EH-catalyzed nucleophilic attack can occur on both carbon atoms, thereof EHs show not only enantioselectivity but also regioselectivity. According to the reaction mechanism, EHs catalyze the ring-opening via a nucleophilic S_N2 attack by an aspartate residue that forms a transient covalent intermediate, and then, an activated water molecule attacks the carbonyl moiety of the ester-intermediate and releases the product diol. Therefore, inversion of configuration could happen when the attack was performed on a more substituted position.

Accordingly, besides the conventional kinetic resolution, EHs can also catalyze the enantioconvergent hydrolysis of racemic epoxides [83–85], once the two epoxide enantiomers are attacked at different positions as shown in Scheme 1. In contrast to traditional kinetic resolution, in which only 50 % theoretical yield is available, 100 % theoretical yield could be obtained in an enantioconvergent process.

3.1 Styrene Oxide and Its Derivatives

Enantiopure styrene oxide and its derivatives are precursors of various pharmaceuticals and liquid crystal materials. They possess a benzylic carbon atom, which facilitates the formation of a carbo-cation stabilized by the adjacent aromatic ring. As a result, attack at the benzylic carbon is electronically, though sterically impeded. Thus, mixed regiochemical pathways (i.e., attack at both oxirane carbon atoms) are particularly easy within this group of substrates, and an enantioconvergent process could be expected.

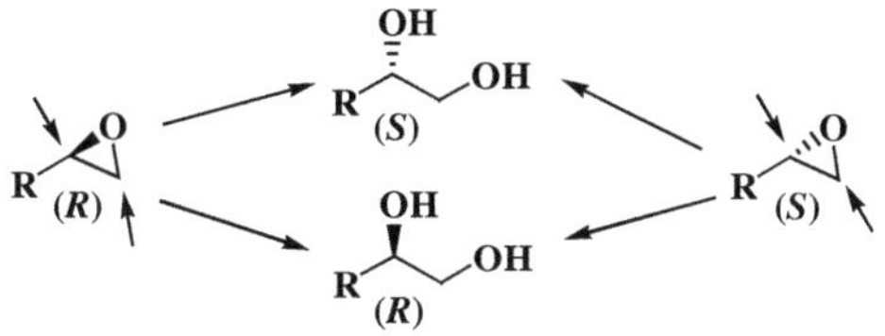

Scheme 1 Enantioconvergent hydrolysis of racemic epoxides

3.1.1 (*R*)-1-Phenyl-1,2-ethanediol

Recombinant *Pichia pastoris* expressing *Rhodotorula glutinis* EH shows high hydrolytic activity toward (*R*)-styrene oxide. Kinetic resolution of styrene oxide was conducted in a styrene oxide–aqueous biphasic system with a substrate load of 526 mM using whole cells of recombinant *P. pastoris*. Optically pure (*S*)-styrene oxide (>98 % *ee*) was obtained with 36 % yield [86]. After reaction optimization, (*S*)-styrene oxide of 98 % *ee* was formed with 41 % yield from 1.8 M *rac*-styrene oxide at pH 8.0, 4 °C in the presence of 40 % (*v*/*v*) Tween-20 and 5 % (*v*/*v*) glycerol [87].

Several enantioconvergent processes have been developed for the preparation of (*R*)-phenyl-1,2-ethanediol, including the combination of EHs from *Aspergillus niger* and *Bacillus sulfurescens* [88]; *Agrobacterium radiobacter* and *Solanum tuberosum* [89]; *A. niger* or *R. glutinis* and *Caulobacter crescentus* [90]; *C. crescentus*, and marine fish *Mugil cephalus* [91]. In these processes, the EHs from *A. niger*, *A. radiobacter*, *R. glutinis,* and *M. cephalus* preferentially attack the terminal carbon of the oxirane ring with retention of the stereochemistry, and (*R*)-diol was formed. On the other hand, the EHs from *B. sulfurescens* and *C. crescentus* attack the benzylic carbon of the (*S*)-enantiomer, which also leads to the formation of (*R*)-diol.

3.1.2 (*R*)-*p*-Nitro Styrene Oxide

(*R*)-*p*-Nitro styrene oxide [(*R*)-*p*NSO] and its corresponding diol are important precursors of (*R*)-Nifenalol, a β-adrenergic blocker with antianginal and antiarrhythmic properties.

A. niger EH (*An*EH) can catalyze the kinetic resolution of *p*NSO, affording (*S*)-*p*NSO (96 % *ee*, 38 % yield) and (*R*)-diol (66 % *ee*, 49 % yield). After recrystallization, enantiopure (*R*)-diol was obtained with 32 % yield. In order to overcome the limitation of 50 % theoretical yield in the resolution process, acid hydrolysis of the remaining (*S*)-*p*NSO was investigated for the preparation of the corresponding (*R*)-diol with inversion of configuration at the stereogenic benzylic carbon atom and realized the desymmetric transformation of *rac*-*p*NSO into (*R*)-diol [92]. The produced (*R*)-diol was cyclized to (*R*)-*p*NSO and then condensated with isopropylamine, affording (*R*)-Nifenalol. The reaction details are shown in Fig. 10.

To obtain the (*R*)-diol with high optical purity, accurate tuning of these two consecutive reactions was necessary since the final *ee* of the diol was directly dependent upon the conversion ratio. According to calculation results based on the enantiomeric ratio, the acid hydrolysis of the remaining (*S*)-*p*NSO was initiated when ee_S reached 95 %, and (*R*)-diol was formed in 90 % yield with 83 % *ee*, and after recrystallization, optically pure (*R*)-diol with 73 % yield was afforded.

Two EHs which could catalyze the enantioconvergent hydrolysis of *p*NSO were discovered from mung bean. Interestingly, these two EHs showed complementary enantioselectivities but with identical regioselectivity. Both EHs can catalyze the complete conversion of *p*NSO to (*R*)-diol. By using the mung bean crude powder as

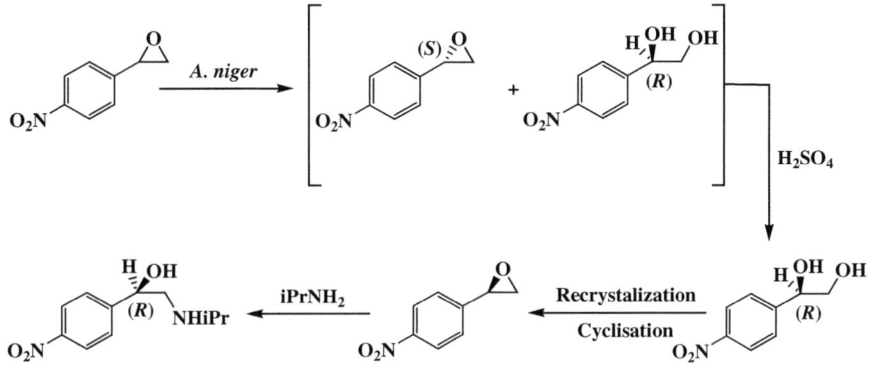

Fig. 10 Chemoenzymatic synthesis of (*R*)-Nifenalol

biocatalyst, optically active (*R*)-diol was produced with 82.4 % *ee* and 83.5 % yield, and after recrystallization, enantiopure (*R*)-diol was obtained with an overall yield of 68.7 % [93]. The crude EH was then immobilized by diatomite adsorption, and Tween-80 was introduced for better substrate dispersion. After process optimization, the *ee* of (*R*)-diol was increased to 84.7 % [94]. An enantioconvergent EH (*Vr*EH1) has been cloned from mung bean, which shows opposite regioselectivity toward (*S*)-*p*NSO (83 % to C_α) in contrast to (*R*)-*p*NSO (87 % to C_β) [95].

3.1.3 Chlorostyrene Oxides

Enantiopure chlorostyrene oxides (CSOs) and their corresponding diols are important building blocks for the synthesis of a series of biologically active molecules, including β_3-adrenergic receptor agonists SR 58611A or AJ-9677, and an effective *N*-methyl-*D*-aspartic acid receptor antagonist eliprodil, as shown in Fig. 11.

Recombinant *S. tuberosum* EH (*St*EH) showed high enantioselectivity and complementary regioselectivity to the enantiomers of 3- and 4-CSO. For both epoxides, the (*S*)-enantiomer was preferentially attacked at the (benzylic) more

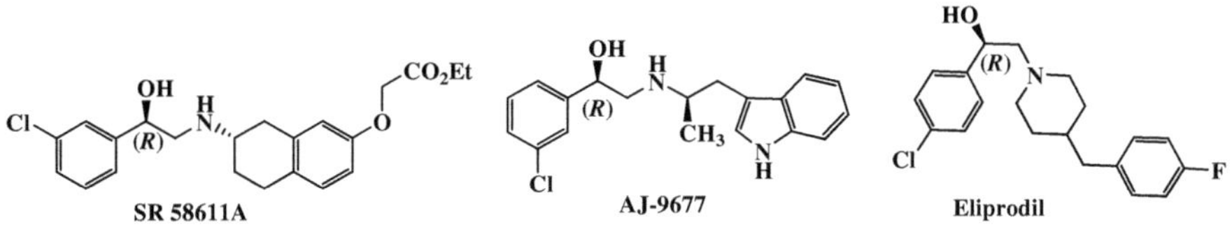

Fig. 11 Pharmaceuticals synthesized from enantiopure chlorostyrene oxides

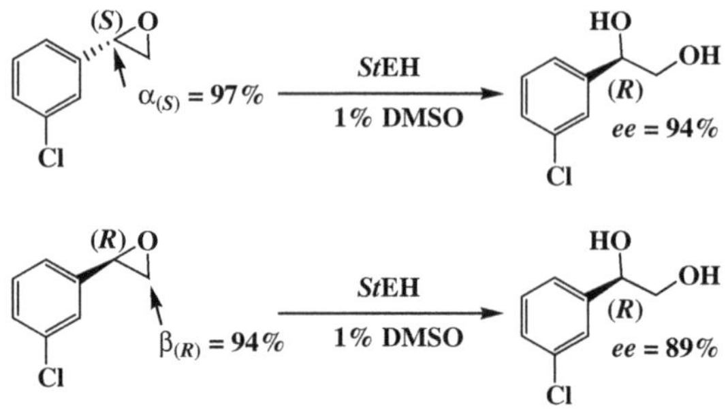

Fig. 12 Enantioconvergent hydrolysis of *rac*-3-chlorostyrene oxide by *St*EH

substituted carbon atom (97 %), whereas the (*R*)-antipode was attacked at the (terminal) less substituted carbon atom (94 %), as exemplified in Fig. 12. Enantioconvergent hydrolysis of 3- and 4-CSOs was performed, giving the corresponding (*R*)-diols with 91 and 74 % *ee*, respectively, after complete conversion of the substrates.

A preparative-scale biohydrolysis of *rac*-3-CSO was performed. For the purpose of minimizing the spontaneous hydrolysis of the 3-CSO and favor the stability of *St*EH, the reaction was performed at 20 °C. After complete conversion of the substrate, the homogeneous reaction mixture was filtered through an ultrafiltration membrane, and the recovered enzyme was reused. Totally, nine batches were performed at a substrate concentration of 10 g/L, the product was pooled, affording (*R*)-diol with 97 % *ee* and an isolated yield of 88 % [96].

*An*EH also shows high enantioselectivity to the 4-CSO and preferentially hydrolyzes the (*R*)-4-CSO, providing (*R*)-diols. Hydrolytic kinetic resolution of *rac*-4-chlorostyrene oxide was performed in heptane with an initial a_w of 0.9 [97]. Both the hydrolytic kinetic resolution efficiency and operational stability of *An*EH were found to be modest to excellent in various binary organic solvent mixtures of heptane and dioxane [98].

Since *St*EH and *An*EH showed complementary enantioselectivities, an enantioconvergent process for the enzymatic hydrolysis of 4-chlorostyrene oxide using a sequential bienzymatic strategy was adopted to realize the ideal 100 % yield of (*R*)-diol. In order to prevent the significant spontaneous hydrolysis, the enzymatic hydrolysis was conducted at 0 °C. *E*-value of about 100 was observed for both enzymes in a kinetic resolution process. As high as 2 M substrate could be efficiently resolved by *An*EH, while for *St*EH, substrate concentration above 200 mM was deleterious to the enzyme activity. Considering the *St*EH was an enantioconvergent EH and appeared to be more sensitive to inhibition by the (*R*)-diol formed, *St*EH was added first to completely transformed (*S*)-4-CSO before the addition of *An*EH. By this way, 0.2 M 4-chlorostyrene oxide was converted to (*R*)-diol with 96 % *ee* and 93 % yield [99]. The formed (*R*)-diol was a precursor for the preparation of (*R*)-eliprodil, as shown in Fig. 13.

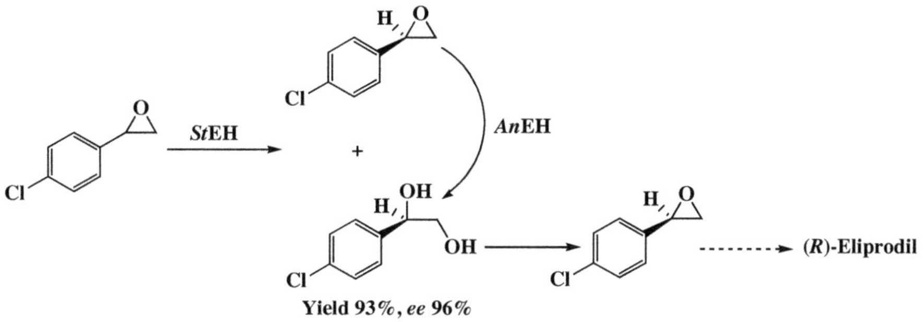

Fig. 13 Chemobienzymatic preparation of (*R*)-eliprodil

3.2 Pyridyloxirane

Enantiopure 2-, 3-, 4-pyridyloxirane are valuable chiral synthons, but they cannot be produced with high optical purity and yield using the conventional metal-based catalysts.

Among 14 EH collections, *An*EH was found to be the best choice for the preparative-scale resolution of 2-pyridyloxirane [100]. Enantiopure (*S*)-2-pyridyloxirane (>99 % *ee*) was obtained in 43 % yield [101]. Gram-scale preparation of (*S*)-2-, 3-, and 4-pyridyloxirane was carried out by hydrolytic kinetic resolution with *An*EH at a substrate concentration of 10 g/L (82 mM) in plain water, and (*S*)-pyridyloxiranes were afforded in a nearly enantiopure form (*ee* > 98 %) [102].

A. radiobacter EH (*Ar*EH) can also catalyze the resolution of 2-, 3- and 4-pyridyloxiranes. An active-site mutation (Tyr215Phe) was introduced into the *Ar*EH yielding a more suitable catalyst for kinetic resolution, and 127 mM (15.4 g/L) of (*S*)-2-pyridyloxirane was obtained in a preparative scale from 300 mM racemic substrate by the *Ar*EH mutant [103].

3.3 Glycidyl Azide

Glycidyl azide is a key chiral C3 epoxide, wherein oxirane carbon atoms are highly active to many nucleophiles and azide is a precursor of amine. The optically pure glycidyl azides are important synthons for the synthesis of vicinal amino alcohols such as (*S*)-atenolol, (*R*)-carnitine, and synthetic antibiotic linezolid, as shown in Fig. 14. EH from *A. niger* showed high enantioselectivity toward *rac*-glycidyl azide and (*R*)-glycidyl azide was produced with 98 % *ee* and 40 % yield (*E* = 21) [104].

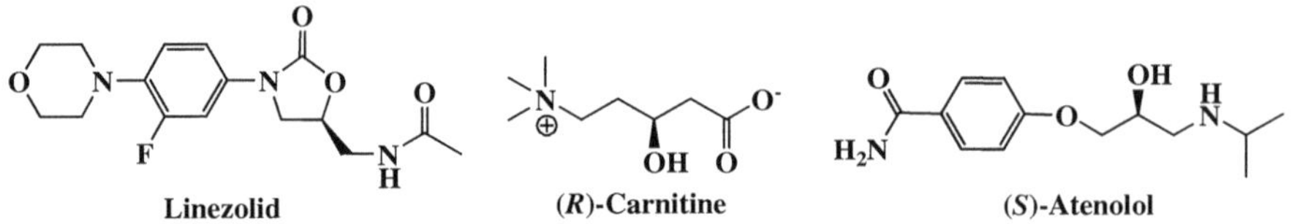

Fig. 14 Pharmaceuticals prepared from optically pure glycidyl azides

3.4 Epichlorohydrin and Aryl Glycidyl Ether

3.4.1 Epichlorohydrin

Enantiopure epichlorohydrins (ECHs) are important C3 chiral building blocks for the synthesis of pharmaceuticals, pesticides, and many other chemicals. For example, (*R*)-ECH serves as a key chiral intermediate for the synthesis of β-blocker drugs, such as metoprolol and alprenolol [105]. While (*S*)-ECH can be used as a precursor for atorvastatins, which is the top-selling cholesterol-lowering drug with global sales of 10 billion US dollars.

EHs with high enantioselectivity to epichlorohydrin are scarce. EHs from *A. niger* [106–108], *Rhodosporidium tortiloides* [109], and *Novosphingobium aromaticivorans* [110] prefer to hydrolyze the (*R*)-ECH, affording (*S*)-ECH, while EHs from *R. glutinis* [111, 112] and *A. radiobacter* [113, 114] prefer the (*S*)-ECH.

ECH is unstable in the aqueous medium. To overcome the spontaneous hydrolysis of ECH in aqueous buffer, organic solvents with little water [107, 108] and biphasic system [113, 114] were employed for the reaction. By using the recombinant *E. coli* whole cells expressing EH from *A. radiobacter* AD1 as biocatalyst, isooctane–aqueous (7:3) biphasic system was used for the reaction and 574 mM ECH was converted, producing (*R*)-ECH with 99.3 % *ee* and 37.5 % yield (analytical yield) [113].

3.4.2 Aryl Glycidyl Ether

Aryl glycidyl ethers are important precursors for the preparation of many β-blocker drugs with a suffix of "lol," including propranolol, metoprolol, and atenolol, as shown in Fig. 15.

Due to the flexible property of the chiral center in aryl glycidyl ethers, few EHs show high enantioselectivity toward these substrates, and most of them are (*S*)-preferred, such as those from *A. niger* [115], *A. radiobacter* [116], and *Trichosporon loubierii* [117, 118]. Protein engineering was employed to increase the enantioselectivity of EH for the enantioselective hydrolysis of aryl glycidyl ethers. After one round of error-prone polymerase chain reaction (epPCR), the *E*-value of *An*EH was increased from 4.6 to 10.8. The improved variant contained three amino acid substitutions, and two of them were spatially far away from the catalytically

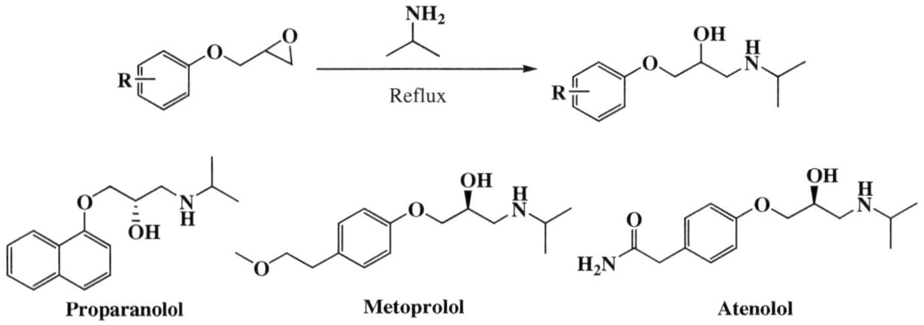

Fig. 15 Preparation of β-blocker drugs from (*R*)-aryl glycidyl ethers

active center [119]. Moreover, the EH was evolved by iterative combinatorial active-site saturation test (CAST), leading to significantly improved enantioselectivity (enantiomeric ratio *E* from 4.6 to 115) for phenyl glycidyl ether (PGE) [120]. The enantioselectivity of EH from *A. radiobacter* was enhanced using epPCR and DNA shuffling, eight mutants showed significantly improved enantioselectivity (up to 13-fold) toward *p*-nitro-PGE and three other epoxides [121].

A bacterial strain, *Bacillus megaterium* ECU1001, was isolated from soil samples by using PGE as sole carbon and energy source [122], which preferentially hydrolyze the (*R*)-PGE, affording (*S*)-epoxide and (*R*)-diol with high enantioselectivity ($E = 47.8$) [123]. The *E*-value (enantiomeric ratio) was increased to 69.3 by using surfactant tween-80 as additive to help disperse the water-insoluble substrate [124]. Isooctane–aqueous biphasic system was employed to overcome the low solubility and instability of PGE in the aqueous phase, and *E*-value was further increased to 94. Resolution of 90.1 g/L PGE (based on isooctane phase) was carried out, affording enantiopure (*S*)-PGE with a yield (analytical yield) of 44.5 % [125, 126].

An (*R*)-enantioselective epoxide hydrolase (*Bm*EH) was cloned from *B. megaterium* ECU1001 [127], high-to-excellent enantioselectivities ($E > 200$) were achieved in the bioresolution of PGE, *ortho*-substituted PGEs, and *meso*-nitro PGEs using the recombinant *Bm*EH. The crystal structure of *Bm*EH was resolved recently. By analyzing the active site of *Bm*EH, two residues (Met145 and Phe128) were identified as potential hot spots for enhancing the *Bm*EH activity toward the bulky substrates. After site-directed mutation of the two predicted hot spots, the activity of the *Bm*EH was improved by 6–430-folds toward nine typical β-blocker precursors [128].

3.4.3 Benzyl Glycidyl Ether

Optically pure benzyl glycidyl ether (BGE) plays an important role in the synthesis of numerous drugs and natural products. For example, (*S*)-BGE is an intermediate of (+)-cryptocarya diacetate, a natural product used in the treatment of headaches, morning sickness, and cancer pulmonary diseases. (*R*)-BGE is an intermediate for

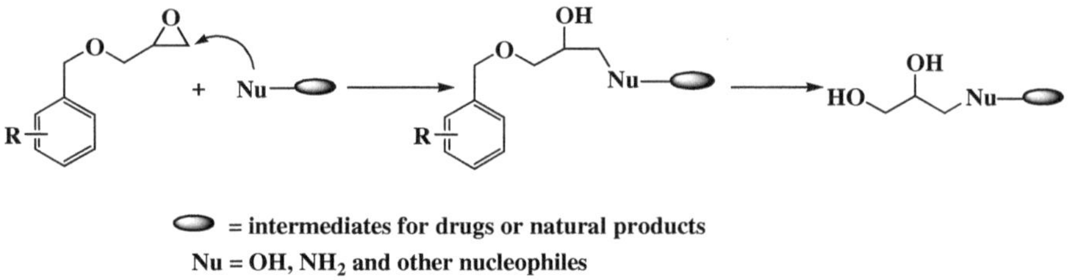

Fig. 16 Application of optically active BGE and its derivates in the synthesis of complex compounds

the synthesis of the anti-tumor and anti-leukemic drug, Synargentoide A. Optically active derivatives of BGE, such as methylbenzyl glycidyl ethers (MBGE) and dimethoxylbenzyl glycidyl ethers (DMBGE), are important intermediates for the synthesis of more complex compounds [129] (Fig. 16). As a result, synthesis of optically pure BGE and its derivatives has received considerable interest.

Due to the linear structure adjacent to the chiral center, EHs with high enantioselectivity to these compounds are quite limited. EHs from *Rhodotorula* sp. and *A. niger* preferred (*S*)-BGE [130–132], while the EH from *Rhodococcus fascians* M022 preferred the (*R*)-enantiomer [131], the *E*-values of these EHs were lower than 10. *Talaromyces flavus* containing a constitutive EH showed relatively high enantioselectivity to BGE, with an *E*-value of 13. By using the whole cells as biocatalyst, (*R*)-BGE (96 % *ee*) was obtained [133].

The enantioselectivity of enzymatic resolution of BGE could be significantly enhanced with a methyl substitution at the 2-position of BGE. For example, whole cells of *Rhodococcus ruber* SM 1789 showed high enantioselectivity ($E > 200$) to *rac*-2-methyl-BGE, and (*R*)-2-methyl-BGE was obtained with 98 % *ee* and 43 % yield, and the (*R*)-diol was produced with 97 % *ee* and 44 % yield [134]. Furthermore, an enantioconvergent process was performed via a tandem reaction of *Rhodococcus* sp. CBS 717.73 EH-catalyzed kinetic resolution of 2-methyl-BGE and an acid-catalyzed hydrolysis of the remaining (*R*)-2-methyl-BGE with inversion of the configuration, furnishing (*R*)-diols as the sole product in 97 % *ee* and 78 % yield.

3.5 Cascade Reactions

Cascade reactions are green and promising process for organic synthesis since it can avoid the usually metal catalysts involved protecting and deprotecting steps and costly intermediate isolation process, thus making the process cost-effective for target molecule synthesis [135]. Recombinant *E. coli* cells coexpressing styrene monooxygenase and enantioconvergent EH have been constructed for efficient enantioselective dihydroxylation of various terminal aryl olefins [136]. Using styrene and its phenyl-substituted derivatives as substrates, (*S*)-vicinal diols were produced with medium-to-excellent enantiopurities by the monooxygenase coexpressed with *Sphingomonas* sp. EH, while (*R*)-vicinal diols were formed with the monooxygenase coexpressed with *St*EH. This type of cascade biocatalysis provides an attractive alternative to Sharpless dihydroxylation, accepting *cis*-alkene and offering enantioselective *trans*-dihydroxylation.

Recently, the recombinant *E. coli* cells expressing *A. radiobacter* halohydrin dehalogenase (HHDH) and *Ar*EH, respectively, were immobilized by adsorption onto perlite and used for the preparation of (*R*)-epichlorohydrin from 1,3-dichloro-2-propanol in a cascade reaction. In the first step, racemic epichlorohydrin was produced with 73 % yield, and the final yield of enantiopure (*R*)-epichlorohydrin reached 25.1 % from 10 mM 1,3-dichloro-2-propanol [137].

4 Deracemization with Nitrilases

Nitrile compounds are ubiquitous in nature mainly in the form of cyanoglycosides, cyanolipids, ricinine, and phenylacetonitrile, etc. [138]. They can be used for the manufacture of a series of polymers or as feedstock, solvents, and extractants in chemical industry or pesticides in agriculture. They are also very important intermediates for the synthesis of a variety of pharmaceuticals, agrochemicals, and fine chemicals because of their broad chemical versatility [139].

Nitrilases (EC 3.5.5.1) and nitrile hydratases (EC 4.2.1.84) are two classes of important nitrile-converting enzymes, the former directly hydrolyze nitriles into the corresponding carboxylic acids and NH_3 in a single step, while the later first convert nitriles into amides, which are then transformed into carboxylic acids and NH_3 by amidases (EC 3.5.1.4) (Scheme 2) [140]. In recent years, nitrilase-mediated biocatalysis has attracted substantial interest from both academia and industry since it can be performed under mild reaction conditions combined with excellent selectivity (chemoselectivity, regioselectivity, and enantioselectivity). The enzymatic approach is significantly superior to traditional chemical methods that usually require harsh reaction conditions such as high temperatures, strongly acidic, or basic environment [141–143].

This section attempts to describe the use of nitrilase for the synthesis of pharmaceuticals, agrochemicals, fine chemicals, and their building blocks.

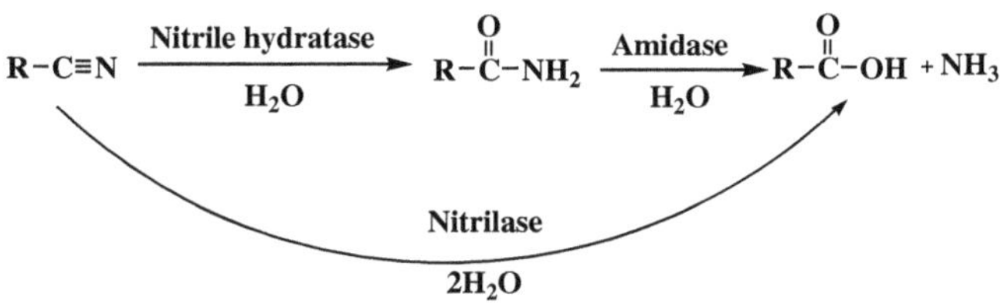

Scheme 2 Pathways for nitrile hydrolysis

4.1 Optically Pure α-Hydroxy Carboxylic Acid and Its Derivatives

4.1.1 Optically Pure Mandelic Acid

Enantiopure α-hydroxy acid and its derivatives serve as important chiral synthons for the synthesis of various pharmaceuticals, a chiral determination reagent, and a resolving reagent, for example, (*R*)-mandelic acid, is a key intermediate for the synthesis of semisynthetic penicillin, cephalosporin, antitumor agent, and antiobesity drugs [144], while (*S*)-mandelic acid can be used to synthesize the nonsteroidal anti-inflammatory drugs celecoxib and deracoxib [145]. Great effort has been paid on the development of nitrilase-catalyzed synthesis of optically pure mandelic acid from racemic mandelonitrile recently. The reaction is usually performed at slightly alkaline pH conditions, in which (*R*)-selective nitrilase preferentially hydrolyzes (*R*)-mandelonitrile to (*R*)-mandelic acid, whereas the unreacted (*S*)-mandelonitrile is spontaneously racemized in situ under the alkaline conditions and the newly formed (*R*)-mandelonitrile is used for the hydrolysis over again, thereby allowing the reaction to be proceeded in a DKR manner and affording 100 % theoretical yield (Fig. 17) [144].

Recently, a nitrilase producing strain *Alcaligenes* sp. ECU0401 was isolated from soil samples in our laboratory, which showed excellent enantioselectivity (>99.9 % *ee*) toward *R*-mandelonitrile [146]. The nitrilase gene was then cloned and overexpressed in *E. coli*, resulting in about 160-fold enhancement in nitrilase expression [147], and the enzyme production was further increased up to 19,000 U/L (50-fold improvement) by optimization of culture conditions and glycerol feeding [148]. Using the recombinant *E. coli* cells as catalyst, totally 520 mM (79 g/L) (*R*)-mandelic acid could be produced from 600 mM mandelonitrile in a fed-batch reaction and the space time yield (STY) reached 108 g (product) L^{-1} d^{-1} [149]. To relieve substrate inhibition, a toluene–water biphasic reaction system was adopted and the

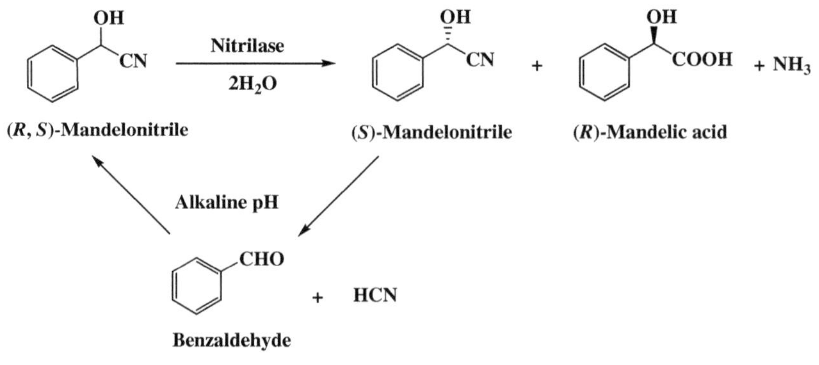

Fig. 17 Nitrilase-catalyzed synthesis of (*R*)-mandelic acid from racemic mandelonitrile in a dynamic kinetic resolution manner

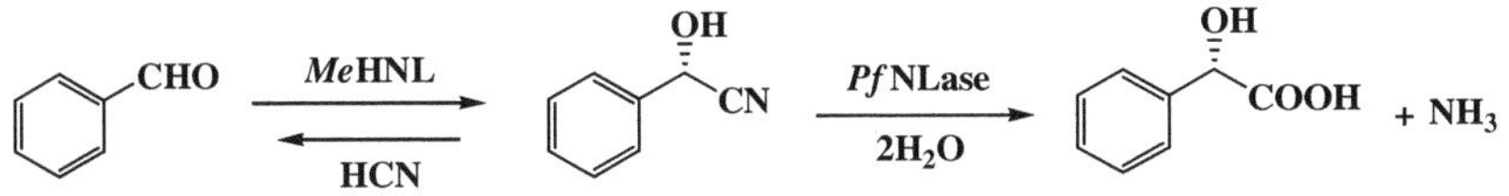

Fig. 18 Enzymatic cascade procedure for the synthesis of (*S*)-mandelic acid using *S*-selective hydroxynitrile lyase (*Me*HNL) and non-selective nitrilase in tandem

STY was increased to 352 g (product) L^{-1} d^{-1}. Furthermore, the recombinant *E. coli* cells were immobilized in calcium alginate to facilitate product isolation and biocatalyst recovery; finally, 110.7 g (*R*)-mandelic acid was obtained by recycling the immobilized biocatalyst in a 2-L-stirred tank reactor, giving a catalyst productivity of 13.8 g (product) g^{-1} (cells) [150]. Xue et al. developed an integrated bioprocess for the efficient production of (*R*)-mandelic acid with the immobilized *Alcaligenes faecalis* ZJUTB10 in a packed bed bioreactor which was incorporated with an in situ product recovery system to overcome product inhibition. This reaction system was very stable and gave a productivity of 8.87 mM/h in 16 h of reaction; totally, 550 mmol of (*R*)-mandelic acid with excellent enantiomeric excess (>99 %) was accumulated after 80 h of reaction [151]. Currently, BASF and Mitsubishi are producing (*R*)-mandelic acid from racemic mandelonitrile at several tons per year [74].

Since most mandelonitrile hydrolases are *R*-selective, it is not applicable to synthesize (*S*)-mandelic acid directly from racemic mandelonitrile. To overcome this limitation, Baum et al. developed a bienzymatic cascade reaction system including an (*S*)-hydroxynitrile lyase from *Manihot esculenta* and a non-selective arylacetonitrilase from *Pseudomonas fluorescens* EBC191 for the synthesis of (*S*)-mandelic acid from benzaldehyde and cyanide (Fig. 18). An aqueous–ionic liquid biphasic system was adopted to alleviate the inhibitory effect of benzaldehyde on nitrilase activity; this system allowed to convert up to 700 mM benzaldehyde in the ionic liquid phase with a product yield of 87–100 %. Unfortunately, the nitrilase also showed nitrile hydratase activity; therefore, (*S*)-mandeloamide was formed as a by-product in about 50 % of the total product [152]. A third enzyme, an amidase from *Rhodococcus erythopolis,* was then incorporated to the bienzymatic cascade system, and all three enzymes were co-immobilized in cross-linked enzyme aggregates, allowing the production of (*S*)-mandelic acid in 90 % yield and >99 % enantiomeric purity without any by-product [153].

4.1.2 Optically Pure *o*-Chloromandelic Acid

(*R*)-*o*-Chloromandelic acid is a key chiral precursor for the synthesis of the platelet aggregation inhibitor, (*S*)-clopidogrel, which is sold under the commercial name of Plavix®. Plavix® is a very important drug in reducing the risk of stroke, heart attack, and death in patients with a previous stroke, unstable angina, heart attack, or peripheral arterial disease caused by blood clots. In 2009, Plavix® has become the

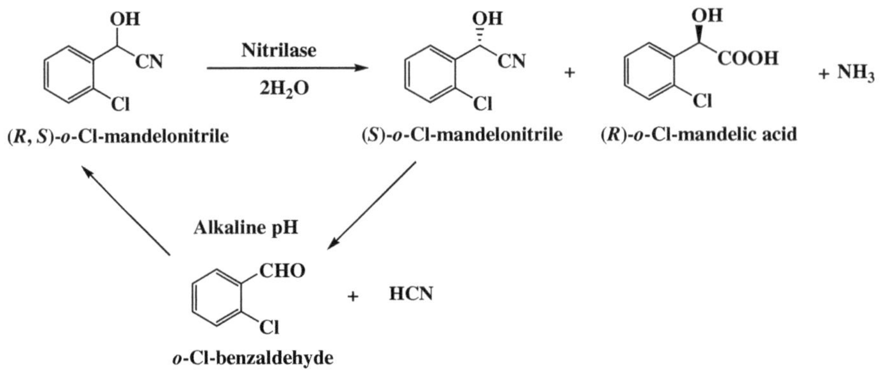

Fig. 19 Synthesis of (*R*)-*o*-chloromandelic acid from *o*-chloromandelonitrile by nitrilase-mediated hydrolysis

second top-selling drug in the world with global sales of over 10 billion US dollars per year. Several enzymatic methods have been developed for the synthesis of (*R*)-*o*-chloromandelic acid; of them, the nitrilase-catalyzed DKE of *o*-chloromandelonitrile represents one of the most promising approaches since under slightly alkaline conditions the unreacted (*S*)-*o*-chloromandelonitrile is spontaneously racemized to its racemate similar to nitrilase-mediated mandelonitrile hydrolysis, thereby affording 100 % theoretical yield (Fig. 19).

In order to explore new nitrilase for *o*-chloromandelonitrile hydrolysis, the data mining strategy based on BLAST was employed using the nitrilase sequence of *Alcaligenes* sp. ECU0401 as the template. Totally, seven nitrilases showing 40–60 % amino acid identities with the template were cloned and expressed in *E. coli*, after screening based on activity and enantioselectivity, a new nitrilase from *Labrenzia aggregata* (LaN) was discovered, which could catalyze the enantioselective hydrolysis of *o*-chloromandelonitrile to (*R*)-*o*-chloromandelic acid with 96.5 % *ee*. To enhance the process efficiency, a toluene–water biphasic reaction system was used to relieve substrate inhibition, in which up to 300 mM *o*-chloromandelonitrile could be completely transformed, giving an isolated yield of 94.5 %, and a space time yield of 154 g (product) L^{-1} d^{-1}, respectively [154].

4.2 Enantiomerically Pure β-Hydroxy Carboxylic Acids

Optically pure β-hydroxy carboxylic acids are key building blocks for the synthesis of natural products, antibiotics, and chiral auxiliaries. Meanwhile, they can also be used for the manufacture of copolyesters in the film, fiber, molding, and coating industry [155]. The versatile application of chiral β-hydroxy carboxylic acids as important synthons has triggered the development of efficient and eco-friendly methodologies for the preparation of enantiomerically pure β-hydroxy carboxylic acids, including enzymatic reduction of β-ketoesters [156], enzymatic resolution of racemic acylated β-hydroxyesters [157], or kinetic resolution of racemic β-hydroxy carboxylic acid esters [158]. Optically pure β-hydroxy carboxylic acids can also be accessed by enantioselective hydrolysis of the easily available β-hydroxy nitriles. However, conventional chemical hydrolysis of nitriles usually requires harsh reaction conditions and elevated temperatures, which in turn result in the undesirable elimination of the functional group carried by the nitriles and lead to by-product formation. Nitrilase-mediated hydrolysis of nitriles represents an attracting alternative since the reaction can be carried out at environmental benign conditions, avoiding the protection and deprotection of functional groups, and most importantly, nitrilases are always highly selective.

4.2.1 Optically Pure 3-Hydroxy-3-phenylpropionate and Its Derivatives

In an attempt to explore the synthetic applicability of nitrilases obtained by genome mining, β-hydroxy nitriles were subjected to hydrolysis by these nitrilases. Both nitrilases (NIT6803 from cyanobacterium *Synechocystis* sp. strain PCC 6803 and bll6402 from *Bradyrhizobium japonicum* strain USDA110) could catalyze the enantioselective hydrolysis of β-hydroxy nitriles to give enantioenriched β-hydroxy carboxylic acids, NIT6803 produced (*S*)-enriched β-hydroxy carboxylic acids, while bll6402 produced (*R*)-enriched β-hydroxy carboxylic acids, but both with low-to-moderate enantioselectivity [159, 160]. To address the enantioselectivity issue, a two-step one-pot process involving carbonyl reductase and nitrilase was developed, in which β-ketonitriles were stereoselectively reduced by carbonyl reductase to afford (*R*)- or (*S*)-β-hydroxy nitriles, which were then hydrolyzed by nitrilase to produce optically active β-hydroxy carboxylic acids in high yields (Fig. 20). Another advantage of this bienzymatic cascade reaction process is that the isolation of the intermediates β-hydroxy nitriles could be avoided, thereby lowering the process cost and minimizing the environmental impact [161].

4.2.2 (*R*)-4-Cyano-3-hydroxybutyric Acid

Lipitor®, the world's top-selling drug, is a cholesterol-lowering drug, and a member of the statin family of HMG-CoA (HMG = 3-hydroxy-3-methylglutaryl) reductase inhibitors. It contains a (3*R*,5*S*)-dihydroxyhexanoate side chain with two chiral centers. The huge market requirement of cholesterol-lowering drugs has stimulated great efforts invested on the efficient and economic synthesis of the chiral side chain, including ketoreductase-catalyzed asymmetric reduction of carbonyl precursors, lipase-promoted kinetic resolution of respective esters, aldolase-mediated carbon–carbon bond-forming reaction of aldehydes, and nitrilase-based desymmetrization of prochiral 3-hydroxyglutaronitrile. Among these methods reported, the nitrilase-catalyzed desymmetrization of 3-hydroxyglutaronitrile to afford (*R*)-4-cyano-3-hydroxybutyric acid shows great advantage since it can be carried

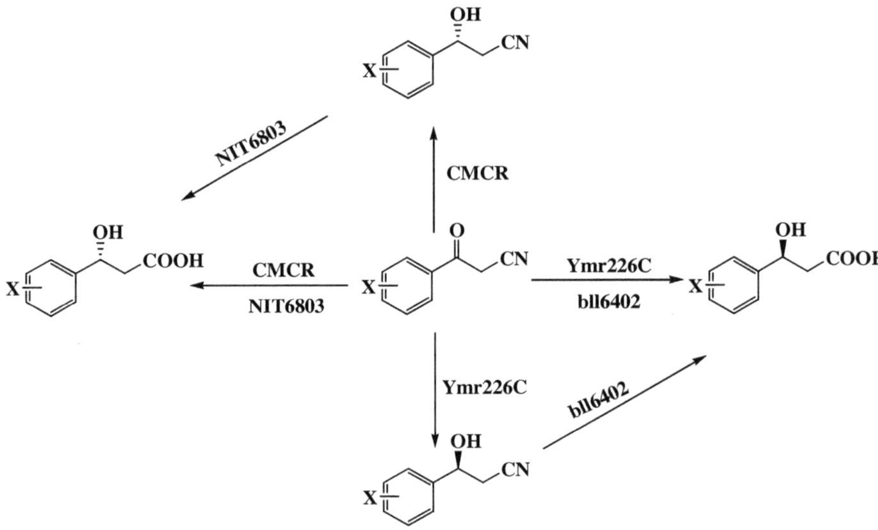

Fig. 20 Two-step one-pot synthesis of optically pure β-hydroxy carboxylic acids

out in 100 % theoretical yield; furthermore, nitrilase is a cofactor-free enzyme, thereby avoiding the supplement of expensive cofactor in the bioreaction process. DeSantis et al. created a nitrilase tool box containing more than 200 new nitrilases by extracting DNA directly from environmental samples collected from different locations of the world [162]. By screening the nitrilase tool box, four enzymes were found to be able to produce (*R*)-4-cyano-3-hydroxybutyric acid from 3-hydroxyglutaronitrile with high conversion (>95 %) and *ee* (>90 %). One of them was then chosen to perform gram-scale (1 g, 240 mM substrate) preparation, affording (*R*)-4-cyano-3-hydroxybutyric acid in 98 % yield and 95 % *ee*, the ethyl ester of which is an important intermediate for the manufacture of cholesterol-lowering drug Lipitor® (Fig. 21).

Unfortunately, the enantioselectivity of the nitrilase dramatically decreased as the substrate concentration increased for the cost-effective production of (*R*)-4-cyano-3-hydroxybutyric acid. To address this problem, a novel directed evolution technology named as the gene site saturation mutagenesis (GSSM) was employed aiming to obtain a nitrilase variant that could convert 3-hydroxyglutaronitrile to (*R*)-4-cyano-3-hydroxybutyric acid with high enantioselectivity at high substrate concentration (3 M). Combined with a high-throughput screening method, they were able to identify a best variant (A190H) which could efficiently transformed

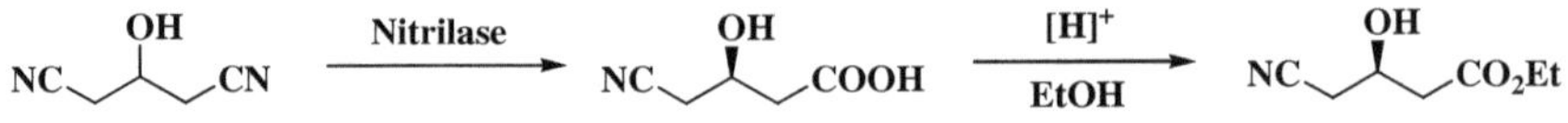

Fig. 21 Nitrilase-catalyzed desymmetrization of 3-hydroxyglutaronitrile to produce (*R*)-4-cyano-3-hydroxybutyric acid

3 M 3-hydroxyglutaronitrile to (*R*)-4-cyano-3-hydroxybutyric acid (96 % isolated yield, 98.5 % *ee*) in 15 h with a space time yield of 619 g (product) L^{-1} d^{-1} [163].

Based on the above results, Bergeron et al. developed a three-step process for the synthesis of (*R*)-4-cyano-3-hydroxybutyric acid starting from the low-cost epichlorohydrin, in which the epichlorohydrin was first cyanided by sodium cyanide to give 3-hydroxyglutaronitrile, which was then subjected to hydrolysis by nitrilase. After 16-h reaction, (*R*)-4-cyano-3-hydroxybutyric acid was produced with 100 % conversion and 99 % *ee* [164].

4.2.3 (*S*)-3-Hydroxybutyric Acid

Lennon et al. devised a strategy that combines toxicity, starvation, and induction studies together with subsequent high-throughput screening method based on a 96-well plate system to rapid identification of bacterial isolates showing nitrilase activity [165]. This strategy enabled the fast screening of 256 novel nitrilase producing bacterial strains toward β-hydroxy nitriles. One of the bacterial strains, which was identified as *Rhodococcus erythropolis* SET1, was found to catalyze deracemization of 3-hydroxybutyronitrile with excellent enantioselectivity. In a preparative-scale reaction, optically pure (*S*)-3-hydroxybutyric acid was successfully produced from 3-hydroxybutyronitrile with 42 % yield and >99.9 % *ee* (Fig. 22).

4.2.4 Ethyl (*R*)-3-Hydroxyglutarate

Optically active (*R*)-ethyl-3-hydroxyglutarate is a key precursor for the synthesis of a potent statin drug Rosuvastatin, which has received great interest in the therapy of patients with coronary artery disease because of its great potential in lowering the level of low-density lipoprotein cholesterol and medical security as compared to other statins. In order to develop a highly efficient and cost-effective method for the synthesis of enantiopure (*R*)-ethyl-3-hydroxyglutarate, a β-hydroxy aliphatic nitriles hydrolase-producing strain, identified as *R. erythropolis* ZJB-0910, was isolated by a colorimetric screening method [166]. Under the optimal reaction conditions using

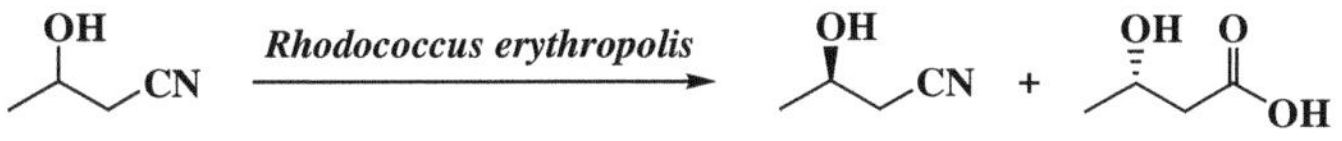

Fig. 22 Enantioselective hydrolysis of 3-hydroxybutyronitrile for the synthesis of optically pure (*S*)-3-hydroxybutyric acid

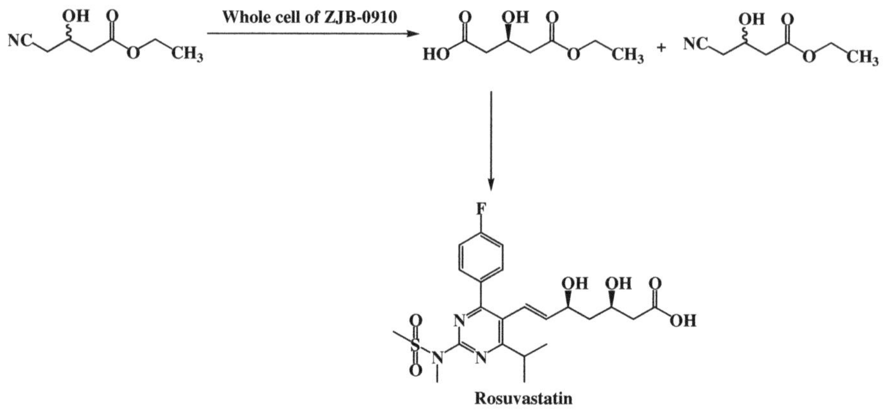

Fig. 23 Enantioselective hydrolysis of racemic ethyl 4-cyano-3-hydroxybutyrate for the synthesis of optically active (*R*)-ethyl-3-hydroxyglutarate

the whole cells of *R. erythropolis* ZJB-0910 as biocatalyst, enantiomerically pure (*R*)-ethyl-3-hydroxyglutarate was produced from racemic ethyl 4-cyano-3-hydroxybutyrate with 46.2 % yield and >99 % *ee* (Fig. 23).

4.3 Enantiomerically Pure γ-Hydroxy Carboxylic Acids

γ-Butyrolactones serve as important intermediates for the synthesis of natural products and pharmaceuticals, and they are also main components of flavors, fragrance, and insect pheromones [167]. γ-Butyrolactones can be produced directly from γ-hydroxy carboxylic acids via lactonization, while γ-hydroxy carboxylic acids can be accessed by nitrilase-catalyzed hydrolysis of γ-hydroxynitriles under mild reaction conditions (Fig. 24). A series of optically active γ-hydroxy carboxylic acids with different side-chain lengths and structures were prepared from their respective γ-hydroxynitriles by commercial nitrilases with moderate-to-high enantioselectivity. The formed γ-hydroxy carboxylic acids can then be transformed into optically pure lactones through lactonization [168].

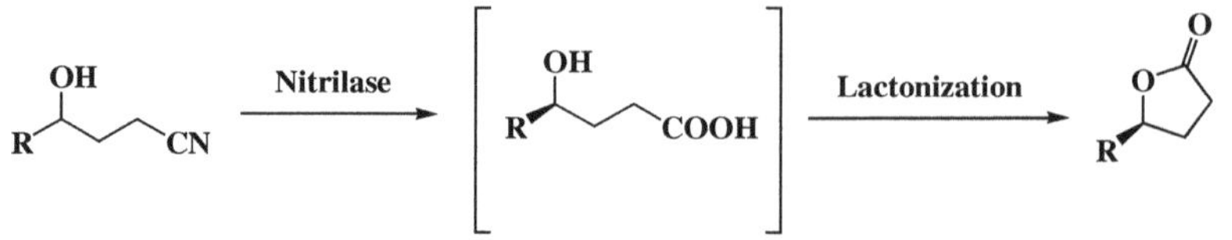

Fig. 24 Nitrilase-mediated hydrolysis of hydroxynitriles to hydroxyacids and subsequent lactonization of hydroxyacids to form lactones

4.4 Important Enantioenriched Cyano Acids

4.4.1 (*S*)-2-Cyano-2-methylpentanoic Acid

(*S*)-2-Cyano-2-methylpentanoic acid can be converted to β-amino acids by chemical reduction of the cyano group, while β-amino acids are key building blocks for the synthesis of a series of biologically active compounds. Therefore, the prerequisite to access the biologically active compounds is the development of an efficient and green synthetic route to produce (*S*)-2-cyano-2-methylpentanoic acid. The well-studied nitriles converting bacterium *Rhodococcus rhodochrous* J1, which produces both nitrilase and nitrile hydratase activity, were chosen as the potential biocatalyst. To prevent potential influence of nitrile hydratase activity on the reaction, the nitrile hydratase activity was removed by cloning the nitrilase gene responsible for the target reaction alone from *R. rhodochrous* J1 and expressing it in *E. coli* JM109. The recombinant *E. coli* cells could enantioselectively hydrolyze 2-methyl-2-propylmalononitrile to form (*S*)-2-cyano-2-methylpentanoic acid (Fig. 25). In a scale-up reaction, 80 g (*S*)-2-cyano-2-methylpentanoic acid was successfully produced from 2-methyl-2-propylmalononitrile with 97 % molar yield and 96 % *ee* after 24-h transformation without any by-product [169].

4.4.2 (3*S*)-3-Cyano-5-methyl Hexanoic Acid

(3*S*)-3-Cyano-5-methyl hexanoic acid is a key precursor for the preparation of a marketed GABA analog, pregabalin (Lyrica® API), which is used for the treatment of neuropathic pain and partial seizures [170]. One approach to synthesize (3*S*)-3-cyano-5-methyl hexanoic acid is the (*S*)-selective nitrilase-mediated regio- and enantioselective hydrolysis of racemic isobutylsuccinonitrile, and the untouched enantiomer (3*R*)-isobutylsuccinonitrile could readily be racemized under basic conditions, allowing to afford (3*S*)-3-cyano-5-methyl hexanoic acid in 100 % theoretical yield (Fig. 26). Ten plant and bacterial nitrilases were cloned from GenBank and their ability to catalyze the target reaction was preliminarily examined by a fluorescent assay based on the NH_3 product and further verified by chiral GC analysis. Five nitrilases showed activity toward isobutylsuccinonitrile with the best one AtNit 1 giving 45 % conversion and 98 % *ee*, while the others are very poor nitrilase toward the target substrate. However, from the view point of synthetic application, the specific activity of AtNit 1 was still too low to serve as an efficient catalyst. Therefore, protein engineering based on error-prone PCR was then

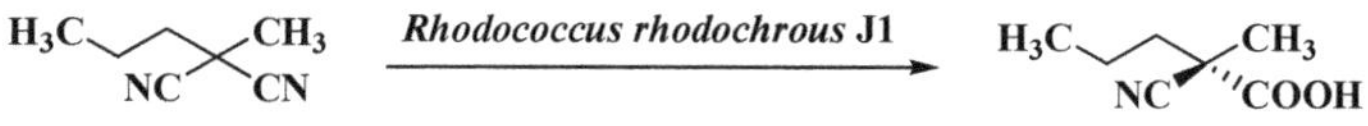

Fig. 25 Enantioselective hydrolysis of 2-methyl-2-propylmalononitrile to produce (*S*)-2-cyano-2-methylpentanoic acid by recombinant *Rhodococcus rhodochrous* J1 nitrilase

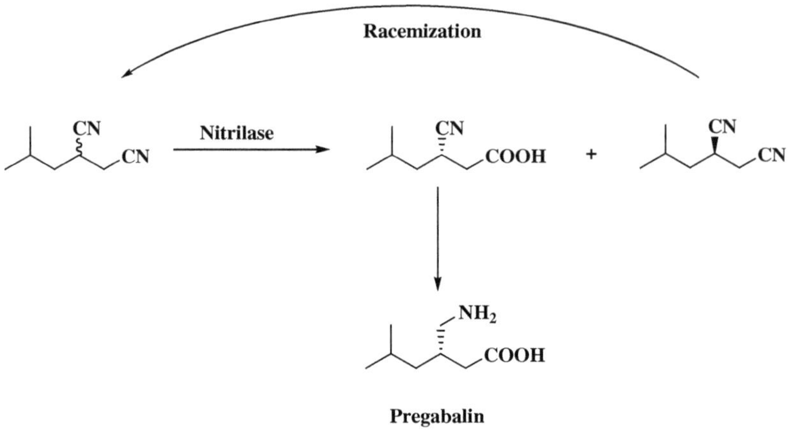

Fig. 26 Nitrilase-catalyzed synthesis of (3*S*)-3-cyano-5-methyl hexanoic acid from racemic isobutylsuccinonitrile

performed to engineer the AtNit 1 aiming to improve its catalytic efficiency. After screening of about 9,962 mutant clones (1 % of the mutant library), a single-mutant C236S was found to show about threefold increase in the activity for the hydrolysis of isobutylsuccinonitrile without affecting its enantioselectivity.

5 Asymmetric Synthesis with Keto Reductases

Chiral alcohols are key building blocks for the synthesis of a variety of biologically active molecules and active pharmaceutical ingredients. Numerous biocatalytic processes, including ketoreductase-catalyzed asymmetric reduction of prochiral carbonyl compounds, EH-mediated hydrolysis of epoxides, lipase-promoted kinetic resolution of esters, HHDH-catalyzed ring-opening of epoxides, and aldolase-based aldehyde condensation, have been developed for the efficient and eco-friendly synthesis of optically pure alcohols [6, 171, 172]. Of these methods, the ketoreductase-catalyzed asymmetric reduction of prochiral ketones represents a promising approach for the synthesis of chiral alcohols since it can be performed under mild reaction conditions with a theoretical yield of 100 % and excellent stereoselectivity. From the view point of industrial application, a promising biocatalytic process is expected to meet the criteria of high substrate concentration (≥100 g/L), low catalyst usage (≤5 g/L), excellent enantioselectivity (>99.5 %), and high volumetric productivity (≥100 g L^{-1} d^{-1}) combined with no or little amount of cofactor consumption (<0.5 g/L) [173].

In this section, the efficient and cost-effective synthesis of chiral alcohols by ketoreductase-catalyzed asymmetric reduction will be discussed.

5.1 Optically Active Ethyl 4-chloro-3-hydroxybutyrate

5.1.1 Ethyl (*S*)-4-chloro-3-hydroxybutyrate

Ethyl (*S*)-4-chloro-3-hydroxybutyrate is an important chiral building block for the synthesis of HMG-CoA reductase inhibitor, which is the active component of the cholesterol-lowering drug Lipitor. An impressive contribution for the synthesis of ethyl (*R*)-4-cyano-3-hydroxybutyrate is presented by Codexis, which involves a two-step process: At first, a ketoreductase coupled with a glucose dehydrogenase (GDH) for cofactor regeneration was employed to asymmetrically reduce ethyl 4-chloroacetoacetate resulting in the formation of ethyl (*S*)-4-chloro-3-hydroxybutyrate, which was then transformed into ethyl (*R*)-4-cyano-3-hydroxybutyrate by HHDH via the formation of an epoxide intermediate (Fig. 27). Protein engineering was adopted to improve the catalyst efficiency and the process is now running at 2,000 L scale [174]. This pioneer work was assigned the Presidential Green Chemistry Challenge Award in 2006.

Besides the work of Codexis, great effort has also been paid on the development of an efficient process to produce ethyl (*S*)-4-chloro-3-hydroxybutyrate based on ketoreductase-catalyzed asymmetric reductions by several different research groups. An acetoacetyl-CoA reductase was discovered from *Ralstonia eutropha* by a bioinformatic-based enzyme-screening method, which could catalyze the highly stereoselective reduction of ethyl 4-chloroacetoacetate to ethyl (*S*)-4-chloro-3-hydroxybutyrate. By using the recombinant *E. coli* cells coexpressing the reductase from *R. eutropha* and a glucose dehydrogenase from *B. subtilis* for the regeneration of NADPH as catalyst, 48.7 g/L ethyl (*S*)-4-chloro-3-hydroxybutyrate was successfully produced from ethyl 4-chloroacetoacetate with an optical purity of 99.8 % *ee* [175]. Recently, Wang et al. reported the cloning and expression of a highly active and stereoselective NADH-dependent reductase from *Streptomyces coelicolor* by genome mining [176]. In a water–toluene biphasic reaction system, up to 600 g/L (3.6 M) ethyl 4-chloro-3-oxobutyrate was asymmetrically reduced to ethyl (*S*)-4-chloro-3-hydroxybutyrate by the recombinant *E. coli* cells using isopropanol as co-substrate for cofactor regeneration, giving a product yield of 93 and >99 % *ee*. The volumetric productivity of the process reached as high as 609 g L^{-1} d^{-1} and a total turnover number of more than 12,000, indicating the great potential of this reductase for industrial application.

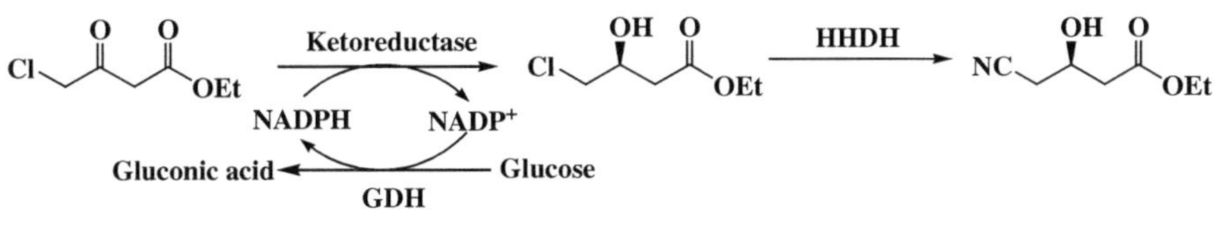

Fig. 27 A two-step biocatalytic process for the synthesis of ethyl (*R*)-4-cyano-3-hydroxybutyrate

5.1.2 Ethyl (*R*)-4-chloro-3-hydroxybutyrate

Optically pure ethyl (*R*)-4-chloro-3-hydroxybutyrate is an intermediate for L-carnitine, and several biocatalytic methods have been developed to produce ethyl (*R*)-4-chloro-3-hydroxybutyrate [177–180]. Nevertheless, the substrate loading used in these processes was relatively low, and the enantioselectivity was unsatisfactory, thereby hindering their practical application. Therefore, there is still an urgent need for novel and highly efficient biocatalyst for the cost-effective production of ethyl (*R*)-4-chloro-3-hydroxybutyrate. A carbonyl reductase gene (*yueD*) was identified from the genome sequence of *Bacillus* sp. ECU0013 and overexpressed in *E. coli*. The recombinant reductase showed activity toward a series of substrates including aromatic ketones, α- and β-keto esters, especially ethyl 4-chloro-3-oxobutyrate. In an aqueous–toluene biphasic system using the recombinant *E. coli* cells coexpressing carbonyl reductase (*yueD*) and glucose dehydrogenase as catalyst, 215 g/L (1.3 M) ethyl 4-chloro-3-oxobutyrate was transformed into ethyl (*R*)-4-chloro-3-hydroxybutyrate by a fed-batch strategy with a product yield of 91.7 and 99.6 % *ee* (Fig. 28) [181].

5.2 *Methyl (R)-o-Chloromandelate*

As the key intermediate for the synthesis of the top-second selling drug Plavix® (clopidogrel bisulfate), methyl (*R*)-*o*-chloromandelate has attracted great interest from both academy and industry. A variety of synthetic methods have been reported for the synthesis of this chiral intermediate, including ketoreductase-catalyzed asymmetric reduction of methyl *o*-chlorobenzoylformate, nitrilase-mediated DKR of *o*-chloromandelonitrile, hydroxynitrile lyase-based hydrocyanation of *o*-chlorobenzaldehyde followed by hydrolysis with nitrilase, lipase/esterase-promoted kinetic resolution of *o*-chloromandelic acid esters, or α-acetoxy-*o*-chlorophenyl acetic acid. Among these methods, the ketoreductase-based asymmetric reduction of methyl *o*-chlorobenzoylformate represents an attractive approach, since the reaction can be carried out at 100 % theoretical yield, and most importantly, the expensive cofactor can readily be regenerated in situ by coupling a glucose dehydrogenase (Fig. 29).

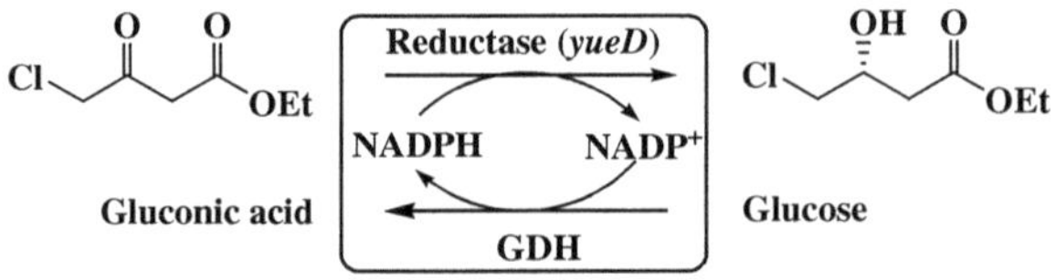

Fig. 28 Ketoreductase-catalyzed asymmetric synthesis of (*R*)-4-chloro-3-hydroxybutyrate

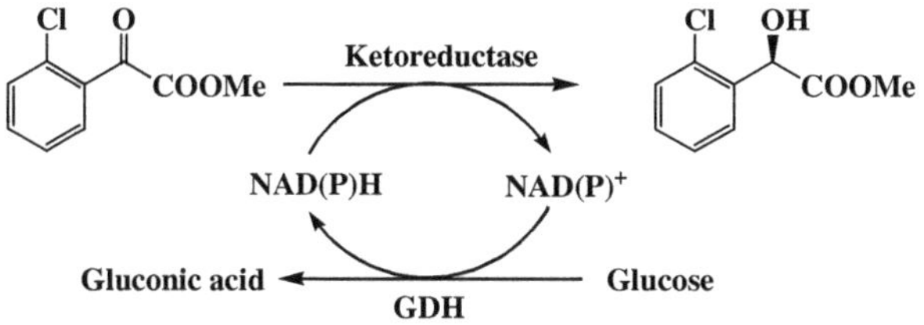

Fig. 29 Stereoselective reduction of methyl *o*-chlorobenzoylformate to methyl (*R*)-*o*-chloromandelate by ketoreductase coupled with glucose dehydrogenase

An efficient biocatalytic reduction process for the synthesis of methyl (*R*)-*o*-chloromandelate was developed by Ema et al., in which a recombinant *E. coli* cell coexpressing a carbonyl reductase (Gre2p) from *Saccharomyces cerevisiae* and a glucose dehydrogenase from *B. megaterium* was utilized for the reduction of methyl *o*-chlorobenzoylformate with externally added $NADP^+$, affording methyl (*R*)-*o*-chloromandelate at 198 g/L [182]. A new carbonyl reductase, CgKR1, which shows high activity and excellent stereoselectivity toward methyl *o*-chlorobenzoylformate, was discovered from *Candida glabrata* by in silico data mining based on sequence homology using Gre2p as the template. Using the crude enzyme of CgKR1 together with glucose dehydrogenase as catalyst, as much as 300 g/L of methyl *o*-chlorobenzoylformate could be stoichiometrically reduced to methyl (*R*)-*o*-chloromandelate with a product yield of 87 and 98.7 % *ee*. The volumetric productivity of this process reached as high as 700 g L^{-1} d^{-1}, suggesting a great potential for practical application [183]. Recently, a more efficient carbonyl reductase, YtbE, was identified from a tool box of carbonyl reductases cloned from *Bacillus* sp. ECU0013 and coexpressed with a glucose dehydrogenase in *E. coli*. Up to 500 g/L of methyl *o*-chlorobenzoylformate could be completely converted to optically pure methyl (*R*)-*o*-chloromandelate in an aqueous–ethyl caprylate biphasic reaction system by the recombinant *E. coli* cells with an isolated yield of 88 and >99 % *ee*, affording a volumetric productivity of 812 g L^{-1} d^{-1}. It is noteworthy that no any external cofactor was added during the biocatalytic process, which will significantly reduce the production cost [184].

5.3 Ethyl 2-hydroxy-4-phenylbutyrate

5.3.1 Ethyl (*R*)-2-hydroxy-4-phenylbutyrate

Ethyl (*R*)-2-hydroxy-4-phenylbutyrate is a key chiral synthon for the preparation of a variety of angiotensin-converting enzyme (ACE) inhibitors, while the latter is widely used for the treatment of hypertension and congestive heart failure. Although a number of methods for the preparation of ethyl (*R*)-2-hydroxy-4-phenylbutyrate have been investigated [185–187], the ketoreductase-catalyzed

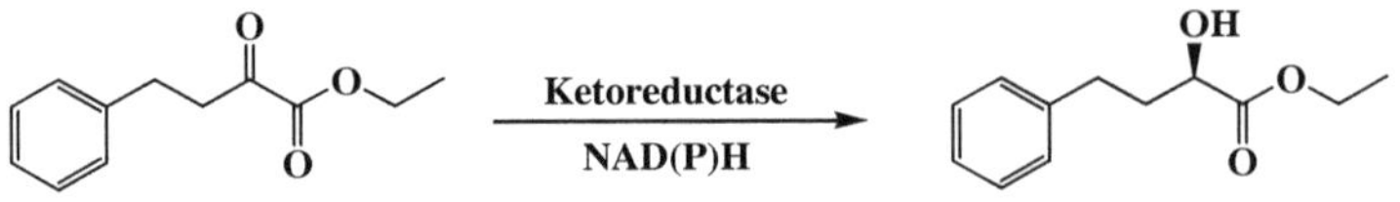

Fig. 30 Synthesis of ethyl (*R*)-2-hydroxy-4-phenylbutyrate by asymmetric reduction of ethyl 2-oxo-4-phenylbutyrate with reductase

asymmetric synthesis from ethyl 2-oxo-4-phenylburyrate represents an attractive alternative due to its high theoretical product yield, excellent stereoselectivity, and environmental friendliness (Fig. 30). Therefore, several microorganisms including *Candida boidinii* CIOC21, *Candida krusei* SW2026, *Pichia angusta*, and *S. cerevisiae*, have been employed for the synthesis of ethyl (*R*)-2-hydroxy-4-phenylbutyrate [188–191]. However, they share some common limitations such as low substrate loading, the requirement of external cofactor addition, and/or inadequate stereoselectivity that hinder their industrial application.

To discover more efficient catalyst for the practical synthesis of ethyl (*R*)-2-hydroxy-4-phenylbutyrate, a new reductase, CgKR2, was identified as the most promising catalyst candidate from 13 recombinant reductases obtained by genome mining due to its highest activity and stereoselectivity. Using the recombinant *E. coli* cells expressing CgKR2 and glucose dehydrogenase as catalyst, as much as 206 g/L (1 M) of ethyl 2-oxo-4-phenylbutyrate was completely reduced to ethyl (*R*)-2-hydroxy-4-phenylbutyrate with 84 % isolated yield and >99 % *ee*, the volumetric productivity reached 700 g L^{-1} d^{-1}, which is significantly higher than the highest value reported so far in literature. It should be noted that during the biocatalytic process, no cofactor was externally added, thereby greatly lowering the production cost [192]. The reductase, CgKR1, discovered by Ma et al. also showed very high activity in the reduction of ethyl 2-oxo-4-phenylbutyrate, and up to 412 g/L (2 M) of substrate could be stoichiometrically converted to the target product but with a little lower enantioselectivity (98.1 % *ee*) [183]. Recently, an aqueous–octanol biphasic reaction system combined with fed-batch strategy was developed by Ni et al., in which 330 g/L (1.6 M) ethyl 2-oxo-4-phenylbutyrate was successfully reduced to ethyl (*R*)-2-hydroxy-4-phenylbutyrate with 99.5 % *ee* using the recombinant *E. coli* cells coexpressing a reductase gene and a glucose dehydrogenase gene from *B. subtilis* as catalyst [193].

5.3.2 Ethyl (*S*)-2-hydroxy-4-phenylbutyrate

Similar to the synthesis of ethyl (*R*)-2-hydroxy-4-phenylbutyrate, its (*S*)-enantiomer can also be accessed through asymmetric reduction of ethyl 2-oxo-4-phenylbutyrate by reductases with reverse enantioselectivity. An impressive work on the efficient production of ethyl (*S*)-2-hydroxy-4-phenylbutyrate was presented by Ni et al., in which an extremely high substrate concentration (620 g L^{-1}, equal to 3 M) could be completely transformed by the recombinant *E. coli* cells coexpressing a reductase

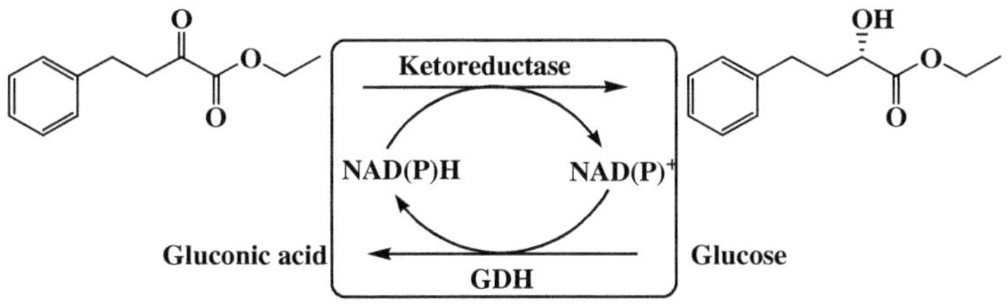

Fig. 31 Preparation of ethyl (*R*)-2-hydroxy-4-phenylbutyrate by asymmetric reduction of ethyl 2-oxo-4-phenylburyrate by reductase coupled with glucose dehydrogenase for cofactor regeneation

(FabG) identified from *Bacillus* sp. by genome mining and a glucose dehydrogenase in the absence of expensive cofactor, affording ethyl (*S*)-2-hydroxy-4-phenylbutyrate with 91 % isolated yield and >99 % *ee* (Fig. 31) [194].

5.4 Optically Active β-Hydroxynitriles

Optically pure β-hydroxynitriles find widespread application in the synthesis of various biologically active compounds and pharmaceuticals. For example, they are important precursors of the popular serotonin/norepinephrine reuptake inhibitors under the commercial name of fluoxetine, atomoxetine, and nisoxetine, which are widely used for the treatment of inception and disorders, including anxiety, alcoholism, bulimia, chronic pain, migraine headaches, sleep and memory disorders, and urinary incontinence [195]. A number of methods have been utilized for the synthesis of chiral β-hydroxynitriles including chemical routes and enzymatic approaches; however, the chemical methods usually require expensive heavy metals as catalyst, which will result in environmental pollution and toxicity issues, and the enantioselectivity is also unsatisfactory [196, 197]. While the enzymatic kinetic resolution approaches are limited by the low theoretical yield of 50 % [198, 199]. Therefore, the ketoreductase-catalyzed asymmetric reduction of β-ketonitriles offers an interesting alternative, in which both enantiomers are accessible in 100 % theoretical yield using biocatalysts with reverse enantioselectivity (Fig. 32).

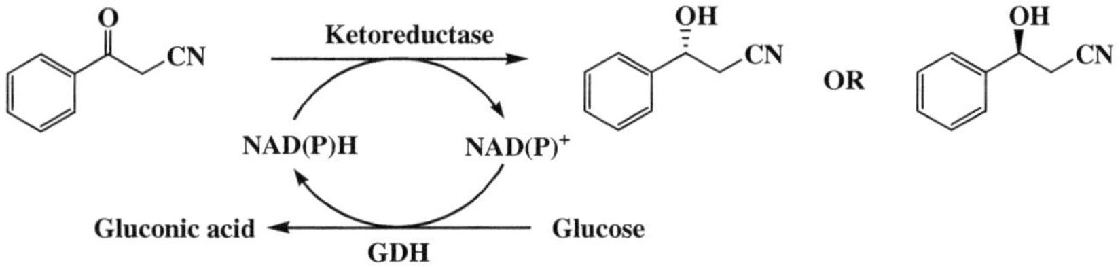

Fig. 32 Synthesis of (*R*)- or (*S*)-β-hydroxynitriles by asymmetric reduction of benzoylacetonitrile using reductase with reverse enantioselectivity

Optically active (*R*)- or (*S*)-β-hydroxynitriles were produced with excellent enantioselectivity and yield from β-ketonitriles by using a carbonyl reductase (CMCR) from *Candida magnoliae* or an alcohol dehydrogenase (Ymr226c) from *S. cerevisiae* in its isolated form, respectively, without any α-ethylated by-product, which is often observed in the whole cell biocatalysis [161]. Recently, Xu et al. reported the development of a biphasic system, in which recombinant *E. coli* cells coexpressing a data-mined carbonyl reductase *Dh*CR from *Debaryomyces hansenii* or *Cg*CR from *C. glabrata,* and a glucose dehydrogenase for in situ cofactor regeneration were employed for the asymmetric reduction of benzoylacetonitrile, as much as 145 g/L substrate based on the organic phase was completely transformed into (*R*)- or (*S*)-β-hydroxynitriles with >99 % *ee*, and no any α-ethylated by-product was detected [200].

5.5 Optically Active Aryl Halohydrins

Optically pure aryl halohydrins play an important role in the synthesis of a variety of pharmaceutical relevant compounds, such as β-blockers, β-lactam antibiotics, and chiral biphosphines. [201]. To date, numerous work on the synthesis of enantioenriched aryl halohydrins especially those with biocatalyst have been extensively investigated (Fig. 33). For example, Xie et al. demonstrated the preparation of (*R*)-aryl halohydrin in >99 % *ee* from α-chloroacetophenone using crude enzyme of Adzuki bean with a space time yield of 61.6 g L^{-1} d^{-1} [202], while the (*S*)-aryl halohydrin was produced from α-chloroacetophenone at a substrate concentration of 144 g/L using the alcohol dehydrogenase (LsADH) from *Leifsonia* sp. strain S749 with a space time yield of 104 g L^{-1} d^{-1} and a total turn over number of 935 [203]. Recently, a carbonyl reductase tool box was developed by genome data mining, after screening based on specific activity and substrate tolerance using α-chloroacetophenone as a model substrate, a robust carbonyl reductase (KtCR) from *Kluyveromyces thermotolerans* was discovered. In a biocatalytic reduction process using the recombinant *E. coli* cells expressing KtCR as biocatalyst, up to 154 g/L (1 M) of α-chloroacetophenone was asymmetrically reduced to (*S*)-aryl halohydrin, giving an isolated yield of 92 and >99 % *ee* [204].

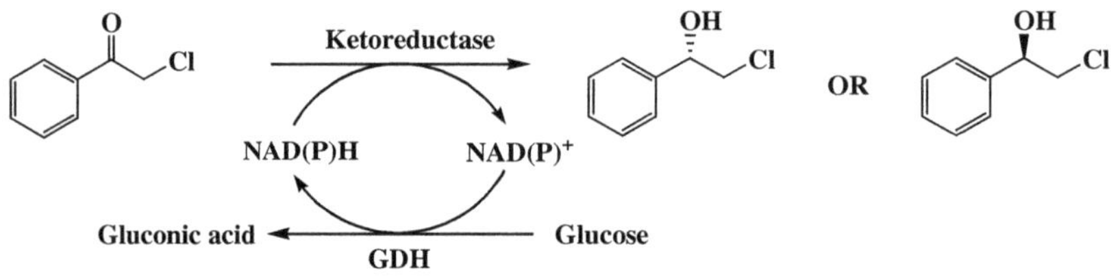

Fig. 33 Synthesis of (*R*)- or (*S*)-2-chloro-1-phenylethanol via asymmetric reduction of α-chloroacetophenone using reductase with reverse enantioselectivity

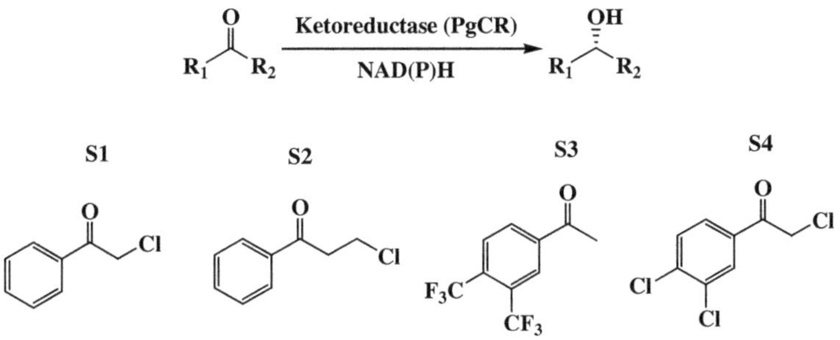

Fig. 34 Asymmetric reduction of aryl-halogenated ketones with ketoreductase (PgCR)

To further explore the synthetic potential of this reductase tool box, a series of aryl-halogenated ketones were subjected to reduction using the reductases from this tool box. Another reductase, PgCR from *Pichia guilliermondii* NRRL Y-324, was found to show broad substrate spectrum, including aryl ketones, aliphatic ketones, and ketoesters. Four aryl-halogenated ketones were then asymmetrically reduced by the isolated reductase coupled with a NADPH regeneration system in a semi-preparative scale (Fig. 34). All of the four aryl halohydrins were produced with excellent enantioselectivity (>99 %) and isolated yields (>80 %) [205].

5.6 *(R)-3-Quinuclidinol*

Enantiomerically pure 3-quinuclidinol is an important chiral intermediate with widespread applications, for example, (*R*)-3-quinuclidinol is a precursor for the synthesis of talsaclidine, revatropate, and cevimeline [206], while (*S*)-3-quinuclidinol is a very promising chiral synthon for serotonin receptor antagonist drugs and new anticholinergic drugs [207]. Therefore, a number of methods including chemical routes and enzymatic approaches have been developed for the preparation of optically pure 3-quinuclidinol. Of these methods reported, the ketoreductase-catalyzed asymmetric reduction of 3-quinuclidinone has attracted most attention since it can be carried out in 100 % theoretical yield, and by coupling a glucose dehydrogenase for in situ cofactor regeneration, the process cost could be significantly reduced (Fig. 35).

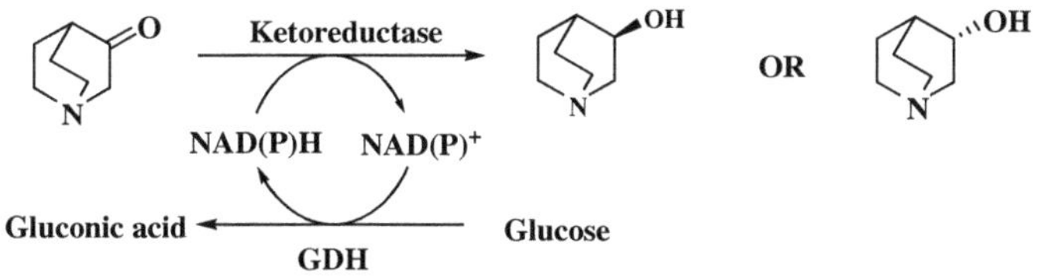

Fig. 35 Synthesis of optically pure 3-quinuclidinol via asymmetric reduction of 3-quinuclidinone with ketoreductase

A recombinant *E. coli* cells coexpressing the reductase from *Rhodotorula rubra* and a glucose dehydrogenase were employed by Uzura et al. for the conversion of 100 g/L (618 mM) of 3-quinuclidinone to (*R*)-3-quinuclidinol in 98.6 % yield and >99.9 % *ee* [208]. (*S*)-3-Quinuclidinol in 92 % yield and >99 % *ee* was formed by asymmetric reduction of 3-quinuclidinone using an *S*-selective reductase producing strain *R. erythropolis* WY1406 [209]. A promising process for the efficient preparation of (*R*)-3-quinuclidinol was demonstrated by Zhang et al. recently [210]. In this process, a new reductase (*Ar*QR) was identified from *A. radiobacter* ECU2556 by screening the laboratory stock microorganisms, which showed high activity and excellent stereoselectivity in the asymmetric reduction of 3-quinuclidinone. For the cofactor regeneration, the reductase and a glucose dehydrogenase from *B. megaterium* was then coexpressed in *E. coli*, and the resultant recombinant *E. coli* cells were utilized for the synthesis of (*R*)-3-quinuclidinol from 3-quinuclidinone. Up to 242 g/L of substrate could be stoichiometrically reduced to the target product with >99 % *ee* and a space–time yield of 916 g L^{-1} d^{-1}, indicating its great potential for practical application.

6 Chiral Amine Synthesis with Amine Transaminases

Optically pure amines and amino acids are usually used as active pharmaceutical ingredients, and also as resolving agents to obtain enantiomerically pure carboxylic acids. Traditional chemical methods to prepare chiral amines or amino acids usually require the involvement of expensive transition metal catalysts and unavoidably result in environmental pollutions, while the eco-friendly enzymatic routes represent a promising alternative [211–213]. Till date, several enzymatic approaches for the preparation of optically active amines and amino acids including lipase, amidase, monoamine oxidase, amine dehydrogenase, and amine transaminase have been developed; among these methods reported, the amine transaminase-catalyzed kinetic resolution of racemic amines or asymmetric synthesis from the corresponding prochiral ketones represents one of the most promising approaches. Especially, the amine transaminase-mediated asymmetric synthesis seems more advantageous, since it can afford 100 % theoretical yield.

In this section, the asymmetric synthesis of chiral amines catalyzed by amine transaminases will be discussed.

6.1 Sitagliptin

So far the most successful application of amine transaminase for organic synthesis is the large-scale manufacture of antidiabetic compound sitagliptin developed by Codexis and Merck & Co together (Fig. 36) [214]. A substrate walking, modeling, site saturation mutagenesis, and directed evolution strategy were adopted to tailor

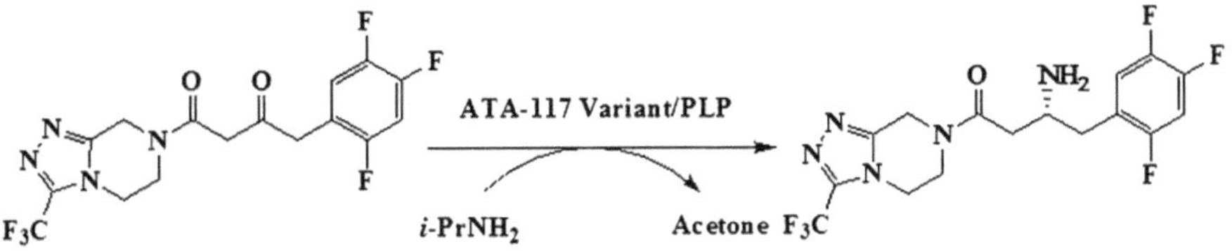

Fig. 36 Biocatalytic synthesis of sitagliptin mediated by an amine transaminase (ATA-117) variant

an amine transaminase (ATA-117) originally inactive toward prositagliptin ketone for practical application in a manufacturing setting. Structural homology model analysis and docking studies revealed that the enzyme could not bind the target substrate due to steric hindrance in the small binding pocket and potentially unfavorable interactions in the large binding pocket. Therefore, a truncated methyl ketone analog of prositagliptin ketone was applied to the screening of the site saturation libraries of residues lining the large pocket of the active site, resulting in a variant (S223P) with 11-fold activity improvement toward the methyl ketone analog. In the second library construction, residues potentially interact with the trifluorophenyl group and those selected from structural considerations were subjected to saturation mutagenesis based on S223P, which gave the first detectable transaminase activity toward the target substrate. The variant active toward the target substrate was then used for the second round of evolution, in which beneficial mutations from both small and large binding pockets were combined. A variant with 75-fold increased activity toward prositagliptin ketone was found after screening of the library. In order to tolerate the harsh reaction conditions such as high substrate/co-substrate loading, high organic solvent concentration, and high reaction temperatures, the best variant from the second round was subjected to another 9 rounds of evolution, and process-like conditions were applied to the screening steps; specifically, the substrate concentration was increased from 2 to 100 g/L, the *i*PrNH$_2$ concentration from 0.5 to 1 M, DMSO from 5 to 50 %, the pH from 7.5 to 8.5, and the temperature from 22 to 45 °C. Finally, a best variant containing 27 mutations was obtained, which met the required process target and could converted 200 g/L prositagliptin ketone to sitagliptin with >99.95 % *ee* and 92 % yield.

Interestingly, the engineered amine transaminase also showed broad substrate specificity, and some chiral amines including various trifluoromethyl-substituted amines as well as phenylethylamines with electron-rich substituents and pyrrolidines that cannot be prepared via traditional reduction amination previously can now be produced by the amine transaminase variant with excellent stereoselectivity (Fig. 37).

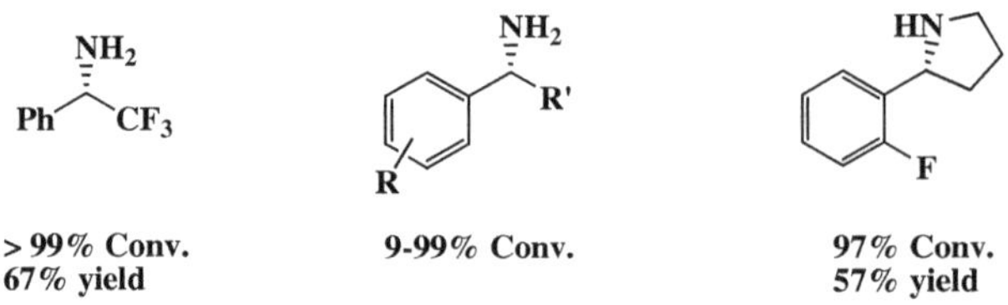

Fig. 37 Chiral amines and pyrrolidine prepared by ATA-117 variants

6.2 sec-Butylamine

To obtain optically active (*R*)-sec-butylamine, a biocatalytic resolution process was developed, in which whole cells of recombinant *E. coli* expressing an (*S*)-selective amine transaminase from *B. megaterium* SC6394 were employed for the resolution of racemic sec-butylamine using pyruvate as amino acceptor and pyridoxal phosphate as a cofactor (Fig. 38) [215]. This process was performed in 15-L scale, in which 750 g of racemic substrate was resolved to afford 585 g (*R*)-sec-butylamine·1/2H_2SO_4 salt (46.6 % isolated yield, 99.2 % *ee*).

The bottleneck of the above-mentioned resolution process is the theoretical yield of 50 %; thereof, an asymmetric synthesis process would be a useful alternative for practical application. The ω-transaminase (ATA-117)-catalyzed asymmetric synthesis of optically pure (*R*)-sec-butylamine from 2-butanone using D- or L-alanine as amino donor (ATA-117) was developed by Kroutil et al. [216]. In order to shift the reaction equilibrium to full conversion, lactate dehydrogenase was applied to remove the by-product pyruvate. Under the optimized reaction conditions, 50 mM 2-butanone could be converted with 98 % conversion and >99 % *ee* (Fig. 39).

6.3 Other Chiral Amines

Although amine transaminase-catalyzed asymmetric synthesis theoretically provides a 100 % yield of the product, the existence of the reaction equilibrium that favors the substrate over the product will result in incomplete conversion. One solution for the problem is the use of excess amino donor; the other strategy is the introduction of a second reaction to further transform the by-product. For example, Hohne et al. reported the use of lactate dehydrogenase and pyruvate decarboxylase for the removal of the by-product pyruvate in the asymmetric synthesis of several

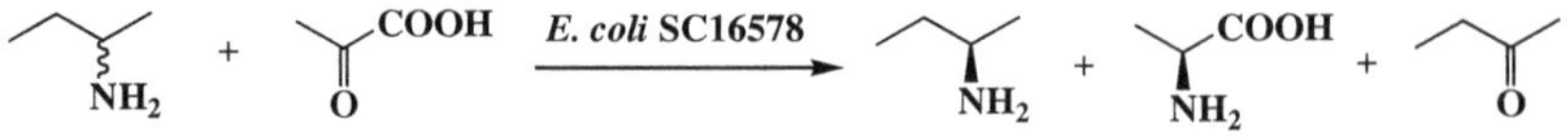

Fig. 38 Resolution of sec-butylamine by transaminase-catalyzed transamination

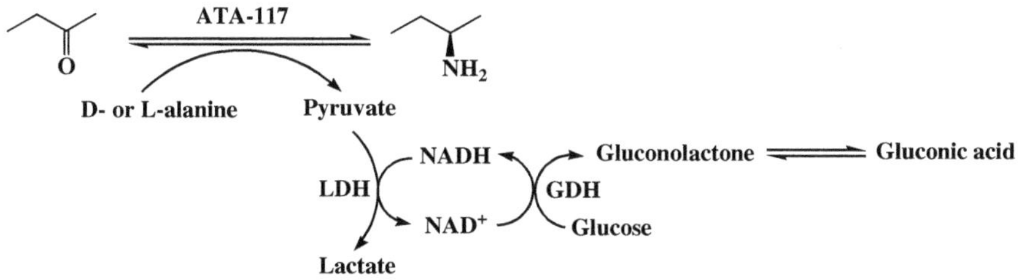

Fig. 39 Asymmetric synthesis of sec-butylamine with ω-transaminase. Lactate dehydrogenase was employed to remove the by-product pyruvate and shifted the equilibrium to the product direction

optically active amines, both resulted in significantly higher conversation rate and pyruvate decarboxylase showed slightly better results (Fig. 40) [217]. Recently, an efficient single-enzymatic cascade for the asymmetric synthesis of chiral 1-phenylethylamine and its derivatives employing 3-aminocyclohexa-1,5-dienecarboxylic acid as the amino donor was developed by Berglund et al. [218], in which the by-product ketone was spontaneously transformed into 3-hydroxybenzoic acid, pushing the equilibrium to the desired direction, allowing a theoretical yield of 100 % (Fig. 41).

Unlike various (*S*)-selective amine transaminases reported, the number of (*R*)-selective enzymes are relatively scarce. To find more (*R*)-selective amine transaminases for (*R*)-amines synthesis, an in silico strategy for enzyme identification was developed [13]. They first analyzed the structural information of respective enzymes to assess the possibility for the evolution of an (*R*)-selective amine transaminases; then, a prediction of key amino acids need to be changed was made; an annotation algorithm based on key motifs was adopted to exclude unwanted enzyme activities; after database search, protein sequences fulfilled the predicted criteria were identified and cloned from synthetic genes; at the last step,

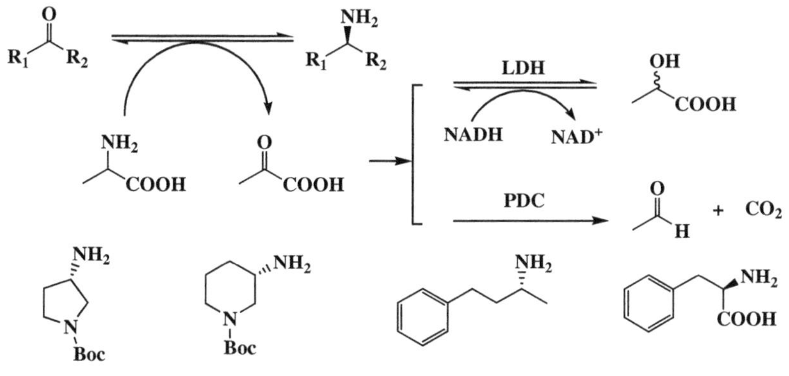

Fig. 40 Asymmetric synthesis of chiral amines by combining transaminase and lactate dehydrogenase or pyruvate decarboxylase

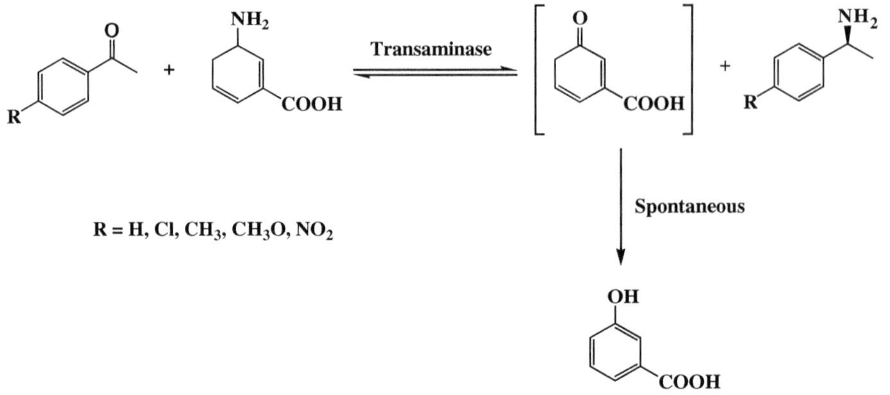

Fig. 41 Asymmetric synthesis of chiral 1-phenylethylamines and its derivatives by ω-transaminase. The employment of 3-aminocyclohexa-1,5-dienecarboxylic acid drove the reaction to full conversion via spontaneous transformation of the product ketone to 3-hydroxybenzoic acid

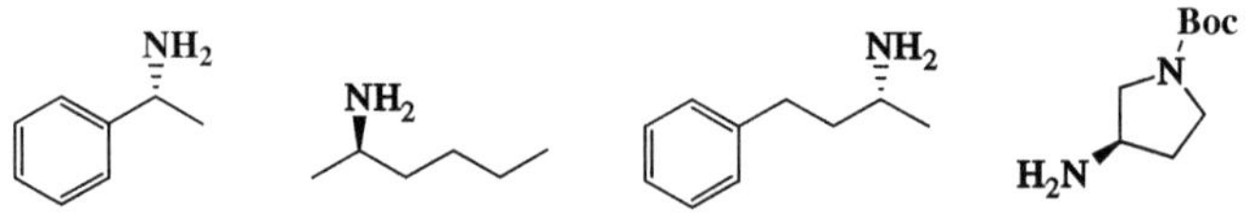

Fig. 42 Chiral amines synthesized by the data-mined (*R*)-selective amine transaminases

after protein expression of respective genes, the resultant enzymes were subjected to activity and selectivity investigation. Finally, up to 17 (*R*)-selective amine transaminases that could catalyze the synthesis of several (*R*)-amines with excellent optical purity were discovered by this in silico approach (Fig. 42). This is also a nice example for the data mining of novel enzymes.

7 Perspectives

In the past years, we have seen a big step forward in the application of biocatalysis for the synthesis of a series of optically pure chiral chemicals both in laboratory and industry with the fast growth of various powerful biocatalysts discovered. Although the genomic database offers a large pool of potential biocatalyst resources and provides great opportunities for the discovery of novel and robust biocatalyst, protein engineering including rational design, semi-rational design, random mutagenesis, and de novo enzyme design combined with reliable high-throughput screening strategy is required to further tailor the enzyme to meet specific requirement for industrial application. The development of suitable expression vectors and host strains is an important factor influencing the availability of robust biocatalyst in large quantity with acceptable cost. Furthermore, immobilization of biocatalyst on appropriate carrier using proper methods or supplement of suitable additives is crucial for enzyme storage, transportation, and application in large-scale transformation. Additionally, process engineering such as the use of aqueous–organic biphasic reaction system, designing of suitable bioreactors, in situ product recovery, and continuous operations is beneficial for process efficiency and volumetric productivity.

It is also noteworthy that cascade reactions involving two or more enzymatic reaction steps or chemoenzymatic process in one pot have emerged as a promising strategy for the preparation of chiral pharmaceutical building blocks without the need for intermediate isolation, thus making the process cost-effective for target molecular synthesis. Finally, collaboration between biologists, organic chemists, and engineers is expected to promote the biocatalysis technology to be a first choice approach for the eco-friendly, highly efficient and cost-effective synthesis of various chiral chemicals in large scale in the near future.

Acknowledgments This work was financially supported by the National Natural Science Foundation of China (Nos. 21276082 and 21406067), Ministry of Science and Technology, People's Republic of China (Nos. 2011CB710800 and 2011AA02A210), Shanghai Commission of Science and Technology (No. 11431921600), the Fundamental Research Funds for the Central Universities (No. 222201314016), and the China Postdoctoral Science Foundation (No. 2014M560309).

References

1. Bommarius AS, Riebel BR (eds) (2004) Biocatalysis: fundamentals and application. Wiley-VCH, Weinheim
2. Liese A, Seelbach K, Wandrey C (eds) (2006) Industrial biotransformations, 2nd edn. Wiley-VCH, Weinheim
3. Woodley JM (2008) New opportunities for biocatalysis: making pharmaceutical processes greener. Trends Biotechnol 22:321–327
4. Tao JH, Xu JH (2009) Biocatalysis in development of green pharmaceutical processes. Curr Opin Chem Biol 13:43–50
5. Zheng GW, Xu JH (2011) New opportunities for biocatalysis: driving the synthesis of chiral chemicals. Curr Opin Biotechnol 22:784–792
6. Ni Y, Xu JH (2012) Biocatalytic ketone reduction: A green and efficient access to enantiopure alcohols. Biotechnol Adv 30:1279–1288
7. Bornscheuer UT, Huisman GW, Kazlauskas RJ, Lutz S, Moore JC, Robins K (2012) Engineering the third wave of biocatalysis. Nature 485:185–194
8. Huisman GW, Collier SJ (2013) On the development of new biocatalytic processes for practical pharmaceutical synthesis. Curr Opin Chem Biol 17:284–292
9. Ogawa J, Shimizu S (1999) Microbial enzymes: new industrial applications from traditional screening methods. Trends Biotechnol 17:13–21
10. Asano Y (2002) Overview of screening for new microbial catalysts and their uses in organic synthesis—selection and optimization of biocatalysts. J Biotechnol 94:65–72
11. Behrens GA, Hummel A, Padhi SK, Schatzle S, Bornscheuer UT (2011) Discovery and protein engineering of biocatalysts for organic synthesis. Adv Synth Catal 353:2191–2215

12. Davids T, Schmidt M, Bottcher D, Bornscheuer UT (2013) Strategies for the discovery and engineering of enzymes for biocatalysis. Curr Opin Chem Biol 17:215–220
13. Hohne M, Schatzle S, Jochens H, Robins K, Bornscheuer UT (2010) Rational assignment of key motifs for function guides in silico enzyme identification. Nat Chem Biol 6:807–813
14. Gong JS, Lu ZM, Li H, Zhou ZM, Shi JS, Xu ZH (2013) Metagenomic technology and genome mining: emerging areas for exploring novel nitrilases. Appl Microbiol Biotechnol 97:6603–6611
15. Reetz MT (2002) Lipases as practical biocatalysts. Curr Opin Chem Biol 6:145–150
16. Bornscheuer UT (2002) Microbial carboxyl esterases: classification, properties and application in biocatalysis. Fems Microbiol Rev 26:73–81
17. Kirk O, Borchert TV, Fuglsang CC (2002) Industrial enzyme applications. Curr Opin Biotechnol 13:345–351
18. Panda T, Gowrishankar BS (2005) Production and applications of esterases. Appl Microbiol Biotechnol 67:160–169
19. Fojan P, Jonson PH, Petersen MTN, Petersen SB (2000) What distinguishes an esterase from a lipase: a novel structural approach. Biochimie 82:1033–1041
20. Crepin V, Faulds C, Connerton I (2004) Functional classification of the microbial feruloyl esterases. Appl Microbiol Biotechnol 63:647–652
21. Jaeger KE, Reetz MT (1998) Microbial lipases form versatile tools for biotechnology. Trends Biotechnol 16:396–403
22. Jaeger KE, Eggert T (2002) Lipases for biotechnology. Curr Opin Biotechnol 13:390–397
23. Hasan F, Shah AA, Hameed A (2006) Industrial applications of microbial lipases. Enzyme Microb Technol 39:235–251
24. Stecher H, Faber K (1997) Biocatalytic deracemization techniques: dynamic resolutions and stereoinversions. Synthesis 1:1–16
25. Pàmies O, Bäckvall JE (2004) Chemoenzymatic dynamic kinetic resolution. Trends Biotechnol 22:130–135
26. Ebbers EJ, Ariaans GJ, Houbiers JP, Bruggink A, Zwanenburg B (1997) Controlled racemization of optically active organic compounds: prospects for asymmetric transformation. Tetrahedron 53:9417–9476
27. Simeo Y, Kroutil W, Faber K (2007) Biocatalytic deracemization: dynamic resolution, stereoinversion, enantioconvergent processes, and cyclic deracemization. CRC Press, Boca Raton
28. Martín-Matute B, Bäckvall JE (2007) Dynamic kinetic resolution catalyzed by enzymes and metals. Curr Opin Chem Biol 11:226–232
29. Larissegger-Schnell B, Glueck SM, Kroutil W, Faber K (2006) Enantio-complementary deracemization of (±)-2-hydroxy-4-phenylbutanoic acid and (±)-3-phenyllactic acid using lipase-catalyzed kinetic resolution combined with biocatalytic racemization. Tetrahedron 62:2912–2916
30. Larissegger-Schnell B, Kroutil W, Faber K (2005) Chemo-enzymatic synthesis of (*R*)- and (*S*)-2-hydroxy-4-phenylbutanoic acid via enantio-complementary deracemization of (±)-2-hydroxy-4-phenyl-3-butenoic acid using a racemase-lipase two-enzyme system. Synlett 12:1936–1938
31. Wu WH, Akoh CC, Phillips RS (1996) Lipase-catalyzed stereoselective esterification of *dl*-menthol in organic solvents using acid anhydrides as acylating agents. Enzyme Microb Technol 18:536–539
32. Xu JH, Kawamoto T, Tanaka A (1995) Efficient kinetic resolution of *dl*-menthol by lipase-catalyzed enantioselective esterification with acid anhydride in fed-batch reactor. Appl Microbiol Biotechnol 43:402–407
33. Omata T, Iwamoto N, Kimura T, Tanaka A, Fukui S (1981) Stereoselective hydrolysis of *dl*-menthyl succinate by gel-entrapped *Rhodotorula minuta var. texensis* cells in organic solvent. Eur. J. Appl Microbiol Biotechnol 11:199–204

34. Vorlová S, Bornscheuer UT, Gatfield I, Hilmer JM, Bertram HJ, Schmid RD (2002) Enantioselective hydrolysis of *dl*-menthyl benzoate to *l*-(-)-menthol by recombinant *Candida rugosa* lipase Lip1. Adv Synth Catal 344:1152–1155
35. Zheng GW, Yu HL, Zhang JD, Xu JH (2009) Enzymatic production of *l*-menthol by a high substrate concentration tolerable esterase from newly isolated *Bacillus subtilis* ECU0554. Adv Synth Catal 351:405–414
36. Zheng GW, Pan J, Yu HL, Ngo-Thi MT, Li CX, Xu JH (2010) An efficient bioprocess for enzymatic production of *l*-menthol with high ratio of substrate to catalyst using whole cells of recombinant *E. coli*. J Biotechnol 150:108–114
37. Zheng GW, Yu HL, Li CX, Pan J, Xu JH (2011) Immobilization of *Bacillus subtilis* esterase by simple cross-linking for enzymatic resolution of *dl*-menthyl acetate. J Mol Catal B-Enzym 70:138–143
38. Tian X, Zheng GW, Li CX, Wang ZL, Xu JH (2011) Enantioselective production of (*S*)-1-phenyl-1,2-ethanediol from dicarboxyesters by recombinant *Bacillus subtilis* esterase. J Mol Catal B-Enzym 73:80–84
39. Zheng GW, Liu XY, Zhang ZJ, Tian P, Lin GQ, Xu JH (2013) Separation of enantiopure *m*-substituted 1-phenylethanols in high space-time yield using *Bacillus subtilis* esterase. RSC Adv 3:20446–20449
40. Romano D, Ferrario V, Molinari F, Gardossi L, Montero JMS, Torre P, Converti A (2006) Kinetic resolution of (R, S)-1, 2-O-isopropylideneglycerol by esterification with dry mycelia of moulds. J Mol Catal B-Enzym 41:71–74
41. Molinari F, Solange Cavenago K, Romano A, Romano D, Gandolfi R (2004) Enantioselective hydrolysis of (*R,S*)-isopropylideneglycerol acetate with *Kluyveromyces marxianus*. Tetrahedron: Asymmetry 15:1945–1947
42. Dröge MJ, Bos R, Quax WJ (2001) Paralogous gene analysis reveals a highly enantioselective 1,2-*O*-isopropylideneglycerol caprylate esterase of *Bacillus subtilis*. Eur J Biochem 268:3332–3338
43. Godinho LF, Reis CR, Rozeboom HJ, Dekker FJ, Dijkstra BW, Poelarends GJ, Quax WJ (2012) Enhancement of the enantioselectivity of carboxylesterase A by structure-based mutagenesis. J Biotechnol 158:36–43
44. Liu AMF, Somers NA, Kazlauskas RJ, Brush TS, Zocher F, Enzelberger MM, Bornscheuer UT, Horsman GP, Mezzetti A, Schmidt-Dannert C (2001) Mapping the substrate selectivity of new hydrolases using colorimetric screening: lipases from *Bacillus thermocatenulatus* and *Ophiostoma piliferum*, esterases from *Pseudomonas fluorescens* and *Streptomyces diastatochromogenes*. Tetrahedron: Asymmetry 12:545–556
45. Aragozzini F, Maconi E, Scolastico S (1991) Enantioselective hydrolysis of benzoyl 1, 2-*O*-isopropylideneglycerol by *Bacillus coagulans*. Appl Microbiol Biotechnol 34:433–435
46. Molinari F, Brenna O, Valenti M, Aragozzini F (1996) Isolation of a novel carboxylesterase from *Bacillus coagulans* with high enantioselectivity toward racemic esters of 1, 2-*O*-isopropylideneglycerol. Enzyme Microb Technol 19:551–556
47. Romano D, Falcioni F, Mora D, Molinari F, Buthe A, Ansorge-Schumacher M (2005) Enhanced enantioselectivity of *Bacillus coagulans* in the hydrolysis of 1,2-*O*-isopropylidene glycerol esters by thermal knock-out of undesired enzymes. Tetrahedron: Asymmetry 16:841–845
48. Godinho LF, Reis CR, Tepper PG, Poelarends GJ, Quax WJ (2011) Discovery of an *Escherichia coli* esterase with high activity and enantioselectivity toward 1, 2-*O*-isopropylideneglycerol esters. Appl Environ Microbiol 77:6094–6099
49. Godinho LF, Reis CR, van Merkerk R, Poelarends GJ, Quax WJ (2012) An esterase with superior activity and enantioselectivity towards 1,2-*O*-isopropylideneglycerol esters obtained by protein design. Adv Synth Catal 354:3009–3015
50. Kourist R, de María PD, Bornscheuer UT (2008) Enzymatic synthesis of optically active tertiary alcohols: expanding the biocatalysis toolbox. ChemBioChem 9:491–498
51. Kourist R, Bornscheuer UT (2011) Biocatalytic synthesis of optically active tertiary alcohols. Appl Microbiol Biotechnol 91:505–517

52. Coope JF, Main BG (1995) Biocatalytic resolution of a tertiary quinuclidinol ester using pig liver esterase. Tetrahedron: Asymmetry 6:1393–1398
53. Henke E, Pleiss J, Bornscheuer UT (2002) Activity of lipases and esterases towards tertiary alcohols: insights into structure-function relationships. Angew Chem Int Ed 41:3211–3213
54. Hari Krishna S, Persson M, Bornscheuer UT (2002) Enantioselective transesterification of a tertiary alcohol by lipase A from *Candida antarctica*. Tetrahedron: Asymmetry 13:2693–2696
55. Henke E, Bornscheuer UT, Schmid RD, Pleiss J (2003) A molecular mechanism of enantiorecognition of tertiary alcohols by carboxylesterases. ChemBioChem 4:485–493
56. Oh KH, Nguyen GS, Kim EY, Kourist R, Bornscheuer UT, Oh TK, Yoon JH (2012) Characterization of a novel esterase isolated from intertidal flat metagenome and its tertiary alcohols synthesis. J Mol Catal B-Enzym 80:67–73
57. Rehdorf J, Behrens G, Nguyen GS, Kourist R, Bornscheuer UT (2012) *Pseudomonas putida* esterase contains a GGG(A)X-motif confering activity for the kinetic resolution of tertiary alcohols. Appl Microbiol Biotechnol 93:1119–1126
58. Bassegoda A, Fillat A, Pastor FIJ, Diaz P (2013) Special *Rhodococcus* sp. CR-53 esterase Est4 contains a GGG(A)X-oxyanion hole conferring activity for the kinetic resolution of tertiary alcohols. Appl Microbiol Biotechnol 97:8559–8568
59. Heinze B, Kourist R, Fransson L, Hult K, Bornscheuer UT (2007) Highly enantioselective kinetic resolution of two tertiary alcohols using mutants of an esterase from *Bacillus subtilis*. Protein Eng Des Sel 20:125–131
60. Bartsch S, Kourist R, Bornscheuer UT (2008) Complete inversion of enantioselectivity towards acetylated tertiary alcohols by a double mutant of a *Bacillus subtilis* esterase. Angew Chem Int Ed 47:1508–1511
61. Bassegoda A, Nguyen GS, Schmidt M, Kourist R, Diaz P, Bornscheuer UT (2010) Rational protein design of *Paenibacillus barcinonensis* esterase EstA for kinetic resolution of tertiary alcohols. ChemCatChem 2:962–967
62. Shaw NM, Robins KT, Kiener A (2003) Lonza: 20 years of biotransformations. Adv Synth Catal 345:425–435
63. Wang YS, Zheng RC, Xu JM, Liu ZQ, Cheng F, Feng ZH, Liu LL, Zheng YG, Shen YC (2010) Enantioselective hydrolysis of (*R*)-2,2-dimethylcyclopropane carboxamide by immobilized cells of an *R*-amidase-producing bacterium, *Delftia tsuruhatensis* CCTCC M 205114, on an alginate capsule carrier. J Ind Microbiol Biotechnol 37:503–510
64. Wang Q, Yang F, Du H, Hossain MM, Bennett D, Grubisha DS (1998) The synthesis of *S*-(+)-2,2-dimethylcyclopropane carboxylic acid: a precursor for cilastatin. Tetrahedron: Asymmetry 9:3971–3977
65. Liang FY, Huang J, He JY, Wang P (2012) Improved enantioselective hydrolysis of racemic ethyl-2,2-dimethylcyclopropanecarboxylate catalyzed by modified Novozyme 435. Biotechnol Bioproc Eng 17:952–958
66. Liu CH, Pan J, Ye Q, Xu JH (2013) Enzymatic production of Cilastatin intermediate via highly enantioselective hydrolysis of methyl (±)-2,2-dimethylcyclopropane carboxylate using newly isolated *Rhodococcus* sp. ECU1013. Appl Microbiol Biotechnol 97:7659–7667
67. Cantele F, Restelli A, Riva S, Tentorio D, Villa M (2001) Enzymatic kinetic resolution of methyl 3-phenylglycidate by transesterification with amino alcohols. Adv Synth Catal 343:721–725
68. Nair R, Kasi D, Muthuramalingam S (2004) Enzymatic resolution of (±)-methyl trans-3-phenylglycidate. WO Patent 2004087932
69. Zheng GJ, Yuan QP, Yang L, Zhang X, Wang JJ, Sun WR (2006) Preparation of (2*S*,3*R*)-methyl-3-phenylglycidate using whole cells of *Pseudomonas putida*. J Mol Catal B-Enzym 43:133–136
70. Zhou DJ, Pan J, Yu HL, Zheng GW, Xu JH (2013) Target-oriented discovery of a new esterase-producing strain *Enterobacter* sp. ECU1107 for whole cell-catalyzed production of (2*S*,3*R*)-3-phenylglycidate as a chiral synthon of Taxol. Appl Microbiol Biotechnol 97:6293–6300

71. Martinez CA, Hu S, Dumond Y, Tao J, Kelleher P, Tully L (2008) Development of a chemoenzymatic manufacturing process for pregabalin. Org Process Res Dev 12:392–398
72. Zheng RC, Wang TZ, Fu DJ, Li AP, Li XJ, Zheng YG (2013) Biocatalytic synthesis of chiral intermediate of pregabalin with high substrate loading by a newly isolated *Morgarella morganii* ZJB-09203. Appl Microbiol Biotechnol 97:4839–4847
73. Zheng RC, Li AP, Wu ZM, Zheng JY, Zheng YG (2012) Enzymatic production of (*S*)-3-cyano-5-methylhexanoic acid ethyl ester with high substrate loading by immobilized *Pseudomonas cepacia* lipase. Tetrahedron: Asymmetry 23:1517–1521
74. Groger H (2001) Enzymatic routes to enantiomerically pure aromatic α-hydroxy carboxylic acids: a further example for the diversity of biocatalysis. Adv Synth Catal 343:547–558
75. Wong CH, Chen ST, Hennen WJ, Bibbs JA, Wang YF, Liu JL, Pantoliano MW, Whitlow M, Bryan PN (1990) Enzymes in organic synthesis: use of subtilisin and a highly stable mutant derived from multiple site-specific mutations. J Am Chem Soc 112:945–953
76. Ma JB, Wu L, Guo F, Gu JL, Tang XL, Jiang L, Liu J, Zhou JH, Yu HW (2013) Enhanced enantioselectivity of a carboxyl esterase from *Rhodobacter sphaeroides* by directed evolution. Appl Microbiol Biotechnol 97:4897–4906
77. Fuganti C, Rosell CM, Servi S, Tagliani A, Terreni M (1992) Enantioselective recognition of the phenacetyl moiety in the pen G acylase catalysed hydrolysis of phenylacetate esters. Tetrahedron: Asymmetry 3:383–386
78. Ju X, Yu HL, Pan J, Wei DZ, Xu JH (2010) Bioproduction of chiral mandelate by enantioselective deacylation of α-acetoxyphenylacetic acid using whole cells of newly isolated *Pseudomonas* sp. ECU1011. Appl Microbiol Biotechnol 86:83–91
79. Ma BD, Yu HL, Pan J, Liu JY, Ju X, Xu JH (2013) A thermostable and organic-solvent tolerant esterase from *Pseudomonas putida* ECU1011: catalytic properties and performance in kinetic resolution of α-hydroxy acids. Bioresour Technol 133:354–360
80. Basavaiah D, Krishna PR (1995) Synthesis of chiral α-aryl-α-hydroxyacetic acids: substituent effects in pig-liver acetone powder (PLAP) induced enantioselective hydrolysis. Tetrahedron 51:2403–2416
81. Campbell RF, Fitzpatrick K, Inghardt T, Karlsson O, Nilsson K, Reilly JE, Yet L (2003) Enzymatic resolution of substituted mandelic acids. Tetrahedron Lett 44:5477–5481
82. Ma BD, Kong XD, Yu HL, Zhang ZJ, Dou S, Xu YP, Ni Y, Xu JH (2014) Increased catalyst productivity in α-hydroxy acids resolution by esterase mutation and substrate modification. ACS Catal 4:1026–1031
83. Lee EY, Shuler ML (2007) Molecular engineering of epoxide hydrolase and its application to asymmetric and enantioconvergent hydrolysis. Biotechnol Bioeng 98:318–327
84. Lee EY (2008) Epoxide hydrolase-mediated enantioconvergent bioconversions to prepare chiral epoxides and alcohols. Biotechnol Lett 30:1509–1514
85. Schober M, Faber K (2013) Inverting hydrolases and their use in enantioconvergent biotransformations. Trends Biotechnol 31:468–478
86. Lee EY, Yoo SS, Kim HS, Lee SJ, Oh YK, Park S (2004) Production of (*S*)-styrene oxide by recombinant *Pichia pastoris* containing epoxide hydrolase from *Rhodotorula glutinis*. Enzyme Microb Technol 35:624–631
87. Yoo SS, Park S, Lee EY (2008) Enantioselective resolution of racemic styrene oxide at high concentration using recombinant *Pichia pastoris* expressing epoxide hydrolase of *Rhodotorula glutinis* in the presence of surfactant and glycerol. Biotechnol Lett 30:1807–1810
88. Pedragosa-Moreau SAA, Furstoss R (1996) Microbiological transformations 32. Use of epoxide hydrolase mediated biohydrolysis as a way to enantiopure epoxides and vicinal diols: application to substituted styrene oxides. Tetrahedron 52:4593–4606
89. Cao L, Lee JT, Chen W, Wood TK (2006) Enantioconvergent production of (*R*)-1-phenyl-1,2-ethanediol from styrene oxide by combining the *Solanum tuberosum* and an evolved *Agrobacterium radiobacter* AD1 epoxide hydrolases. Biotechnol Bioeng 94:522–529
90. Hwang S, Choi CY, Lee EY (2008) One-pot biotransformation of racemic styrene oxide into (*R*)-1,2-phenylethandiol by two recombinant microbial epoxide hydrolases. Biotechnol Bioprocess Eng 13:453–457

91. Kim HS, Lee OK, Hwang S, Kim BJ, Lee EY (2008) Biosynthesis of (*R*)-phenyl-1,2-ethanediol from racemic styrene oxide by using bacterial and marine fish epoxide hydrolases. Biotechnol Lett 30:127–133
92. Pedragosa-Moreau S, Morisseau C, Baratti J, Zylber J, Archelas A, Furstoss R (1997) Microbiological transformations 37. An enantioconvergent synthesis of the β-blocker (*R*)-Nifenalol using a combined chemoenzymatic approach. Tetrahedron 53:9707–9714
93. Xu W, Xu JH, Pan J, Gu Q, Wu XY (2006) Enantioconvergent hydrolysis of styrene epoxides by newly discovered epoxide hydrolases in mung bean. Org Lett 8:1737–1740
94. Ju X, Pan J, Xu JH (2008) Enantioconvergent hydrolysis of *p*-nitrostyrene oxide catalyzed by mung bean epoxide hydrolase. Chin J Catal 29:696–700
95. Zhu QQ, He WH, Kong XD, Fan LQ, Zhao J, Li SX, Xu JH (2014) Heterologous overexpression of *Vigna radiata* epoxide hydrolase in *Escherichia coli* and its catalytic performance in enantioconvergent hydrolysis of *p*-nitrostyrene oxide into (*R*)-*p*-nitrophenyl glycol. Appl Microbiol Biotechnol 98:207–218
96. Monterde MI, Lombard M, Archelas A, Cronin A, Arand M, Furstoss R (2004) Enzymatic transformations. Part 58: enantioconvergent biohydrolysis of styrene oxide derivatives catalysed by the *Solanum tuberosum* epoxide hydrolase. Tetrahedron: Asymmetry 15:2801–2805
97. Karboune S, Archelas A, Baratti J (2006) Properties of epoxide hydrolase from *Aspergillus niger* for the hydrolytic kinetic resolution of epoxides in pure organic media. Enzyme Microb Technol 39:318–324
98. Karboune S, Archelas A, Baratti JC (2010) Free and immobilized *Aspergillus niger* epoxide hydrolase-catalyzed hydrolytic kinetic resolution of racemic *p*-chlorostyrene oxide in a neat organic solvent medium. Process Biochem 45:210–216
99. Manoj KM, Archelas A, Baratti J, Furstoss R (2001) Microbiological transformations. Part 45: a green chemistry preparative scale synthesis of enantiopure building blocks of Eliprodil: elaboration of a high substrate concentration epoxide hydrolase-catalyzed hydrolytic kinetic resolution process. Tetrahedron 57:695–701
100. Genzel Y, Archelas A, Broxterman QB, Schulze B, Furstoss R (2002) Microbiological transformations 50: selection of epoxide hydrolases for enzymatic resolution of 2-, 3-or 4-pyridyloxirane. J Mol Catal B-Enzym 16:217–222
101. Genzel Y, Archelas A, Broxterman QB, Schulze B, Furstoss R (2000) Microbiological transformations. Part 46: preparation of enantiopure (*S*)-2-pyridyloxirane via epoxide hydrolase-catalysed kinetic resolution. Tetrahedron: Asymmetry 11:3041–3044
102. Genzel Y, Archelas A, Broxterman QB, Schulze B, Furstoss R (2001) Microbiological transformations. 47. A step toward a green chemistry preparation of enantiopure (*S*)-2-, -3-, and -4-pyridyloxirane via an epoxide hydrolase catalyzed kinetic resolution. J Org Chem 66:538–543
103. Genzel Y, Archelas A, Spelberg JHL, Janssen DB, Furstoss R (2001) Microbiological transformations. Part 48: Enantioselective biohydrolysis of 2-, 3-and 4-pyridyloxirane at high substrate concentration using the *Agrobacterium radiobacter* AD1 epoxide hydrolase and its Tyr215Phe mutant. Tetrahedron 57:2775–2779
104. Chen L, Shen HL, Wei C, Zhu Q (2013) Bioresolution of (*R*)-glycidyl azide by *Aspergillus niger* ZJUTZQ208: a new and concise synthon for chiral vicinal amino alcohols. Appl Microbiol Biotechnol 97:2609–2616
105. Jung SH, Linh PT, Lim HK, Kim HJ, Kim KH, Kang JS (2000) Enantioselective preparation of metoprolol and its major metabolites. Arch Pharm Res 23:226–229
106. Yildirim D, Tukel SS, Alptekin O, Alagoz D (2013) Immobilized *Aspergillus niger* epoxide hydrolases: cost-effective biocatalysts for the preparation of enantiopure styrene oxide, propylene oxide and epichlorohydrin. J Mol Catal B-Enzym 88:84–90
107. Choi WJ, Lee EY, Yoon SJ, Yang ST, Choi CY (1999) Biocatalytic production of chiral epichlorohydrin in organic solvents. J Biosci Bioeng 88:339–341
108. Jin HX, Hu ZC, Zheng YG (2012) Enantioselective hydrolysis of epichlorohydrin using whole *Aspergillus niger* ZJB-09173 cells in organic solvents. J Biosci 37:695–702

109. Liu ZQ, Zhang LP, Cheng F, Ruan LT, Hu ZC, Zheng YG, Shen YC (2011) Characterization of a newly synthesized epoxide hydrolase and its application in racemic resolution of (R, S)-epichlorohydrin. Catal Commun 16:133–139
110. Woo JH, Hwang YO, Kang JH, Lee HS, Kim SJ, Kang SG (2010) Enantioselective hydrolysis of racemic epichlorohydrin using an epoxide hydrolase from *Novosphingobium aromaticivorans*. J Biosci Bioeng 110:295–297
111. Kim HS, Lee JH, Park S, Lee EY (2004) Biocatalytic preparation of chiral epichlorohydrins using recombinant *Pichia pastoris* expressing epoxide hydrolase of *Rhodotorula glutinis*. Biotechnol Bioprocess Eng 9:62–64
112. Lee EY (2007) Enantioselective hydrolysis of epichlorohydrin in organic solvents using recombinant epoxide hydrolase. J Ind Eng Chem 13:159–162
113. Zou SP, Yan HW, Hu ZC, Zheng YG (2013) Enzymatic resolution of epichlorohydrin catalyzed by whole cells in an organic solvent/buffer biphasic system. Chin J Catal 34:1339–1347
114. Jin HX, Liu ZQ, Hu ZC, Zheng YG (2013) Biosynthesis of (*R*)-epichlorohydrin at high substrate concentration by kinetic resolution of racemic epichlorohydrin with a recombinant epoxide hydrolase. Eng Life Sci 13:385–392
115. Choi WJ, Park HJ, Lee EY, Choi CY (1998) Kinetic resolution for optically active epoxides by microbial enantioselective hydrolysis. Biotechnol Tech 12:225–228
116. Spelberg JHL, Rink R, Kellogg RM, Janssen DB (1998) Enantioselectivity of a recombinant epoxide hydrolase from *Agrobacterium radiobacter*. Tetrahedron: Asymmetry 9:459–466
117. Xu Y, Xu JH, Pan J, Tang YF (2004) Biocatalytic resolution of glycidyl aryl ethers by *Trichosporon loubierii*: cell/substrate ratio influences the optical purity of (*R*)-epoxides. Biotechnol Lett 26:1217–1221
118. Xu Y, Xu JH, Pan J, Zhao L, Zhang SL (2004) Biocatalytic resolution of nitro-substituted phenoxypropylene oxides with *Trichosporon loubierii* epoxide hydrolase and prediction of their enantiopurity variation with reaction time. J Mol Catal B-Enzym 27:155–159
119. Reetz MT, Torre C, Eipper A, Lohmer R, Hermes M, Brunner B, Maichele A, Bocola M, Arand M, Cronin A, Genzel Y, Archelas A, Furstoss R (2004) Enhancing the enantioselectivity of an epoxide hydrolase by directed evolution. Org Lett 6:177–180
120. Reetz MT, Bocola M, Wang LW, Sanchis J, Cronin A, Arand M, Zou JY, Archelas A, Bottalla AL, Naworyta A, Mowbray SL (2009) Directed evolution of an enantioselective epoxide hydrolase: uncovering the source of enantioselectivity at each evolutionary stage. J Am Chem Soc 131:7334–7343
121. van Loo B, Spelberg JHL, Kingma J, Sonke T, Wubbolts MG, Janssen DB (2004) Directed evolution of epoxide hydrolase from *A. radiobacter* toward higher enantioselectivity by error-prone PCR and DNA shuffling. Chem Biol 11:981–990
122. Tang YF, Xu JH, Ye Q, Schulze B (2001) Enantioselective hydrolysis of phenyl glycidyl ether catalyzed by newly isolated *Bacillus megaterium* ECU1001. Chin J Catal 22:1–2
123. Tang YF, Xu JH, Ye Q, Schulze B (2001) Biocatalytic preparation of (*S*)-phenyl glycidyl ether using newly isolated *Bacillus megaterium* ECU1001. J Mol Catal B-Enzym 13:61–68
124. Gong PF, Xu JH, Tang YF, Wu HY (2003) Improved catalytic performance of *Bacillus megaterium* epoxide hydrolase in a medium containing Tween-80. Biotechnol Prog 19:652–654
125. Gong PF, Xu JH (2002) Efficient production of enantiopure (*S*)-glycidyl phenyl ether by an epoxide hydrolase from *Bacillus megaterium* ECU1001 in an organic-aqueous two-phase system. Chin J Catal 23:299–300
126. Gong PF, Xu JH (2005) Bio-resolution of a chiral epoxide using whole cells of *Bacillus megaterium* ECU1001 in a biphasic system. Enzyme Microb Technol 36:252–257
127. Zhao J, Chu YY, Li AT, Ju X, Kong XD, Pan J, Tang Y, Xu JH (2011) An unusual (*R*)-selective epoxide hydrolase with high activity for facile preparation of enantiopure glycidyl ethers. Adv Synth Catal 353:1510–1518

128. Kong XD, Ma Q, Zhou JH, Zeng BB, Xu JH (2014) A smart library of epoxide hydrolase variants with best hits for synthesis of (*S*)-β-blocker precursors. Angew Chem Int Ed 53:6641–6644
129. Gaunt MJ, Yu J, Spencer JB (1998) Rational design of benzyl-type protecting groups allows sequential deprotection of hydroxyl groups by catalytic hydrogenolysis. J Org Chem 63:4172–4173
130. Botes AL, Weijers CAGM, Botes PJ, van Dyk MS (1999) Enantioselectivities of yeast epoxide hydrolases for 1,2-epoxides. Tetrahedron: Asymmetry 10:3327–3336
131. Kotik M, Brichac J, Kyslik P (2005) Novel microbial epoxide hydrolases for biohydrolysis of glycidyl derivatives. J Biotechnol 120:364–375
132. Kotik M, Kyslik P (2006) Purification and characterisation of a novel enantioselective epoxide hydrolase from *Aspergillus niger* M200. Biochim Biophys Acta 1760:245–252
133. Wei C, Chen YY, Shen HL, Wang S, Chen L, Zhu Q (2012) Biocatalytic resolution of benzyl glycidyl ether and its derivates by *Talaromyces flavus*: effect of phenyl ring substituents on enantioselectivity. Biotechnol Lett 34:1499–1503
134. Hellstrom H, Steinreiber A, Mayer SF, Faber K (2001) Bacterial epoxide hydrolase-catalyzed resolution of a 2,2-disubstituted oxirane: optimization and upscaling. Biotechnol Lett 23:169–173
135. O'Reilly E, Iglesias C, Turner NJ (2014) Monoamine oxidase-ω-transaminase cascade for the deracemisation and dealkylation of amines. ChemCatChem 6:992–995
136. Wu SK, Chen YZ, Xu Y, Li AT, Xu QS, Glieder A, Li Z (2014) Enantioselective trans-dihydroxylation of aryl olefins by cascade biocatalysis with recombinant *Escherichia coli* coexpressing monooxygenase and epoxide hydrolase. ACS Catal 4:409–420
137. Jin HX, Liu ZQ, Hu ZC, Zheng YG (2013) Production of (*R*)-epichlorohydrin from 1,3-dichloro-2-propanol by two-step biocatalysis using haloalcohol dehalogenase and epoxide hydrolase in two-phase system. Biochem Eng J 74:1–7
138. Banerjee A, Sharma R, Banerjee UC (2002) The nitrile-degrading enzymes: current status and future prospects. Appl Microbiol Biotechnol 60:33–44
139. Martinkova L, Mylerova V (2003) Synthetic application of nitrile-converting enzymes. Curr Org Chem 7:1279–1295
140. Singh R, Sharma R, Tewari N, Geetanjali, Rawat DS (2006) Nitrilase and its application as a 'green' catalyst. Chem Biodivers 3:1279–1287
141. Martinkova L, Kren V (2010) Biotransformations with nitrilases. Curr Opin Chem Biol 14:130–137
142. Gong JS, Lu ZM, Li H, Shi JS, Zhou ZM, Xu ZH (2012) Nitrilases in nitrile biocatalysis: recent progress and forthcoming research. Microb Cell Fact 11:142–159
143. Wang MX (2005) Enantioselective biotransformations of nitriles in organic synthesis. Top Catal 35:117–130
144. Yamamoto K, Oishi K, Fujimatsu I, Komatsu K (1991) Production of (*R*)-(-)-mandelic acid from mandelonitrile by *Alcaligenes faecalis* ATCC 8750. Appl Environ Microbiol 57:3028–3032
145. Mateo C, Chmura A, Rustler S, van Rantwijk F, Stolz A, Sheldon RA (2006) Synthesis of enantiomerically pure (*S*)-mandelic acid using an oxynitrilase-nitrilase bienzymatic cascade: a nitrilase surprisingly shows nitrile hydratase activity. Tetrahedron: Asymmetry 17:320–323
146. He YC, Xu JH, Xu Y, Ouyang LM, Pan J (2007) Biocatalytic synthesis of (*R*)-(-)-mandelic acid from racemic mandelonitrile by a newly isolated nitrilase-producer *Alcaligenes* sp. ECU0401. Chin Chem Lett 18:677–680
147. Zhang ZJ, Xu JH, He YC, Ouyang LM, Liu YY (2011) Cloning and biochemical properties of a highly thermostable and enantioselective nitrilase from *Alcaligenes* sp. and its potential for (*R*)-(-)-mandelic acid production. Bioprocess Biosyst Eng 34:315–322
148. Liu JF, Zhang ZJ, Li AT, Pan J, Xu JH (2011) Significantly enhanced production of recombinant nitrilase by optimization of culture conditions and glycerol feeding. Appl Microbiol Biotechnol 89:665–672

149. Zhang ZJ, Xu JH, He YC, Ouyang LM, Liu YY, Imanaka T (2010) Efficient production of (*R*)-(-)-mandelic acid with highly substrate/product tolerant and enantioselective recombinant *Alcaligenes* sp. nitrilase. Process Biochem 45:887–891
150. Zhang ZJ, Pan J, Liu JF, Xu JH, He YC, Liu YY (2011) Significant enhancement of (*R*)-mandelic acid production by relieving substrate inhibition of recombinant nitrilase in toluene–water biphasic system. J Biotechnol 152:24–29
151. Xue YP, Xu M, Chen HS, Liu ZQ, Wang YJ, Zheng YG (2013) A novel integrated bioprocess for efficient production of (*R*)-(−)-mandelic acid with immobilized *Alcaligenes faecalis* ZJUTB10. Org Process Res Dev 17:213–220
152. Baum S, van Rantwijk F, Stolz A (2012) Application of a recombinant *Escherichia coli* whole-cell catalyst expressing hydroxynitrile lyase and nitrilase activities in ionic liquids for the production of (*S*)-mandelic acid and (*S*)-mandeloamide. Adv Synth Catal 354:113–122
153. Chmura A, Rustler S, Paravidino M, van Rantwijk F, Stolz A, Sheldon RA (2013) The combi-CLEA approach: enzymatic cascade synthesis of enantiomerically pure (*S*)-mandelic acid. Tetrahedron: Asymmetry 24:1225–1232
154. Zhang CS, Zhang ZJ, Li CX, Yu HL, Zheng GW, Xu JH (2012) Efficient production of (*R*)-*o*-chloromandelic acid by deracemization of *o*-chloromandelonitrile with a new nitrilase mined from *Labrenzia aggregata*. Appl Microbiol Biotechnol 95:91–99
155. Zhu D, Ankati H, Mukherjee C, Yang Y, Biehl ER, Hua L (2007) Asymmetric reduction of β-ketonitriles with a recombinant carbonyl reductase and enzymatic transformation to optically pure β-hydroxy carboxylic acids. Org Lett 9:2561–2563
156. Zhu D, Mukherjee C, Rozzell JD, Kambourakis S, Hua L (2006) A recombinant ketoreductase tool-box. Assessing the substrate selectivity and stereoselectivity toward the reduction of β-ketoesters. Tetrahedron 62:901–905
157. Xu C, Yuan C (2005) *Candida Rugosa* lipase-catalyzed kinetic resolution of β-hydroxy-β-arylpropionates and δ-hydroxy-δ-aryl-β-oxopentanoates. Tetrahedron 61:2169–2186
158. Boaz NW (1992) Enzymatic hydrolysis of ethyl 3-hydroxy-3-phenylpropanoate: observations on an enzyme active-site model. J Org Chem 57:4289–4292
159. Mukherjee C, Zhu D, Biehl ER, Hua L (2006) Exploring the synthetic applicability of a cyanobacterium nitrilase as catalyst for nitrile hydrolysis. Eur J Org Chem 2006:5238–5242
160. Kamila S, Zhu D, Biehl ER, Hua L (2006) Unexpected stereorecognition in nitrilase-catalyzed hydrolysis of β-hydroxy nitriles. Org Lett 8:4429–4431
161. Ankati H, Zhu D, Yang Y, Beihl ER, Hua L (2009) Asymmetric synthesis of both antipodes of β-hydroxy nitriles and β-hydroxy carboxylic acids via enzymatic reduction or sequential reduction/hydrolysis. J Org Chem 74:1658–1662
162. DeSantis G, Zhu Z, Greenberg WA, Wong K, Chaplin J, Hanson SR, Farwell B, Nicholson LW, Rand CL, Weiner DP, Robertson DE, Burk MJ (2002) An enzyme library approach to biocatalysis: development of nitrilases for enantioselective production of carboxylic acid derivative. J Am Chem Soc 124:9024–9025
163. DeSantis G, Wong K, Farwell B, Chatman K, Zhu Z, Tomlinson G, Huang H, Tan X, Bibbs L, Chen P, Kretz K, Burk MJ (2003) Creation of a productive, highly enantioselective nitrilase through gene site saturation mutagenesis (GSSM). J Am Chem Soc 125:11476–11477
164. Bergeron S, Chaplin DA, Edwards JH, Ellis BSW, Hill CL, Holt-Tiffin K, Knight JR, Mahoney T, Osborne AP, Ruecroft G (2006) Nitrilase-catalyzed desymmetrization of 3-hydroxyglutaronitrile: preparation of a statin side chain intermediate. Org Process Res Dev 10:661–665
165. Coady TM, Coffey LV, O'Reilly C, Owens EB, Lennon CM (2013) A high throughput screening strategy for the assessment of nitrile-hydrolyzing activity towards the production of enantiopure β-hydroxy acids. J Mol Catal B-Enzym 97:150–155
166. Dong HP, Liu ZQ, Zheng YG, Shen YC (2010) Novel biosynthesis of (*R*)-ethyl-3-hydroxyglutarate with (*R*)-enantioselective hydrolysis of racemic ethyl 4-cyano-3-hydroxybutyrate by *Rhodococcus erythropolis*. Appl Microbiol Biotechnol 87:1335–1345

167. Brown HC, KulKarni SV, Racherla US (1994) Chiral synthesis via organoboranes. 39. A facile synthesis of γ-substituted-γ-butyrolactones in exceptionally high enantiomeric purity. J Org Chem 59:365–369
168. Pollock JA, Clark KM, Martynowicz BJ, Pridgeon MG, Rycenga MJ, Stolle KE, Taylor SK (2007) A mild synthesis of lactones via enantioselective hydrolysis of hydroxynitriles. Tetrahedron: Asymmetry 18:1888–1892
169. Yoshida T, Mitsukura K, Mizutani T, Nakashima R, Shimizu Y, Kawabata H, Nagasawa T (2013) Enantioselective synthesis of (*S*)-2-cyano-2-methylpentanoic acid by nitrilase. Biotechnol Lett 35:685–688
170. Xie Z, Feng J, Garcia E, Bernett M, Yazbeck D, Tao J (2006) Cloning and optimization of a nitrilase for the synthesis of (3*S*)-3-cyano-5-methyl hexanoic acid. J Mol Catal B-Enzym 41:75–80
171. Huisman GW, Liang J, Krebber A (2010) Practical chiral alcohol manufacture using ketoreductases. Curr Opin Chem Biol 14:122–129
172. Magano J, Dunetz JR (2012) Large-scale carbonyl reductions in the pharmaceutical industry. Org Process Res Dev 16:1156–1184
173. Hollmann F, Arends IWCE, Holtmann D (2011) Enzymatic reductions for the chemist. Green Chem 13:2285–2313
174. Ma SK, Gruber J, Davis C, Newman L, Gray D, Wang A, Grate J, Huisman GW, Sheldon RA (2010) A green-by-design biocatalytic process for atorvastatin intermediate. Green Chem 12:81–86
175. Yamamoto H, Matsuyama A, Kobayashi Y (2003) Synthesis of ethyl (*S*)-4-chloro-3-hydroxybutyrate using *fabG*-homologues. Appl Microbiol Biotechnol 61:133–139
176. Wang LJ, Li CX, Ni Y, Zhang J, Liu X, Xu JH (2011) Highly efficient synthesis of chiral alcohols with a novel NADH-dependent reductase from *Streptomyces coelicolor*. Bioresour Technol 102:7023–7028
177. Kataoka M, Yamamoto K, Kawabata H, Wada M, Kita K, Yanase H, Shimizu S (1999) Stereoselective reduction of ethyl 4-chloro-3-oxobutanoate by *Escherichia coli* transformant cells coexpressing the aldehyde reductase and glucose dehydrogenase genes. Appl Microbiol Biotechnol 51:486–490
178. Suzuki T, Idogaki H, Kasai N (1999) Dual production of highly pure methyl (*R*)-4-chloro-3-hydroxybutyrate and (*S*)-3-hydroxy-γ-butyrolactone with *Enterobacter* sp. Enzyme Microb Technol 24:13–20
179. Yamamoto H, Matsuyama A, Kobayashi Y (2002) Synthesis of ethyl (*R*)-4-chloro-3-hydroxybutanoate with recombinant *Escherichia coli* cells expressing (*S*)-specific secondary alcohol dehydrogenase. Biosci Biotechnol Biochem 66:481–483
180. Yu MA, Wei YM, Zhao L, Jiang L, Zhu XB, Qi W (2007) Bioconversion of ethyl 4-chloro-3-oxobutanoate by permeabilized fresh brewer's yeast cells in the presence of allyl bromide. J Ind Microbiol Biotechnol 34:151–156
181. Ni Y, Li CX, Wang LJ, Zhang J, Xu JH (2011) Highly stereoselective reduction of prochiral ketones by a bacterial reductase coupled with cofactor regeneration. Org Biomol Chem 9:5463–5468
182. Ema T, Ide S, Okita N, Sakai T (2008) Highly efficient chemoenzymatic synthesis of methyl (*R*)-*o*-chloromandelate, a key intermediate for clopidogrel, via asymmetric reduction with recombinant *Escherichia coli*. Adv Synth Catal 350:2039–2044
183. Ma HM, Yang LL, Ni Y, Zhang J, Li CX, Zheng GW, Yang HY, Xu JH (2012) Stereospecific reduction of methyl *o*-chlorobenzoylformate at 300 g L^{-1} without additional cofactor using a carbonyl reductase mined from *Candida glabrata*. Adv Synth Catal 354:1765–1772
184. Ni Y, Pan J, Ma HM, Li CX, Zhang J, Zheng GW, Xu JH (2012) Bioreduction of methyl *o*-chlorobenzoylformate at 500 g L^{-1} without external cofactors for efficient production of enantiopure clopidogrel intermediate. Tetrahedron Lett 53:4715–4717
185. Liese A, Kragl U, Kierkels H, Schulze B (2002) Membrane reactor development for the kinetic resolution of ethyl 2-hydroxy-4-phenylburate. Enzyme Microb Technol 30:673–681

186. Chadha A, Manohar M, Soundararajan T, Lokeswari TS (1996) Asymmetric reduction of 2-oxo-4-phenylbutanoic acid ethyl ester by *Daucus carota* cell cultures. Tetrahedron: Asymmetry 7:1571–1572
187. Lin WQ, He Z, Jing Y, Cui X, Liu H, Mi AQ (2001) A practical synthesis of ethyl (*R*)-and (*S*)-2-hydroxy-4-phenylbutanoate and D-homophenylalanine ethyl ester hydrochloride from L-malic acid. Tetrahedron: Asymmetry 12:1583–1587
188. Chen YZ, Lin H, Xu XY, Xia SW, Wang LX (2008) Preparation the key intermediate of angiotensin-converting enzyme (ACE) inhibitors: high enantioselective production of ethyl (*R*)-2-hydroxy-4-phenylbutyrate with *Candida boidinii* CIOC21. Adv Synth Catal 350:426–430
189. Zhang W, Ni Y, Sun ZH, Zheng P, Lin WQ, Zhu P, Ju NF (2009) Biocatalytic synthesis of ethyl (*R*)-2-hydroxy-4-phenylbutyrate with *Candida krusei* SW2026: a practical process for high enantiopurity and product titer. Process Biochem 44:1270–1275
190. Lacerda PSB, Ribeiro JB, Leite SGF, Ferrara MA, Coelho RB, Bon EPS, Silvalima EL, Antunes OAC (2006) Microbial reduction of ethyl 2-oxo-4-phenylbutyrate. Searching for (*R*)-enantioselectivity. New access to the enalapril like ACE inhibitors. Tetrahedron: Asymmetry 17:1186–1188
191. Kaluzna I, Andrew AA, Bonilla M, Martzen MR, Stewart JD (2002) Enantioselective reductions of ethyl 2-oxo-4-phenylbutyrate by *Saccharomyces cerevisiae* dehydrogenases. J Mol Catal B-Enzym 17:101–105
192. Shen ND, Ni Y, Ma HM, Wang LJ, Li CX, Zheng GW, Zhang J, Xu JH (2012) Efficient synthesis of a chiral precursor for angiotensin-converting enzyme (ACE) inhibitors in high space-time yield by a new reductase without external cofactors. Org Lett 14:1982–1985
193. Ni Y, Su YN, Li HD, Zhou JY, Sun ZH (2013) Scalable biocatalytic synthesis of optically pure ethyl (*R*)-2-hydroxy-4-phenylbutyrate using a recombinant *E. coli* with high catalyst yield. J Biotechnol 168:493–498
194. Ni Y, Li CX, Zhang J, Shen ND, Bornscheuer UT, Xu JH (2011) Efficient reduction of ethyl 2-oxo-4-phenylbutyrate at 620 g L^{-1} by a bacterial reductase with broad substrate spectrum. Adv Synth Catal 353:1213–1217
195. Hammond RJ, Poston BW, Ghiviriga I, Feske BD (2007) Biocatalytic synthesis towards both antipodes of 3-hydroxy-3-phenylpropanenitrile a precursor to fluoxetine, atomoxetine and nisoxetine. Tetrahedron Lett 48:1217–1219
196. Liu PN, Gu PM, Wang F, Tu YQ (2004) Efficient heterogeneous asymmetric transfer hydrogenation of ketones using highly recyclable and accessible silica-immobilized Ru-TsDPEN catalyst. Org Lett 6:169–172
197. Vazquez-Villa H, Reber S, Ariger MA, Carreira EM (2011) Iridium diamine catalysts for the asymmetric transfer hydrogenation of ketones. Angew Chem Int Ed 50:8979–8981
198. Kamal A, Ramesh Khanna GB, Ramu R (2002) Chemoenzymatic synthesis of both enantiomers of fluoxetine, atomoxetine and nisoxetine: lipase catalyzed resolution of 3-aryl-3-hydroxypropanenitrile. Tetrahedron: Asymmetry 13:2039–2051
199. Hann EC, Sigmund AE, Fager SK, Cooling FB, Gavagan JE, Ben-Bassat A, Chauhan S, Payne MS, Hennessey SM, DiCosimo R (2003) Biocatalytic hydrolysis of 3-hydroxyalkanenitriles to 3-hydroxyalkanoic acids. Adv Synth Catal 345:775–782
200. Xu GC, Yu HL, Zhang ZJ, Xu JH (2013) Stereocomplementary bioreduction of β-ketonitrile without ethylated byproduct. Org Lett 15:5408–5411
201. Seisser B, Lavandera I, Faber K, Lutje Spelberg JH, Kroutil W (2007) Stereo-complementary two-step cascades using a two-enzyme system leading to enantiopure epoxides. Adv Synth Catal 349:1399–1404
202. Xie Y, Xu JH, Lu WY, Lin GQ (2009) Adzuki bean: a new resource of biocatalyst for asymmetric reduction of aromatic ketones with high stereoselectivity and substrate tolerance. Bioresour Technol 100:2463–2468
203. Itoh N, Isotani K, Nakamura M, Inoue K, Isogai Y, Makino Y (2012) Efficient synthesis of optically pure alcohols by asymmetric hydrogen-transfer biocatalysis: application of engineered enzymes in a 2-propanol-water medium. Appl Microbiol Biotechnol 93:1075–1085

204. Xu GC, Yu HL, Zhang XY, Xu JH (2012) Access to optically active aryl halohydrins using a substrate-tolerant carbonyl reductase discovered from *Kluyveromyces thermotolerans*. ACS Catal 2:2566–2571
205. Xu GC, Yu HL, Xu JH (2013) Facile access to chiral alcohols with pharmaceutical relevance using a ketoreductase newly mined from *Pichia guilliermondii*. Chin J Chem 31:349–354
206. Nomoto F, Hirayama Y, Ikunaka M, Inoue T, Otsuka K (2003) A practical chemoenzymatic process to access (*R*)-quinuclidin-3-ol on scale. Tetrahedron: Asymmetry 14:1871–1877
207. Fancelli D, Caccia C, Fomaretto MG, McArthur R, Severino D, Vaghi F, Varasi M (1996) Serotoninergic 5-HT_3 and 5-HT_4 receptor activities of dihydrobenzofuran carboxylic acid derivatives. Bioorg Med Chem Lett 6(3):263–266
208. Uzura A, Nomoto F, Sakoda A, Nishimoto Y, Kataoka M, Shimizu S (2009) Stereoselective synthesis of (*R*)-3-quinuclidinol through asymmetric reduction of 3-quinuclidinone with 3-quinuclidinone reductase of *Rhodotorula rubra*. Appl Microbiol Biotechnol 83:617–626
209. Wang Y, Li JJ, Wu QQ, Zhu DM (2013) Microbial stereospecific reduction of 3-quinuclidinone with newly isolated *Nocardia* sp. and *Rhodococcus erythropolis*. J Mol Catal B-Enzym 88:14–19
210. Zhang WX, Xu GC, Huang L, Pan J, Yu HL, Xu JH (2013) Highly efficient synthesis of (*R*)-3-quinuclidinol in a space–time yield of 916 g L^{-1} d^{-1} using a new bacterial reductase *Ar*QR. Org Lett 15(19):4917–4919
211. Hohne M, Bornscheuer UT (2009) Biocatalytic routes to optically active amines. ChemCatChem 1:42–51
212. Koszelewski D, Tauber K, Faber K, Kroutil W (2010) ω-Transaminase for the synthesis of non-racemic α-chiral primary amines. Trend Biotechnol 28:324–332
213. Mathew S, Yun H (2012) ω-Transaminases for the production of optically pure amines and unnatural amino acids. ACS Catal 2:993–1001
214. Savile CK, Janey JM, Mundorff EC, Moore JC, Tam S, Jarvis WR, Colbeck JC, Krebber A, Fleitz FJ, Brands J, Devine PN, Huisman GW, Hughes GJ (2010) Biocatalytic asymmetric synthesis of chiral amines from ketones applied to sitagliptin manufacture. Science 329:305–309
215. Hanson RL, Davis BL, Chen Y, Goldberg SL, Parker WL, Tully TP, Montana MA, Patel RN (2008) Preparation of (*R*)-amines from racemic amines with an (*S*)-amine transaminase from *Bacillus megaterium*. Adv Synth Catal 350:1367–1375
216. Koszelewski D, Lavandera I, Clay D, Rozzell D, Kroutil W (2008) Asymmetric synthesis of optically pure pharmacologically relevant amines employing ω-transaminases. Adv Synth Catal 350:2761–2766
217. Hohne M, Kuhl S, Robins K, Bornscheuer UT (2008) Efficient asymmetric synthesis of chiral amines by combining transaminase and pyruvate decarboxylase. ChemBioChem 9:363–365
218. Wang B, Land H, Berglund P (2013) An efficient single-enzymatic cascade for asymmetric synthesis of chiral amines catalyzed by ω-transaminase. Chem Commun 49:161–163

Adv Biochem Eng Biotechnol (2016) 155: 107–140
DOI 10.1007/10_2014_294

Published Online: 11 January 2015

Construction of *Escherichia Coli* Cell Factories for Production of Organic Acids and Alcohols

Pingping Liu, Xinna Zhu, Zaigao Tan, Xueli Zhang and Yanhe Ma

Abstract Production of bulk chemicals from renewable biomass has been proved to be sustainable and environmentally friendly. *Escherichia coli* is the most commonly used host strain for constructing cell factories for production of bulk chemicals since it has clear physiological and genetic characteristics, grows fast in minimal salts medium, uses a wide range of substrates, and can be genetically modified easily. With the development of metabolic engineering, systems biology, and synthetic biology, a technology platform has been established to construct *E. coli* cell factories for bulk chemicals production. In this chapter, we will introduce this technology platform, as well as *E. coli* cell factories successfully constructed for production of organic acids and alcohols.

Keywords *Escherichia coli* · Bulk chemicals · Cell factories · Metabolic engineering · Synthetic biology

Contents

P. Liu · X. Zhu · Z. Tan · X. Zhang · Y. Ma
Tianjin Institute of Industrial Biotechnology, Chinese Academy of Sciences,
Tianjin, China

P. Liu · X. Zhu · Z. Tan · X. Zhang (✉)
Key Laboratory of Systems Microbial Biotechnology, Chinese Academy of Sciences,
32 West 7th Ave, Tianjin Airport Economic Area, Tianjin 300308, China
e-mail: zhang_xl@tib.cas.cn

Z. Tan
University of Chinese Academy of Sciences, Tianjin, China

1 Introduction

Bulk chemicals are produced at high volume and relatively low cost, which can be either directly used or used as platform chemicals for production of derivatives in the chemical industry [1, 2]. The annual production volumes of bulk chemicals are usually in the range of 1–100 million tons, and their selling prices are less than 2,000 dollars per ton [2]. Traditionally, bulk chemicals are predominantly produced from non-renewable fossil resources via the petrochemical routes [1, 3]. Because of the decrease of global storage, fluctuations of petroleum prices, trade imbalances, and political considerations, the cost for production of bulk chemicals through petrochemical process is increasing [1, 4–10]. In addition, petrochemical processes always consume a lot of energy and cause serious environmental pollutions [9–15].

Production of bulk chemicals from biomass resources by microbial cell factories is an alternative route, which is renewable and environment friendly compared to petrochemical route [11, 16–18]. However, only a few bulk chemicals can be produced by microbial cell factories. On the other hand, although some bulk chemicals can be synthesized by engineered microorganism, the producing capabilities, including titer, yield, productivity, and physiological characteristics, are not good enough to compete with petrochemical routes. Thus, it is important to expand the product range, as well as to improve producing capabilities of cell factories to decrease producing cost for commercialization.

The rapid development of metabolic engineering, systems biology, and synthetic biology has facilitated construction of microbial cell factories for producing bulk chemicals [1, 16, 19, 20]. *Escherichia coli* is the most commonly used host strain for cell factories construction since it has clear physiological and genetic characteristics and can be genetically modified easily [3, 21]. *E. coli* also grows fast in minimal salt medium and can utilize both hexose and pentose in the biomass. Currently, bio-based bulk chemicals produced by *E. coli* include not only those formed by *E. coli* native metabolic pathways such as lactate [22–25] and succinate [26–29], but also those produced by heterologous pathways or totally new synthetic pathways, such as 3-hydroxypropionic acid [30–33], 1,3-propanediol [2], isobutanol [13, 34–36], butanol [13], 1,4-butanediol [37], and alkanes [38, 39]. In this review, we will summarize the technology platforms for construction and optimization of *E. coli* cell factories, as well as representative cases of constructing *E. coli* cell factories for production of organic acids and alcohols.

2 Technology Platform for Construction of *E. coli* Cell Factories

With the developments of systems biology and synthetic biology, a technology platform has been established for the construction of *E. coli* cell factories for bulk chemicals production. This platform includes (1) design of the optimal synthetic pathway, (2) construction of the synthetic pathway, (3) optimization of the synthetic pathway, (4) optimization of the producing capability at the whole cell level, and (5) characterization of the genetic mechanisms (Fig. 1). An initial cell factory can be obtained after the first four steps. The genetic mechanisms identified for high production can be used to further improve the producing capabilities to construct the next-generation cell factories.

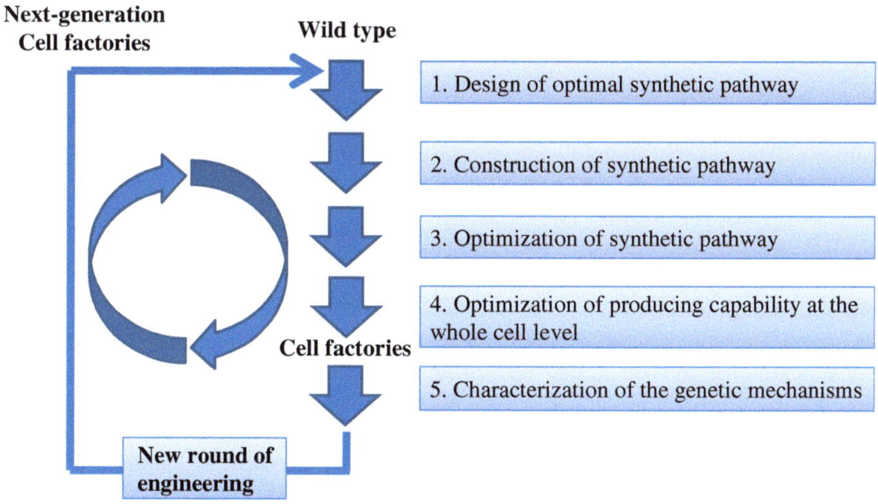

Fig. 1 The technology platform for the construction of *E. coli* cell factories

2.1 Design of the Optimal Synthetic Pathway

E. coli produces mixed acids during the glucose metabolism. In order to produce the target compound, other competitive pathways need to be inactivated. Enough energy supply is necessary to maintain cell growth and metabolism, while appropriate reducing equivalent supply is required to keep redox balanced, especially under anaerobic conditions. On the other hand, most synthetic pathways that convert the designated substrate to target product do not exist in *E. coli*, and sometimes are even not present in nature. Thus, designing the novel synthetic pathway is very important for construction of cell factories. With the development of bioinformatics tools, several genome-scale metabolic network models have been reconstructed for *E. coli* [40–43]. These models can help design the optimal synthetic pathway, discover new engineering target to improve production of target compound, and predict the cellular phenotypes [41]. There have been several reviews describing the developments of metabolic network models and their applications in constructing *E. coli* cell factories [44–49].

With the help of metabolic network models, the optimal synthetic pathway can be designed based on modification of the native metabolic pathways (such as succinate or D-lactate), integration of exogenous reactions by software predication (such as 1,4-Butanediol [12]), genome mining (such as alkanes [20]), and modification of the natural pathway to catalyze unnatural reactions (such as higher-chain alcohols [50, 51]). Different tools used for predication of novel biosynthetic pathways have been reported [52–55]. These tools can not only propose candidate pathways but also supply ranking of the pathways based on different factors employed in the process (such as thermodynamics [56]) to reduce the numbers of pathways to be a reasonable scope for experimental validation [16].

2.2 Construction of the Synthetic Pathway

Gene resources and DNA assembly method are two key factors for construction of synthetic pathway. It is desirable to obtain genes in an easy, quick, and inexpensive way, and to assemble different genes into a complete synthetic pathway with efficient and standard methods.

PCR has been commonly used to obtain target genes. However, original cells having the target genes need to be collected first, which is time-consuming. In addition, many heterologous genes cannot be expressed and translated efficiently in the host strain. With the rapid development of high-throughput chemical synthesis of DNA, achieving gene resources has become independent on original cells. It is also possible to optimize the transcriptional efficiency of target genes by codon optimization. The Church's group developed a microchip-based technology for synthesis of genes and reduced the error rate by ninefold [57]. Using this technology, all 21 genes encoding the proteins of *E. coli* 30S ribosomal subunit were synthesized and the translation efficiency in vitro was optimized through alteration of codon bias. An on-chip gene synthesis technology, including ink-jet printing, isothermal oligonucleotide amplification, and parallel gene assembly, was integrated on a single microchip [58]. By using a mismatch-specific endonuclease for error correction, the error rate was reduced to about 0.19 errors per kb.

Standardized DNA assembly is another limiting technology for construction of the synthetic pathway. Several BioBrick assembly standards [59] have been developed. The DNA unit is flanked by standardized sequences, and the assembly can be achieved by a simple and standardized restriction/ligation method [60, 61]. However, the BioBrick approach still has several disadvantages, such as the remaining of 6-bp scars resulting from each binary BioBrick assembly and the limitation of rearrangement of every intermediate part [62]. Several new technologies, such as sequence- and ligation-independent cloning (SLIC), Gibson isothermal assembly, and circular polymerase extension cloning (CPEC), have been designed, which can supply standardized, scarless, sequence-independent, and multi-part DNA assembly [62–65]. All these technologies are dependent on the 5′ homology sequence flanked at the two ends of DNA part. The biological characteristics and mechanisms for these methods have been reviewed [61].

All methods mentioned above are carried out in vitro. The Zhao group at the University of Illinois at Urbana-Champaign developed a DNA assembly method in vivo [66, 67]. DNA parts with homologous sequences were transformed into *Saccharomyces cerevisiae* and assembled based on the high homologous recombination efficiency of the yeast. The assembled DNA devices were then extracted and transformed into *E. coli* for evaluation or expression. This method is efficient and independent on enzymes in vitro.

DNA integration into chromosome has also been developed to construct a genetically stable strain for industrial production. Homologous recombination based on the λ Red recombinase has been developed [68–70]. The one-step homologous recombination method uses an antibiotic marker for selection, which is flanked by two FLP recognition target (FRT) fragments. The antibiotic marker can be removed by the FLP recombinase, facilitating multiple rounds of genetic engineering. This method can integrate or delete genes quickly but leave a 68-bp FRT scar on chromosome each time, and repeated use of this system has the potential to result in large unintended chromosomal deletions [68, 69, 71]. To facilitate sequential gene manipulations, a two-stage recombination strategy was developed, which was based on the sensitivity of *E. coli* to sucrose when levansucrase (*sacB*) is expressed in cell. In the first recombination, the target chromosomal genes are replaced by a DNA cassette containing an antibiotic marker and the *sacB* gene. In the second recombination step, the antibiotic marker and the *sacB* gene are removed by selection for the resistance to sucrose [69, 72–74].

2.3 Optimization of the Synthetic Pathway

The native and new constructed synthetic pathways are always not efficient. Some enzymes might have low activities and become the rate-limiting steps for the whole pathway. Some toxic intermediates might accumulate in the cell, thus leading to decreased cell growth and flux imbalances. In order to improve the producing capability, several strategies have been developed to optimize the synthetic pathway in three levels.

2.3.1 Optimization of the Synthetic Pathway in a Single Gene Level

Gene overexpression was commonly used to increase the activities of rate-limiting enzymes among the synthetic pathway for the purpose of modulating metabolic fluxes. However, this simple overexpression strategy rarely reached the optimal transcript level and appeared to be unsuccessful in most cases to improve producing capability [75]. Promoter library has been developed as a solution to this all-or-nothing expression strategy, which could provide variable promoters with a wide range of strength for fine-tuning of gene expression [76–80]. Two methods have been developed for creating promoter library. One type was obtained by keeping the conserved −35 and −10 sequences intact and randomizing the surrounding nucleotides [78, 79]. The other type was obtained by mutating the sequence of an existing promoter using error-prone PCR [76, 77].

However, plasmid-based gene expression has several disadvantages for the engineering of genetically stable strains [69]. Plasmid maintenance is a metabolic burden on the host cell, especially for high-copy number plasmids [81], and only few natural unit-copy plasmids have the desirable genetic stability [82]. In addition, only low-copy number plasmids have replication that is timed with the cell cycle, and thus, it is difficult to maintain a consistent copy number in all cells [82]. It is thus desirable to integrate the target genes into chromosome followed by fine-tuning of their expression. With the aid of Red recombination technology [68], promoter libraries were recently constructed directly in the chromosome [83–85], which might be more suitable for modulation of gene expression directly in the chromosome.

Different promoters with varied strengths can be used to control gene transcription precisely to obtain a specific cellular phenotype. One example is the divergence of biomass yield by modulation of phosphoenolpyruvate carboxylase (*ppc*) gene transcription level. When the wild-type promoter of *ppc* was replaced by promoters with varied strengths, there was a positive correlation between the *ppc* transcription level and the biomass yield when the *ppc* transcription level was within a certain range. Excessive *ppc* transcription level led to decreased biomass yield. The promoter library facilitated identifying the optimum transcription level of *ppc* for biomass yield. Another example is the use of promoter library to investigate the relationship between succinate production and PPC or phosphoenolpyruvate carboxykinase (PCK) activity. There was a positive correlation between PCK activity and succinate production. In contrast, there was a positive correlation between PPC activity and succinate production only when PPC activity was within a certain range. Excessive PPC activity decreased the rates of both cell growth and succinate formation [29]. In contrast, plasmid overexpression of *ppc* gene always led to increased succinate production [86], which would mislead our understanding of the relationship between succinate production and PPC activity.

2.3.2 Optimization of the Synthetic Pathway in Multiple Gene Levels

The efficiencies of most synthetic pathways are always not limited by a single rate-limiting reaction [87]. A more broadly accepted opinion is to realize the coordinated expression of multiple genes involved in the synthetic pathway to increase the overall metabolic flux.

A technology for tuning the expression of multiple genes by employing post-transcriptional mechanisms was developed by the Keasling group at the University of California, Berkeley [88]. Libraries of tunable intergenic regions (TIGRs) consisting of several control elements composed by mRNA secondary structures, RNase cleavage sites, and the RBS sequence were constructed and used to differentially change the processes of transcription termination, mRNA stability, and translation initiation [87, 88]. When using this strategy to balance expression of three genes in an operon encoding a heterologous mevalonate biosynthetic pathway, a sevenfold increase of mevalonate production was achieved. Another technology for fine-tuning pathway flux was developed by the same group in 2009 [89]. Synthetic protein scaffolds was built to spatially recruit metabolic enzymes in a specific manner to increase the valid concentration of metabolic intermediates and avoid their accumulation to toxic level. Also, the production levels can be optimized by balancing relative quantities of individual enzymes via changing the number of interaction-domain repeats that locate different enzymes to the synthetic complex. Using this technology, a 77-fold improvement was achieved for the mevalonate biosynthesis.

The Church group at Harvard Medical School developed a powerful tool termed multiple automated genome engineering (MAGE), which can modify many genes in the *E. coli* genome in parallel. This technique was used to optimize the 1-deoxy-D-xylulose-5-phosphate (DXP) synthetic pathway in *E. coli* for improving lycopene production. Twenty-four genes in the DXP pathway were modified simultaneously, and over 4.3 billion variants were created per day. *E. coli* variants with more than fivefold increase in lycopene production were isolated within 3 days [90, 91].

Instead of repeating multiple rounds of gene knockout, synthetic regulatory small RNAs (sRNAs) were designed to finely control gene expression in *E. coli* by the Lee group at Korea Advanced Institute of Science and Technology. Customized synthetic sRNAs were consisted by a scaffold and a target-binding sequence. With utilization of the plasmid-based synthetic sRNA system, one can study the effects of multiple knockdowns on the cell's producing capability in a high-throughput way and simultaneously screen target genes in different *E. coli* strains [92].

2.3.3 Optimization of the Synthetic Pathway Using the Sensor–Regulator System

To precisely control and regulate the heterologous pathway expression due to the change of environment or intracellular conditions, an efficient strategy is to use the sensor–regulator system which can respond to a particular intermediate and stimulate the desired cellular response to enable the cell to efficiently use the cellular resources and improve the producing capability while decrease the accumulation of toxic metabolite [93]. Malonyl-CoA is the rate-limiting precursor involved in the synthetic pathway of several value-added pharmaceuticals and biofuels. By incorporating the *trans*-regulatory protein FapR and the *cis*-regulatory element *fapO* of *Bacillus subtilis*, a hybrid promoter–regulator system was constructed and could respond to a wide range of intracellular malonyl-CoA concentrations in *E. coli* [93]. In another study, the Liao's group designed and engineered a regulatory circuit by recruiting and altering the Ntr regulon which is a global regulatory system to control the pathway expression for lycopene synthesis in *E. coli*. The artificially engineered regulon controlled the gene expression in the lycopene synthetic pathway by sensing the concentration of acetyl phosphate which is the glycolytic pathway hallmark metabolite [94]. Recently, a dynamic sensor–regulator system (DSRS) was developed to dynamically regulate the gene expression in biodiesel biosynthetic pathway by responding to the key intermediate fatty acyl-CoA in *E. coli*. Using this strategy, the fatty acid ethyl ester production was increased threefold compared to that of using constitutive promoters in *E. coli* [95].

2.4 Optimization of Producing Capability at the Whole Cell Level

After optimization of the synthetic pathway, the producing capability of the engineered cell, such as titer, yield, productivity, and physiological characteristics, might still be not good enough for industrial application. The desired cell phenotypes may be affected by factors which are not directly related to the synthetic pathway [96]. In order to obtain an efficient cell factory, the producing capability needs to be optimized further at the whole cell level.

Metabolic evolution was developed by the Ingram group at the University of Florida and has been demonstrated to be an excellent strategy for strain improvement [69]. Synthesis of target product is designed to be the only fermentation pathway to oxidize NADH under anaerobic condition. The cell growth of the engineered cell is coupled with the synthesis of target product, since this is the only way to regenerate NAD^+ for continuous glycolysis to provide ATP for cell growth. This technology has been used widely to improve the production of several bulk chemicals by *E. coli* cell factories, such as D-lactate [97–99], succinate [100–102], and ethanol [103]. In addition, this technology can also be used to improve cell's physiological characteristics, especially tolerance to toxic metabolites or high concentration of target products [104, 105].

Global transcription machinery engineering, or termed gTME, was developed by the Stephanopoulos group at Massachusetts Institute of Technology and has been proved to be a powerful strategy for optimization of a desired phenotype at system level [106–109]. For optimization of a desired phenotype at systems level, gTME has been used to improve ethanol tolerance, lycopene production, and simultaneous tolerance to sodium dodecyl sulfate (SDS) and ethanol. To realize these purposes, one of the components of global cellular transcription machinery (specifically *rpoD* encoding the σ^{70}) in *E. coli* was engineered to globally perturb the transcriptome to help unlock complex phenotypes [109].

Aside from technologies mentioned above, some other tools, such as genome shuffling [110–113] and trackable multiplex recombineering (TRMR) [114], have been designed and used to optimize a target pathway at system level. Great improvement for the chemical production properties has been achieved based on these strategies. It should be noted that high-throughput screening methods are required for the efficient selection.

2.5 Characterization of the Genetic Mechanism

Although metabolic evolution or other global perturbation methods are efficient for improving the producing capability of the engineered cell, the genetic backgrounds of strains obtained by these strategies often remain unclear. Characterization of the genetic mechanisms relative to the improved producing capability is very important. The fast accumulation of omics data, including genomics, transcriptomics, proteomics, metabolomics, and fluxomics, has provided foundation for the understanding of the genetic mechanisms in depth [115–119], which is crucial for further round of engineering to obtain the next-generation cell factories.

Up to now, many *E. coli* cell factories with abilities of producing different bulk chemicals have been constructed and some have been applied in industrial scale. The bio-based bulk chemicals produced by *E. coli* cell factories mainly include organic acids and alcohols, which will be described in detail in the following chapters.

3 Organic Acids

Organic acids have received attractive attentions for their increasing utilization in food industry and great potential as platform chemicals for the manufacture of biodegradable polymers [120, 121]. As an alternative of petroleum-based production, microbial production of organic acids from renewable biomass has been accepted as a feasible process. *E. coli* has been widely engineered to produce organic acids, such as acetate, lactate, pyruvate, 3-hydroxypropionate, succinate, malate, fumarate, glucaric acid, and muconic acid (Table 1).

3.1 D-lactate

As a specialty chemical, D-lactate is widely applied in the food and pharmaceutical industry. A potential huge market for D-lactate is to be combined with L-lactate to produce polylactic acid (PLA), an increasingly attractive biodegradable plastic. The commercial success of PLA will greatly depend on the production cost of the monomers [69]. Wild-type *E. coli* can produce D-lactate in its mixed acid fermentation process (Fig. 2). However, the productivity is low and several undesirable metabolites are produced at the same time. To realize the production of D-lactate in an efficient way, it is necessary to reengineer the metabolic network of *E. coli*.

D-lactate-producing strains were engineered from *E. coli* W3110 by the Ingram's group by inactivating the competitive fermentation pathways, including fumarate reductase (*frdABCD*), alcohol/aldehyde dehydrogenase (*adhE*), and pyruvate formate lyase (*pflB*). A further deletion of the acetate kinase gene (*ackA*) increased the cell mass and lactate productivity. D-lactate production yield of these strains approached the theoretical maximum yield (2 mol/mol glucose) using mineral salts medium [122]. For expanding the substrate range, a cluster of sucrose utilization genes which were characterized and cloned from *E. coli* KO11 were introduced, resulting in production of over 500 mM D-lactate from sucrose [123]. However, these biocatalysts were unable to ferment glucose or sucrose with concentration of up to 10 % completely. Inspired by the construction of ethanol producing strain, a derivative of *E. coli* B was selected as the starting strain for D-lactate production. Based on the growth-based selection (Fig. 3), metabolic evolution was carried out to improve strain performance. The resulting strain SZ194 produced 1.22 M D-lactate with a yield of 1.9 mol/mol using mineral salts medium. The production capability was comparable with lactic acid bacteria [98].

In another study, *E. coli* strain B0013 was engineered for D-lactate production by deletion of acetate kinase and phosphotransacetylase (*ackA-pta*), phosphoenolpyruvate synthase (*pps*), *pflB*, FAD-binding D-lactate dehydrogenase (*dld*), pyruvate oxidase (*poxB*), and *adhE* and *frd* genes. The resulting strain, B0013-070, produced 125 g/L D-lactate [25]. Replacing the *ldhA* promoter with the λ p_R and p_L promoter in strain B0013-070 led to a thermocontrollable strain B0013-070B in which the LDH activity was twofold higher than the parent strain B0013-070 at 42 °C. When the culture temperature reached to 33 °C, the genetic switch would be turned off and strain B0013-070B produced 10 % more biomass under aerobic conditions than stain B0013-070 with trace D-lactate produced. This modification reduced the growth inhibition which was caused by oxygen insufficiency in large-scale fermentation process [23].

Table 1 Production of organic acids by representative *E. coli* cell factories

Chemicals	Strains	Fermentation medium	Fermentation conditions	Fermentation time (h)	Titer (g/L)	Yield (g/g)	References
D-lactate	*E. coli* TG114 (Δ*pflB*, Δ*frd*, Δ*adhE*, Δ*ackA*, Δ*mgsA*)	NBS medium, pH 7.5, 12 % glucose, 1 mM betaine	Batch, anaerobic	41	118	0.98	[98, 99]
	E. coli B0013, Δ*ack-pta*, Δ*pps*, Δ*pflB*, Δ*dld*, Δ*poxB*, Δ*adhE*, Δ*frdA*	Modified M9 medium, glucose[a]	Two-phase fermentation[b]	39	125	0.96[c]	[23]
3-Hydroxypropionate	*E. coli* BL21, pQE-80L-*mcr* pACYCDuet-1r-*accADBC-birA-pntAB*	M9 medium, 100 mM potassium phosphate buffer (pH 7.0), 100 mM glucose	Aerobic, batch	24	1.193	0.016	[33]
	E. coli BL21(DE3), Δ*glpK* Δ*yqhD*, harboring plasmid pELDRR with the *L. brevis DhaB-DhaE* genes and plasmid pCPa72 with the *P. aeruginosa* PSALDH gene	Defined R medium, glycerol	Aerobic, Fed-batch	46	57.3	0.88	[30]
Succinate	*E. coli* K-12, Δ*pfl*, Δ*ldhA*, Δ*ptsG*, pTrc99A-*Rhizobium etli pyc*	Complex medium	Two-phase fermentation	76	99.2	1.1	[143, 149]
	ATCC 8739, Δ*poxB*, Δ*mgsA*, Δ*ldhA*, Δ*adhE*, Δ*ackA*, Δ(*focA-pflB*)	AM1, 100 g/L glucose, 100 mM potassium bicarbonate	Anaerobic, batch	96	83	0.92	[160]
	ATCC 8739, Δ*ptsI*, Δ*ldhA*, Δ*pflB*, *Ppck**-		Anaerobic, batch	96	96	0.89	[163]

(continued)

Table 1 (continued)

Chemicals	Strains	Fermentation medium	Fermentation conditions	Fermentation time (h)	Titer (g/L)	Yield (g/g)	References
	galP, *Ppck*-pck*, Δ*ackA-pta*, *Ppck*-aceAB*, *Ppck*-dcuC*, Δ*mgsA*	AM1 medium, 12 % glucose, 35 mM sodium bicarbonate					
Malate	ATCC 8739, Δ*ldhA*, Δ*pflB*, Δ*ackA*, Δ*adhE*, Δ*fumAC*, Δ*fumB*, Δ*maeB*, Δ*sfcA*, Δ*frdBC*, Δ*mgsA*, Δ*poxB*	NBS medium, 5 % glucose, 100 mM potassium bicarbonate	Two-phase fermentation	72	33.9	1.06	[175]
Fumarate	W3110, Δ*iclR*, Δ*fumC*, Δ*fumA*, Δ*fumB*, Δ*ackA*, Δ*ptsG*, Δ*aspA*, Δ*lacI*, *galP::Ptrc*, harboring pTac15k*ppc*	MR medium supplemented with 20 g/L of glucose, 3 g/L of yeast extract, 2 g/L of Na_2CO_3 and 500 mg/L of L-aspartic acid	Fed-batch	63	28.2	0.389	[190]

[a] Experiments in bioreactor with the initial glucose concentration of 30 g/L, and the residual glucose concentration was maintained above 10 g/L by adding in four batches. The total glucose reached 649.5 g in the end

[b] Aerobic growth and oxygen-limited production

[c] Average lactate yield during oxygen-limited phase

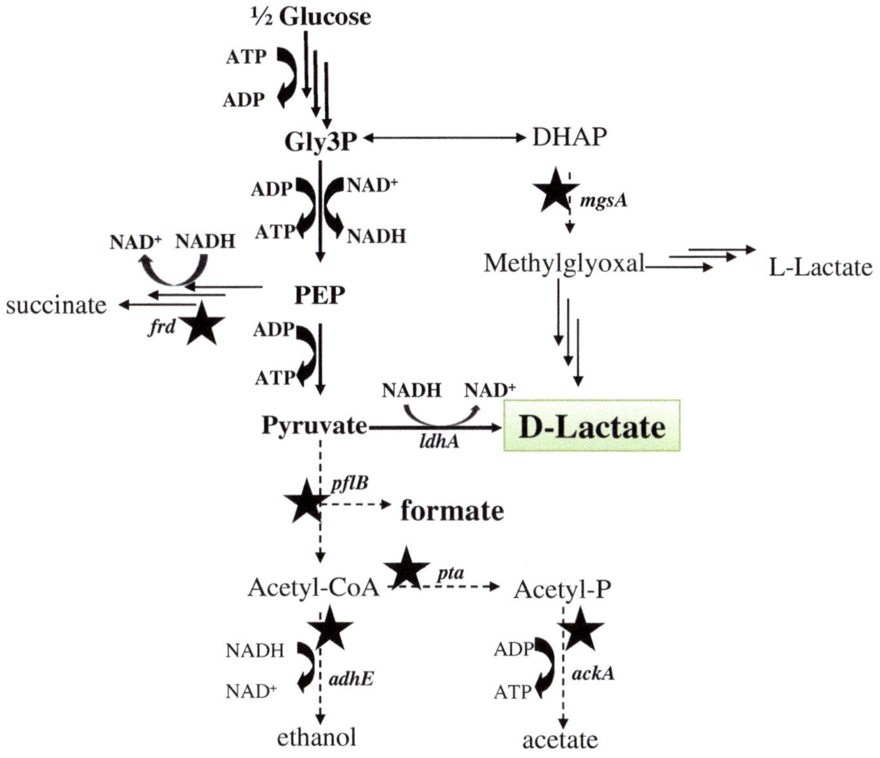

Fig. 2 Construction of *E. coli* cell factories for the production of D-lactate. The *dottled line* and star indicate metabolic reactions which were inactivated to increase production of D-lactate

3.2 3-Hydroxypropionate

3-Hydroxypropionate (3HP), a non-chiral carboxylic acid, has received much attention for its potential applications to produce biodegradable polymer by itself or with other compounds [124, 125]. Additionally, 3HP was an important C3 platform chemical and can be used for the production of various commercially valuable chemicals, such as 1,3-propanediol, acrylic acid, and malonic acid [126]. 3HP has been identified as a metabolic intermediate naturally present in several microorganisms [127–134]. More than a dozen of pathways for 3HP biosynthesis have been proposed based on the natural metabolic pathways or in silico design [135–137]. However, a little fraction of the pathways have been evaluated. The Park's group developed a recombinant *E. coli* strain producing 3HP from glucose involved malonyl-CoA as an intermediate. In this strain, a *mcr* gene encoding the NADPH-dependent malonyl-CoA reductase (MCR) of *Chloroflexus aurantiacus* DSM 635 was introduced into *E. coli*. The recombinant strain produced 0.064 g/L 3HP when cultivated aerobically for 24 h using glucose as the sole carbon source. To improve

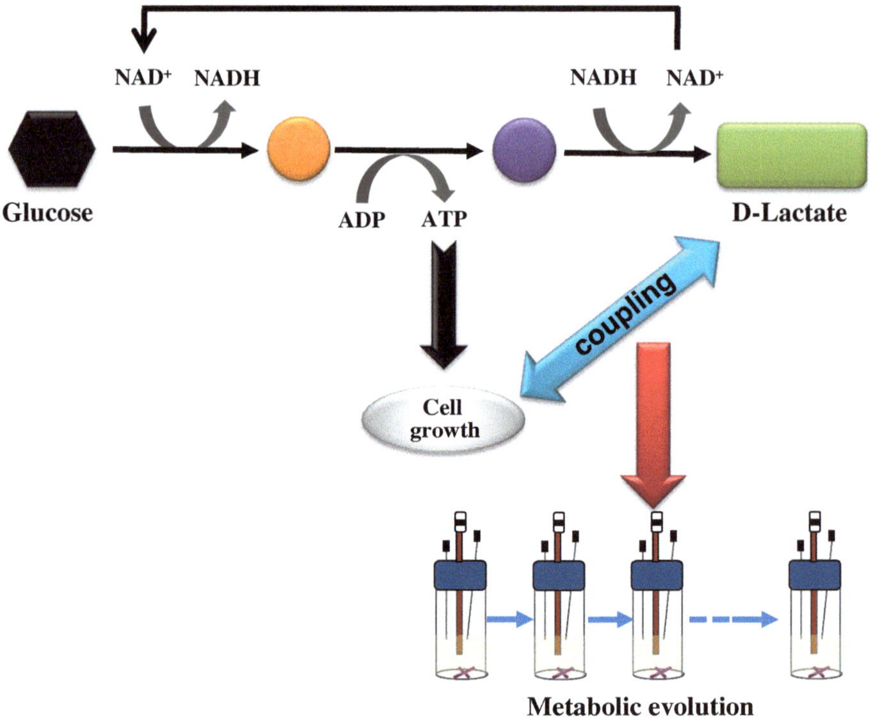

Fig. 3 Metabolic evolution based on the coupling of cell growth and D-lactate production

the 3HP production, the gene cluster *accADBCb* encoding the acetyl-CoA carboxylase and biotinilase of *E. coli* K-12 were overexpressed and this resulted in a twofold improvement in 3HP production. Further genetic modification was carried out to express the gene *pntAB* encoding the membrane-bound transhydrogenase to convert the NADH to NADPH which increased 3HP titer to 0.193 g/L [33].

Compared to producing 3HP from glucose, more studies have been focused on the production of 3HP from glycerol. By heterologous overexpression of the glycerol dehydratase (DhaB) from *Klebsiella pneumobiae* DSM 2026 and aldehyde dehydrogenase (AldH) from *E. coli* K-12 MG1655 in *E. coli* BL21 (DE3), a recombinant *E. coli* strain SH254 was obtained. When fermented aerobically in M9 minimal medium supplemented with glycerol as substrate in shake flask, this strain produced 0.58 g/L 3HP with a yield of 0.48 mol/mol glycerol [138]. Further optimization of the fermentation parameters, such as pH, IPTG concentration, aeration rate, and substrate concentration, led to production of 31 g/L 3HP in 72 h when a fed-batch fermentation process was used [139]. Though the titer of 3HP was improved by optimization of the fermentation parameters, several problems, including the imbalance between DhaB and AldH and instability of DhaB, were still not solved. To overcome these limitations, DhaB and AldH were overexpressed in two compatible plasmids with inducible expression systems and the glycerol

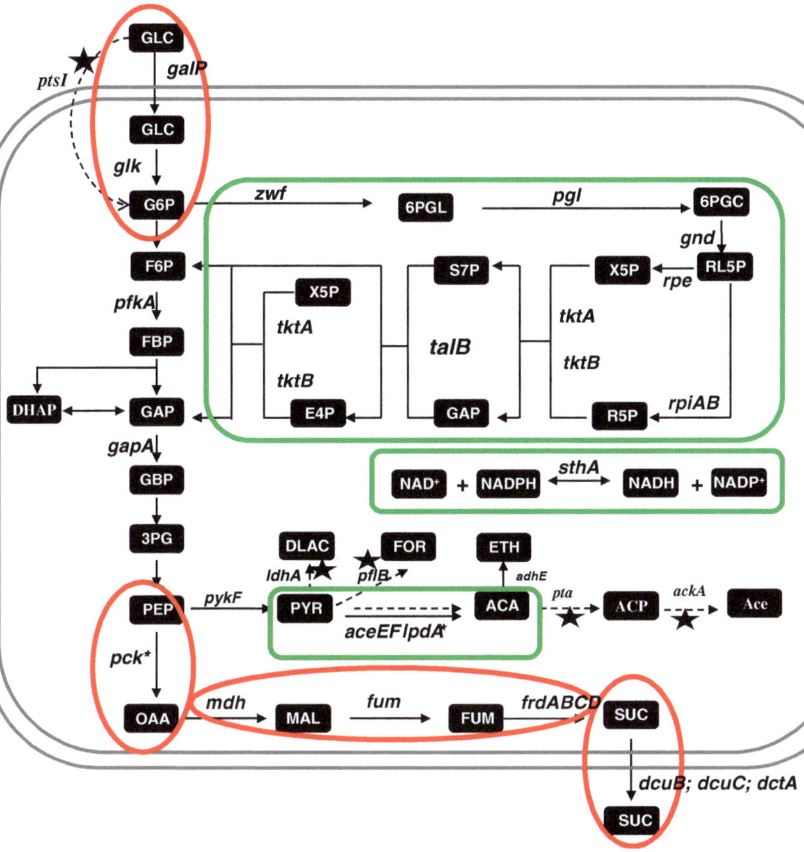

Fig. 4 Genetic mechanisms for high-succinate production in *E. coli* strain HX024. *Green square lines* represent activated metabolic modules (including pentose phosphate, transhydrogenase, and pyruvate dehydrogenase) which are responsible for increased succinate yield. Pentose phosphate pathway was activated to increase the supply of NADPH, which was then converted to NADH through soluble transhydrogenase for succinate production. The sensitivity of pyruvate dehydrogenase to NADH inhibition was eliminated by the *lpdA* gene mutation. Pyruvate dehydrogenase activity increased under anaerobic conditions, which provided additional NADH for succinate production. *Red oval lines* represent activated metabolic modules (including glucose transport, carboxylation, reductive TCA, and succinate transport) which are responsible for increased succinate productivity. *G6P* glucose-6-phosphate, *6PGL* gluconolactone-6-phosphate, *6PGC* 6-phospho gluconate, *Ru5P* ribulose-5-phosphate, *X5P* xylurose-5-phosphate, *R5P* ribose-5-phosphate, *S7P* sedoheptulose-7-phosphate, *E4P* erythrose-4-phosphate, *F6P* fructose- 6-phosphate, *FBP* fructose-1,6-bisphosphate, *GAP* glyceraldehyde-3-phosphate, *DHAP* dihydroxyacetonephosphate, *1,3-BPG* 1,3-bisphosphoglycerate, *3-PG* 3-phosphoglycerate, *2-PG* 2-phosphoglycerate, *PEP* phosphoenolpyruvate, *ACP* acetylphosphate, *Ace* acetate, *DLAC* D-lactate, *FOR* formate, *ETH* ethanol, *OAA* oxaloacetate, *CIT* citrate, *ICIT* isocitrate, *GLO* glyoxylate, *MAL* malate, *FUM* fumarate, *SUC* succinate, NAD^+ oxidized nicotinamide adeninedinucleotide, *NADPH* reducednicotinamide adenine dinucleotidephosphate, *NADH* reduced nicotinamide adeninedinucleotide, and $NADP^+$ oxidized nicotinamide adeninedinucleotide phosphate

dehydratase reactivase (GDR) was expressed at the same time. Then, by using α-ketoglutaric semialdehyde dehydrogenase (KGSADH) from *A. brasilense* to replace the AldH, a recombinant *E. coli* strain SH-BGK1 was constructed which produced 38.7 g/L 3HP aerobically using a fed-batch process [140]. Modulation of glycerol metabolism further increased 3HP titer to 57.3 g/L with a yield of 0.88 g/g glycerol [30]. In another work, based on in silico simulation, two genes *tpiA* (encoding triose phosphate isomerase) and *zwf* (encoding glucose 6-phosphate dehydrogenase) involved in the central metabolism and *yqhD* gene (encoding NADPH-dependent aldehyde reductase) involved in the biosynthetic pathway of the major by-product 1,3-propanediol were identified as the engineering targets to improve 3HP production from glycerol. Deletion of these three genes led to 7.4-fold increase of 3HP titer compared to the parent strain [141].

3.3 Succinate

Succinate can be produced by native *E. coli* as a minor product [56]. In order to produce succinate as the sole product, other competitive pathways need to be eliminated. Strain NZN111 is an engineered *E. coli*, which has lactate dehydrogenase (*ldhA*) and *pflB* inactivated [142–144]. This strain produces undetectable lactate and formate under anaerobic conditions [142, 143]. However, inactivation of these NADH-consuming pathways could also cause redox imbalance within cells, thus leading to decreased cell growth and glucose utilization [143]. Strain NZN111 consumed only 1.8 g/L glucose and produced 1.8 g/L succinate under anaerobic conditions for 44 h [145]. A mutant strain of NZN111 (strain AFP111) was isolated, which recovered cell growth and had increased succinate production under anaerobic conditions. A spontaneous mutation in *ptsG* gene [146–148] of NZN111, which encodes $EIIB^{Glc}$ subunit of phosphoenolpyruvate (PEP): carbohydrate phosphotransferase systems (PTS), was identified to be responsible for the increased cell growth and succinate production [143]. This *ptsG* mutation could enhance PEP precursor supply for succinate synthesis, as well as alleviating glucose repression to the expression of several genes which are crucial to the fermentation. Several genetic manipulations were performed to further improve succinate production of strain AFP111. For instance, overexpression of pyruvate carboxylase (PYC) gene of *Rhizobium etli* in AFP111 increased succinate titer and yield to 99.2 g/L (841 mM) and 1.1 g/g (1.68 mol/mol), respectively, under dual-phase conditions [149].

Under anaerobic conditions, 1 molecule glucose produces 2 molecules NADH through glycolysis, while the production of 1 molecule succinate requires two NADH through the reductive TCA pathway. The maximal succinate yield is only 1 mol/mol glucose, which is much less than the theoretical maximum yield (1.71 mol/mol) [150–153]. In comparison, NADH requirement for succinate synthesis decreases when glyoxylate shunt pathway is utilized. It was calculated that 1.25 molecules NADH was required to synthesize 1 molecule succinate [150]. Glyoxylate shunt pathway is composed of isocitrate lyase (encoded by *aceA*) and malate synthase (encoded by *aceB*) [154]. In the presence of glucose, *aceBAK* operon is strictly repressed by IclR regulator [155–157], and *iclR* deletion was proven to efficiently activate glyoxylate shunt pathway [155]. Sanchez et al. found that deletion of *iclR* in strain SBS550MG (Δ*adhE*, Δ*pflB*, Δ*ack-pta*) harboring pyruvate carboxylase from *Lactococcus lactis* increased its succinate yield to 1.61 mol/mol [150]. Obtaining NADH through formate dehydrogenase is another strategy to improve succinate yield [158]. Blazer et al. reported that overexpression of heterologous NAD^+-dependent formate dehydrogenase from *Candida boidinii* increased succinate yield to 1.74 mol/mol glucose [158]. External formate supplementation further resulted in 6 % increase in succinate yields [158].

Although plenty of successes were obtained in metabolic engineering of *E. coli* to improve succinate production as mentioned above [143, 145, 149, 150, 158, 159], there were several problems that remained to be improved. Many research groups have employed dual-phase fermentation for succinate production, i.e., aerobic growth phase followed by anaerobic fermentation phase [149, 150]. Part of the carbon source is converted to cell mass and carbon dioxide during the aerobic phase, which leads to decreased succinate yield. Supply of dissolved oxygen also increases the energy costs during industrial production. In addition, rich medium is frequently used for fermentation [149, 150], which would increase material costs and downstream purification costs. It is very important to use mineral salts medium and one-step anaerobic process for succinate production. By combining metabolic engineering to inactivate competitive fermentation pathways and metabolic evolution to improve cell growth and succinate production, a high-succinate-producing strain KJ073 was obtained by the Ingram's group which produced 668 mM succinate with a yield of 1.2 mol/mol using mineral salts medium and one-step anaerobic process [160]. The genetic mechanisms for efficient succinate production of strain KJ073 were further identified [101]. PCK activity was increased due to a G-to-A transition at −64 position relative to the ATG start codon of *pck*, which increased the energy supply for cell growth and succinate production under anaerobic condition [101]. In addition, a frame-shift mutation in *ptsI* gene, which encodes the EI component of PTS system [161], was also found in KJ073, which increased PEP precursor supply for succinate production [101]. Reverse metabolic engineering was performed to verify the effects of these two core mutations. After increasing PCK activity and deleting *ptsI* gene in wild-type *E. coli* ATCC 8739, succinate titer and yield increased 3.7- and 4.6-fold compared with parent strain, respectively [162].

Besides inactivating competitive fermentation pathways, increasing energy supply, and increasing precursor supply, the fourth key factor for efficient succinate production is increasing reducing equivalent supply. As mentioned above, activating glyoxylate bypass and recruiting formate dehydrogenase could increase reducing equivalent supply [150, 158]. In addition, two reducing equivalent conserving pathways were identified recently, which could increase succinate yield [163]. By combining metabolic engineering and metabolic evolution, a high-succinate-producing strain HX024 was obtained (Fig. 4), which produced 813 mM succinate with a yield of 1.36 mol/mol using mineral salts medium and one-step anaerobic process [163]. Genetic mechanisms for high yield were then identified through genome sequencing and transcriptome and enzyme assay analysis. Pyruvate dehydrogenase (PDH) activity increased significantly, and sensitivity of PDH to NADH was eliminated by three mutations in LpdA, which is the E3 component of PDH [164–166]. On the other hand, pentose phosphate pathway (PPP) and transhydrogenase SthA [167–169] were activated. More carbon flux could go through the pentose phosphate pathway, thus leading to production of more reducing equivalent in the form of NADPH, which was then converted to NADH through soluble transhydrogenase for succinate production. Reverse metabolic engineering was further performed in the parent strain. Succinate yield increased from 1.12 to 1.5 mol/mol (88 % of theoretical maximum yield) by activating PDH, PPP, and SthA transhydrogenase in combination. It was suggested that the theoretical maximum succinate yield can also be obtained if 85.7 % of the carbon source goes through PPP, using both NADH and NADPH as the reducing equivalents [163]. The other benefit of using the PPP for succinate production is that only half exogenous CO_2 is required, which could reduce the fermentation cost [163].

3.4 Malate

Malate, together with fumarate and succinate, has been identified as one of the 12 most valuable bulk chemicals by the US Department of Energy [126]. It can be produced by several native microorganisms [170–174]. Since converting one molecule pyruvate to one molecule malate only requires one NADH, the theoretical maximum yield for malate production can be 2 mol/mol glucose. Starting from a succinate-producing strain KJ073, the Ingram's group developed an engineered *E. coli* strain for L-malate production [175]. Inactivating fumarase isoenzymes could not convert the succinate-producing strain to produce malate, and the resulting strain still accumulated large amounts of succinate. Fumarate appears to be the immediate precursor for succinate production in a fumarase-negative background. By contrast, it was surprisingly found that inactivation of fumarate reductase alone could reforce the carbon flow into malate production. It was suggested that the thermodynamic equilibrium favors the hydration of fumarate to malate and *E. coli* might have a better malate-transporting capability than fumarate. Inactivation of fumarase and malic enzymes further improve malate production. Strain XZ-T658 was obtained which produced 163 mM malate with a yield of 1.0 mol/mol glucose. When using a two-stage process, 253 mM malate was produced within 72 h and the yield reached 1.42 mol/mol [175].

3.5 Fumarate

Production of fumarate using fermentative process has been studied a century ago [176, 177], and the focus has been concentrated on the *Rhizopus* strains

[176, 178–188]. The best reported strain can produce 126 g/L fumarate with a yield of 0.97 g/g form glucose [189]. Recently, *E. coli* was also engineered by the Lee's group for fumarate production under aerobic conditions [190]. The carbon flux was redirected through the glyoxylate shunt by deletion of the *iclR* gene, while the fumarate production was increased by the deletion of the *fumA*, *fumB,* and *fumC* genes. The engineered strain produced 1.45 g/L fumarate when glucose was used as the substrate. The *ppc* gene was then overexpressed, and the fumarate production increased to 4.09 g/L. To reach better performance, further genetic modifications were carried out including deletion of *arcA* (encoding ArcA transcriptional dual regulator) and *ptsG* genes to increase the oxidative TCA cycle flux, deletion of *aspA* (encoding aspartate ammonia-lyase) to decrease the degradation of fumarate, and replacement of the native promoter of *galP* by a strong *trc* promoter to promote the uptake of glucose. Strain CWF812 was obtained which produced 28.2 g/L fumarate with a yield of 0.389 g/g glucose when fermented in fed-batch for 63 h [190].

3.6 Glucaric Acid

D-Glucaric acid, a compound present in fruits, vegetables, and mammals, has been studied for therapeutic purpose [191–193], and it has potential applications for polymers [126]. The synthetic pathway for D-glucaric acid production from glucose is present naturally in mammals. However, this natural pathway is composed of more than 10 reactions and limits its construction in *E. coli*. To realize the production of D-glucaric acid, the Prather's group designed a synthetic pathway by coexpression of *ino1* encoding *myo*inositol-1-phosphate synthase from *S. cerevisiae*, *miox* encoding *myo*inositol oxygenase from mice, and *udh* encoding the urinate dehydrogenase from *Pseudomonas syringae* in *E. coli*. The resulting strain produced more than 1 g/L of glucaric acid using LB medium with 10 g/L glucose [194]. MIOX was identified as the rate-limiting step in the whole pathway, and its activity was strongly affected by the *myo*inositol concentration. To improve the flux for glucaric acid production, two strategies were carried out. Utilization of protein scaffold to colocalize the three heterologous enzymes in a designable complex resulted in fivefold improvement of glucaric acid titer [195]. On the other hand, protein fusion tags and directed evolution were used to improve MIOX activity, leading to the production of 4.85 g/L glucaric acid from 10.8 g/L *myo*inositol [196].

3.7 Muconic Acid

Muconic acid (MA) is an important unsaturated dicarboxylic acid and has great potential for the production of bioplastics [197–199]. It can also be used as the precursor for the synthesis of important bulk chemicals, such as adipic acid, terephthalic acid, and trimellitic acid [197]. Biosynthesis of muconic acid in *E. coli* has been studied from 1994 by the Frost's group [200]. They described an artificial pathway for muconic acid biosynthesis from glucose by combining the shikimic acid pathway, which is natively present in *E. coli* for aromatic amino acid synthesis, with three heterologous enzymes including 3-dehydroshikimate (DHS) dehydratase, protocatechuic acid (PCA) decarboxylase, and catechol 1,2-dioxygenase (CDO). Inactivation of shikimate dehydrogenase to reduce the DHS consumption and overexpression of transketolase, 3-deoxy-D-arabinoheptulosonate 7-phosphate (DAHP) synthase, and 3-dehydroquinate (DHQ) synthase to increase the availability of DHS were further performed to improve muconic acid production. The resulting strain produced 2.4 g/L muconic acid in a batch fermentation. Deregulation of the feedback inhibition of shikimic acid pathway and overexpression of the critical genes increased muconic acid titer to 38.6 g/L [201]. Optimization of the fermentation process using fed-batch conditions further improved the titer to 59.2 g/L [202].

A novel artificial pathway for MA production in *E. coli* was established by integration of the native tryptophan biosynthetic pathway with a heterologous anthranilate degradation pathway [199]. In this pathway, anthranilate which is an intermediate involved in the native tryptophan biosynthetic pathway was transformed into MA sequentially by anthranilate 1,2-dioxygenase (ADO) from *Pseudomonas aeruginosa* and catechol 1,2-dioxygenase (CDO) from *P. putida*. The MA production was optimized by screening several enzyme candidates and improving the native tryptophan biosynthetic pathway. The resulting strain produced 389 mg/L muconic acid using the modified M9 minimal medium with a mixture carbon sources of glycerol and glucose [199].

Another novel MA synthetic pathway was designed via extending shikimate pathway by introducing the hybrid of a salicylic acid (SA) biosynthetic pathway with its partial degradation pathway [198]. A well-developed phenylalanine-producing strain was first engineered to produce SA by heterologous expression of the isochorismate synthase and isochorismate pyruvate lysate, leading to production of 1.2 g/L of SA. The SA was then converted into MA by introducing salicylate 1-monoxygenase and catechol 1,2-dioxygenase. Optimization of the whole pathway resulted in the production of MA up to 1.5 g/L after 48-h fermentation in shake flasks [198].

3.8 Adipic Acid

As the most important dicarboxylic acid, it is estimated that the market volume of adipic acid is about 2.6 million tons per year in global and an increase of 3–3.5 % will be expected annually [203, 204]. The primary use of adipic acid is as precursor for the production of polyamide nylon-6,6 [200, 203–205]. Traditionally, adipic acid is produced by chemical catalytic pathway in industrial large-scale processes using benzene, an important compound derived from non-renewable fossil resource, as the principal starting compound [200, 203]. To decrease the dependence on fossil feedstock, many efforts have been made in the past years to develop an alternative way to produce adipic acid from renewable biomass resources [11, 200, 201, 203, 205–211].

Although *cis,cis*-muconic acid can be converted to adipic acid by chemical hydrogenation [200, 201], it is still designed to construct cell factories for producing adipic acid directly through glucose fermentation [203, 205, 206, 208, 210]. Recently, the Zhong group at Shanghai Jiao Tong University constructed an artificial adipic acid synthetic pathway in *E. coli* [205]. Acetyl-CoA and succinyl-CoA were condensed to produce the C6 backbone 3-oxoadipyl-CoA, which was then converted to adipic acid sequentially via 3-hydroxyadipyl-CoA, 2,3-dehydroadipyl-CoA, and adipyl-CoA. The six enzymatic steps were catalyzed respectively by the β-ketoadipyl-CoA thiolase (PaaJ) from *E. coli*, 3-hydroxybutyryl-CoA dehydrogenase (Hbd) and crotonase (Crt) from *Clostridium acetobutylicum*, *trans*-enoyl-CoA reductase (Ter) from *Euglena gracilis*, and phosphate butyryltransferase (Ptb) and butyryl kinase (Buk1) from *C. acetobutylicum*. The constructed strain AA1 produced 31 μg/L adipic acid when fermented in minimal R/2 medium supplemented with 10 g/L glucose aerobically at 30°C for 120 h. The adipic acid titer increased to 120 μg/L when replacing Ter with butyryl-CoA dehydrogenase (Bcd) from *C. acetobutylicum*, replacing Hbd with 3-hydroxyacyl-CoA reductase (PaaH1) from *Ralstonia eutropha*, and replacing Crt with the putative enoyl-CoA hydratase (ECH) from *R. eutropha* H16. Supplies of acetyl-CoA and succinyl-CoA precursors were then increased to further improve adipic acid production, resulting in strain AA7 which produced 639 μg/L adipic acid which was about 20-fold higher than that of the starting strain AA1 [205].

4 Alcohols

E. coli cell factories have been constructed for production of a variety of alcohols, such as 1,3-propanediol [19, 212], 1-propanol [213], 1,2-propanediol [214, 215], isopropanol [216], *n*-butanol [217], isobutanol [13], 1,4-butanediol [12], and higher-chain alcohols [13, 51, 218, 219]. Some reviews have been focused on the elucidation of the bio-based production of alcohols using *E. coli* cell factories [2, 3, 217, 220], which will not be described here. This chapter will focus on recently developed cell factories for the production of higher-chain alcohols and 1,4-butanediol.

4.1 Higher-Chain Alcohols

Higher-chain alcohols are attractive biofuel targets because they exhibit higher energy density, lower hygroscopicity, lower vapor pressure, and compatibility with present transportation devices [218]. However, these compounds are not synthesized economically by native organisms [50]. Two different synthetic pathways have been created for the production of these compounds.

By the introduction of the broad-substrate-range 2-keto acid decarboxylases (KDCs) and alcohol dehydrogenases (ADHs) genes, the amino acid synthetic pathways can be redirected to produce higher-chain alcohols from 2-keto acids in *E. coli* [221]. Isobutanol is a representative example. A synthetic pathway for isobutanol production from glucose had been created by the Liao's group through combining branched-chain amino acid synthetic pathway and Ehrlich pathway with 2-keto-isovalerate serving as a precursor [13]. Overexpression of the valine biosynthetic pathway (*ilvIHCD*) and the alcohol producing pathway (*kivD* from *Lactococcus lactis* and *adh2* from *S. cerevisiae*) resulted in production of 1.7 g/L isobutanol. Competitive fermentation pathways, including *adhE*, *ldhA*, *frdAB*, *fnr*, and *pta*, were further deleted to increase the pyruvate supply for the isobutanol production. The resulted strain produced 2.2 g/L isobutanol with a yield of 0.21 g/g glucose. Further improvement was carried out by replacing the native *ilvIH* by the *alsS* gene from *B. subtilis*. AlsS has higher affinity for pyruvate than IlvIH, and the replacement increased isobutanol titer up to 3.7 g/L. With a deletion of the *pflB* gene, the isobutanol titer increased to 22 g/L under microaerobic conditions [13]. When fermented in a 1-L bioreactor instead of the shake flask with in situ isobutanol removal using gas stripping, the isobutanol production could reach a concentration of more than 50 g/L in 72 h [221].

NADPH is the reducing equivalent required for the production of isobutanol. Both keto acid reductoisomerase and alcohol dehydrogenase are NADPH dependent, and two equivalents of NADPH are required for the conversion of pyruvate to isobutanol. In contrast, the common reducing equivalent under anaerobic condition is NADH, which is produced through glycolysis [222]. In order to solve this cofactor imbalance problem, the cofactor specificity of keto acid reductoisomerase and alcohol dehydrogenase enzymes was changed from NADPH to NADH, and theoretical yield was obtained under anaerobic condition [222]. On the other hand, membrane-bound transhydrogenase PntAB and NAD kinase were activated in combination to increase the NADPH supply for improved isobutanol production [35]. Activating these two enzymes increased anaerobic isobutanol yield by 39 % to 0.92 mol/mol glucose [35].

Enzymes involved in the native L-leucine biosynthesis pathway were designed to catalyze the chain elongation, and various 2-keto acids were obtained. Diverse LeuA mutants were generated to suit the different size of the substrates [13, 16, 51, 217, 218]. These 2-keto acids were then successively converted to aldehyde and alcohols in turn by introducing KDCs and ADHs. By this pathway design, one carbon atom was added to the chain in each cycle and a new pool of alcohols can be produced.

A second pathway designed for higher-chain alcohol production was based on acetyl-CoA. Two carbon atoms were added to the chain in each cycle [217, 223–225]. Five reactions were involved in the carbon chain elongation reactions and the production of alcohols, which were catalyzed by the thiolase (AtoB/BktB), dehydrogenase (Hbd/PaaH1), dehydrase (Crt), reductase (Ter), and thioesterase (TesB), respectively. These enzymes were not specific to all compounds with different carbon number. It is critical to find more specific enzymes to increase the efficiency for certain products.

4.2 1,4-Butanediol

As one of the important C4 platform chemicals, 1,4-butanediol (BDO) owns a world market exceeding 1 million tons and is used widely in the manufacture of biopolymers, cosmetics, fine chemicals, and solvents [3, 19]. BDO is predominantly produced from crude oil and natural gas. No biosynthetic pathways have been reported in any natural organisms. The project of constructing an *E. coli* cell factory for BDO production was initiated by Genomatica [12]. All candidate pathways from *E. coli* central metabolites to BDO were elucidated based on the SimPheny Biopathway Predictor software. Rather than by known enzyme reactions, the transformation of functional groups by known chemistry was used as the basis for the Biopathway Predictor algorithm, and this gave a chance to identify novel enzyme activities or to engineer enzymes with specific activities to a particular substrate. As a result, 10,000 pathways for the BDO synthesis from common central metabolites were identified. Then, the proposed pathways were evaluated based on the different factors including maximum theoretical yield, pathway length, number of non-native steps, and thermodynamic feasibility. At last, two pathways for BDO biosynthesis, which involved 4-hydroxybutyrate as the intermediate, were proposed as the highest priority and tested in vivo. A pathway was proved to be potential for BDO production in which BDO was produced from succinate via six enzymatic reactions catalyzed by two *E. coli* native enzymes (succinyl-CoA synthetase and alcohol dehydrogenase) and four heterologous enzymes (CoA-dependent succinate semialdehyde dehydrogenase, 4-hydroxybutyrate dehydrogenase, 4-hydroxybutyryl-CoA transferase, and 4-hydroxybutyryl-CoA reductase). The engineered strain produced over 18 g/L BDO from glucose in 5 days [12]. It could also produce BDO from sucrose, xylose, and biomass-derived mixed sugar streams.

To increase the efficiency of the BDO biosynthetic pathway, butyraldehyde dehydrogenase (Bld) and butanol dehydrogenase from *Clostridium saccharoperbutylacetonicum* were selected and used for BDO production in *E. coli*. Furthermore, random mutagenesis and site-directed mutagenesis were carried out in turn to improve the activity of Bld. The resulted strain could produce BDO with the titer fourfold greater than those of strains expressing the wild-type Bld [37].

5 Perspectives

Production of bulk chemicals by *E. coli* cell factories from renewable biomass resources has been proved to be a sustainable and environment-friendly process to replace the petroleum-based process. Due to the rapid development of metabolic engineering, systems biology, and synthetic biology, great progress has been achieved and many successful *E. coli* cell factories have been constructed. However, there are still several challenges.

Only a few bulk chemicals can be produced biologically by microorganisms. Although several new synthetic pathways have been designed and created, most of the bulk chemicals still do not have biosynthetic pathways. The primary reason is that many chemical reactions do not have natural enzymes. Creation of new enzymes to catalyze the desired chemical reaction by integrating chemistry, protein rational design, and directed evolution will be in demand to full in the gaps existing in the novel biosynthetic pathway [16].

The created synthetic pathway is usually not efficient due to the low catalytic activities of some specific enzymes, especially those new enzymes. It is thus very important to improve the catalytic capabilities of these enzymes so that they are not rate limiting within the whole pathway. In addition, coordinated expression of multiple genes involved in the synthetic pathway is desirable so that there will be no metabolic imbalance problem such as exhaustion of precursor and accumulation of toxic intermediates [226]. Product yield is an important factor to realize low-cost production of bulk chemicals. On the one hand, redox balance is necessary to maintain anaerobic cell growth since the only way to consume the reducing equivalent is through the synthetic pathway of target compound. On the other hand, enough reducing equivalent is required to reach the theoretical maximum yield.

Finally, good physiological characteristics of cell factories are necessary for large-scale industrial production of bulk chemicals [227]. Tolerance to high osmolality and high concentration of target chemicals can increase the final titer and productivity. Tolerance to high temperature can reduce the energy cost and contamination problems. Tolerance to low pH can produce organic acid directly so that complex downstream purification process can be avoided to convert organic acid salt to organic acid. Since modifying single gene usually has no effect on improving physiological characteristics, global perturbation strategies together with high-throughput omics analysis are needed to improve these physiological characteristics and identify the genetic mechanisms so that bulk chemicals can be produced biologically in a cost-comparable way compared to petrochemical process.

Acknowledgments This research was supported by grants from Tianjin Key Technology R&D program of Tianjin Municipal Science and Technology Commission (11ZCZDSY09100, 13ZCZDSY05300), the National Natural Science Foundation of China (31370136), and the Key Deployment Project of the Chinese Academy of Sciences (KGZD-EW-606; KSZD-EW-Z-016-2).

References

1. Vickers C, Klein-Marcuschamer D, Krömer J (2012) Examining the feasibility of bulk commodity production in *Escherichia coli*. Biotechnol Lett 34:585–596
2. Zeng AP, Biebl H (2002) Bulk chemicals from biotechnology: the case of 1,3-propanediol production and the new trends. Adv Biochem Eng Biotechnol 74:239–259

3. Chen X, Zhou L, Tian K et al (2013) Metabolic engineering of *Escherichia coli*: a sustainable industrial platform for bio-based chemical production. Biotechnol Adv 31:1200–1223
4. Keasling JD (2010) Manufacturing molecules through metabolic engineering. Science 330:1355–1358
5. Taher H, Al-Zuhair S, Al-Marzouqi AH et al (2011) A review of enzymatic transesterification of microalgal oil-based biodiesel using supercritical technology. Enzyme Res 2011:468292
6. U.S. Department of Energy EIA (2012) Petroleum marketing monthly. Department of Energy, Washington, DC, DOE Publ. No. EIA-0380(20012/02)
7. U.S. Department of Energy EIA (2006) Annual energy outlook 2006 with projections to 2030. Department of Energy, DOE Publ. No. EIA-0383(2006), Washington, DC
8. U.S. Department of Energy EIA (2007) Annual energy outlook 2007 with projections to 2030. Departmentt of Energy, DOE/EIA-0383(2007), Washington, DC
9. Lee JW, Na D, Park JM et al (2012) Systems metabolic engineering of microorganisms for natural and non-natural chemicals. Nat Chem Biol 8:536–546
10. Demirbas A (2009) Political, economic and environmental impacts of biofuels: a review. 86. Appl Energy 86(1):S108–S117
11. Jang YS, Kim B, Shin JH et al (2012) Bio-based production of C2-C6 platform chemicals. Biotechnol Bioeng 109:2437–2459
12. Yim H, Haselbeck R, Niu W et al (2011) Metabolic engineering of *Escherichia coli* for direct production of 1,4-butanediol. Nat Chem Biol 7:445–452
13. Atsumi S, Hanai T, Liao JC (2008) Non-fermentative pathways for synthesis of branched-chain higher alcohols as biofuels. Nature 451:86–89
14. Jambunathan P, Zhang K (2014) Novel pathways and products from 2-keto acids. Curr Opin Biotechnol 29:1–7
15. Schiweck H, Bär A, Vogel R et al (2000) Sugar Alcohols, in *Ullmann's Encyclopedia of Industrial Chemistry*. Wiley-VCH Verlag GmbH & Co, KGaA
16. Shin JH, Kim HU, Kim DI et al (2013) Production of bulk chemicals via novel metabolic pathways in microorganisms. Biotechnol Adv 31:925–935
17. Jarboe LR, Liu P, Kautharapu KB et al (2012) Optimization of enzyme parameters for fermentative production of biorenewable fuels and chemicals. Comput Struct Biotechnol J 3: e201210005
18. Stephanopoulos G (2012) Synthetic biology and metabolic engineering. ACS Synth Biol 1:514–525
19. Nielsen J, Fussenegger M, Keasling J et al (2014) Engineering synergy in biotechnology. Nat Chem Biol 10:319–322
20. Schirmer A, Rude MA, Li X et al (2010) Microbial biosynthesis of alkanes. Science 329:559–562
21. Yu C, Cao Y, Zou H et al (2011) Metabolic engineering of *Escherichia coli* for biotechnological production of high-value organic acids and alcohols. Appl Microbiol Biotechnol 89:573–583
22. Zhou L, Tian KM, Niu DD et al (2012) Improvement of D-lactate productivity in recombinant *Escherichia coli* by coupling production with growth. Biotechnol Lett 34:1123–1130
23. Zhou L, Niu DD, Tian KM et al (2012) Genetically switched D-lactate production in *Escherichia coli*. Metab Eng 14:560–568
24. Utrilla J, Licona-Cassani C, Marcellin E et al (2012) Engineering and adaptive evolution of *Escherichia coli* for D-lactate fermentation reveals GatC as a xylose transporter. Metab Eng 14:469–476
25. Zhou L, Zuo ZR, Chen XZ et al (2011) Evaluation of genetic manipulation strategies on D-lactate production by *Escherichia coli*. Curr Microbiol 62:981–989
26. Yang J, Wang Z, Zhu N et al (2014) Metabolic engineering of *Escherichia coli* and *in silico* comparing of carboxylation pathways for high succinate productivity under aerobic conditions. Microbiol Res 169:432–440

27. Jiang M, Chen X, Liang L et al (2014) Co-expression of phosphoenolpyruvate carboxykinase and nicotinic acid phosphoribosyltransferase for succinate production in engineered *Escherichia coli*. Enzyme Microb Technol 56:8–14
28. Tang J, Zhu X, Lu J et al (2013) Recruiting alternative glucose utilization pathways for improving succinate production. Appl Microbiol Biotechnol 97:2513–2520
29. Tan Z, Zhu X, Chen J et al (2013) Activating phosphoenolpyruvate carboxylase and phosphoenolpyruvate carboxykinase in combination for improvement of succinate production. Appl Environ Microbiol 79:4838–4844
30. Kim K, Kim SK, Park YC et al (2014) Enhanced production of 3-hydroxypropionic acid from glycerol by modulation of glycerol metabolism in recombinant *Escherichia coli*. Bioresour Technol 156:170–175
31. Jung WS, Kang JH, Chu HS et al (2014) Elevated production of 3-hydroxypropionic acid by metabolic engineering of the glycerol metabolism in *Escherichia coli*. Metab Eng 23:116–122
32. Kwak S, Park YC, Seo JH (2013) Biosynthesis of 3-hydroxypropionic acid from glycerol in recombinant *Escherichia coli* expressing *Lactobacillus brevis dhaB* and *dhaR* gene clusters and *E. coli* K-12 *aldH*. Bioresour Technol 135:432–439
33. Rathnasingh C, Raj SM, Lee Y et al (2012) Production of 3-hydroxypropionic acid via malonyl-CoA pathway using recombinant *Escherichia coli* strains. J Biotechnol 157:633–640
34. Desai SH, Rabinovitch-Deere CA, Tashiro Y et al (2014) Isobutanol production from cellobiose in *Escherichia coli*. Appl Microbiol Biotechnol 98:3727–3736
35. Shi A, Zhu X, Lu J et al (2013) Activating transhydrogenase and NAD kinase in combination for improving isobutanol production. Metab Eng 16:1–10
36. Trinh CT, Li J, Blanch HW et al (2011) Redesigning *Escherichia coli* metabolism for anaerobic production of isobutanol. Appl Environ Microbiol 77:4894–4904
37. Hwang HJ, Park JH, Kim JH et al (2014) Engineering of a butyraldehyde dehydrogenase of *Clostridium saccharoperbutylacetonicum* to fit an engineered 1,4-butanediol pathway in *Escherichia coli*. Biotechnol Bioeng 111:1374–1384
38. Wang W, Lu X (2013) Microbial synthesis of alka(e)nes. Front Bioeng Biotechnol. doi: 10.3389/fbioe.2013.00010
39. Howard TP, Middelhaufe S, Moore K et al (2013) Synthesis of customized petroleum-replica fuel molecules by targeted modification of free fatty acid pools in *Escherichia coli*. Proc Natl Acad Sci U S A 110:7636–7641
40. Riemer SA, Rex R, Schomburg D (2013) A metabolite-centric view on flux distributions in genome-scale metabolic models. BMC Syst Biol 7:33
41. McCloskey D, Palsson BO, Feist AM (2013) Basic and applied uses of genome-scale metabolic network reconstructions of *Escherichia coli*. Mol Syst Biol 9:661
42. Xu Z, Sun X, Sun J (2013) Construction and analysis of the model of energy metabolism in *E. coli*. PLoS One 8:e55137
43. Khodayari A, Zomorrodi AR, Liao JC et al (2014) A kinetic model of *Escherichia coli* core metabolism satisfying multiple sets of mutant flux data. Metab Eng. doi:10.1016/j.ymben.2014.05.014
44. Zomorrodi AR, Lafontaine Rivera JG, Liao JC et al (2013) Optimization-driven identification of genetic perturbations accelerates the convergence of model parameters in ensemble modeling of metabolic networks. Biotechnol J 8:1090–1104
45. Xu Z, Zheng P, Sun J et al (2013) ReacKnock: identifying reaction deletion strategies for microbial strain optimization based on genome-scale metabolic network. PLoS One 8:e72150
46. Chang RL, Andrews K, Kim D et al (2013) Structural systems biology evaluation of metabolic thermotolerance in *Escherichia coli*. Science 340:1220–1223
47. Durot M, Bourguignon P-Y, Schachter V (2009) Genome-scale models of bacterial metabolism: reconstruction and applications. FEMS Microbiol Rev 33:164–190
48. O'Brien EJ, Lerman JA, Chang RL et al (2013) Genome-scale models of metabolism and gene expression extend and refine growth phenotype prediction. Mol Syst Biol 9:693
49. Cho BK, Kim D, Knight EM et al (2014) Genome-scale reconstruction of the sigma factor network in *Escherichia coli*: topology and functional states. BMC Biol 12:4

50. Atsumi S, Liao JC (2008) Metabolic engineering for advanced biofuels production from *Escherichia coli*. Curr Opin Biotechnol 19:414–419
51. Marcheschi RJ, Li H, Zhang K et al (2012) A synthetic recursive "+1" pathway for carbon chain elongation. ACS Chem Biol 7:689–697
52. Planson AG, Carbonell P, Grigoras I et al (2012) A retrosynthetic biology approach to therapeutics: from conception to delivery. Curr Opin Biotechnol 23:948–956
53. Rodrigo G, Carrera J, Prather KJ et al (2008) DESHARKY: automatic design of metabolic pathways for optimal cell growth. Bioinformatics 24:2554–2556
54. Hatzimanikatis V, Li C, Ionita JA et al (2005) Exploring the diversity of complex metabolic networks. Bioinformatics 21:1603–1609
55. Cho A, Yun H, Park JH et al (2010) Prediction of novel synthetic pathways for the production of desired chemicals. BMC Syst Biol 4:35
56. Mavrovouniotis M, Stephanopoulos G (1992) Synthesis of biochemical production routes. Comput Chem Eng 16:605–619
57. Tian J, Gong H, Sheng N et al (2004) Accurate multiplex gene synthesis from programmable DNA microchips. Nature 432:1050–1054
58. Quan J, Saaem I, Tang N et al (2011) Parallel on-chip gene synthesis and application to optimization of protein expression. Nat Biotechnol 29:449–452
59. Shetty RP, Endy D, Knight TF Jr (2008) Engineering BioBrick vectors from BioBrick parts. J Biol Eng 2:5
60. Xu P, Vansiri A, Bhan N et al (2012) ePathBrick: a synthetic biology platform for engineering metabolic pathways in *E. coli*. ACS Synth Biol 1:256–266
61. Ellis T, Adie T, Baldwin GS (2011) DNA assembly for synthetic biology: from parts to pathways and beyond. Integr Biol (Camb) 3:109–118
62. Hillson N (2011) DNA Assembly method standardization for synthetic biomolecular circuits and systems. In: Koeppl H, Setti G, di Bernardo M et al (eds) Design and analysis of biomolecular circuits. Springer, New York, pp 295–314
63. Quan J, Tian J (2014) Circular polymerase extension cloning. Methods Mol Biol 1116:103–117
64. Li MZ, Elledge SJ (2007) Harnessing homologous recombination in vitro to generate recombinant DNA via SLIC. Nat Methods 4:251–256
65. Gibson DG, Young L, Chuang RY et al (2009) Enzymatic assembly of DNA molecules up to several hundred kilobases. Nat Methods 6:343–345
66. Shao Z, Zhao H (2009) DNA assembler, an in vivo genetic method for rapid construction of biochemical pathways. Nucleic Acids Res 37:e16
67. Du J, Yuan Y, Si T et al (2012) Customized optimization of metabolic pathways by combinatorial transcriptional engineering. Nucleic Acids Res 40:e142
68. Datsenko KA, Wanner BL (2000) One-step inactivation of chromosomal genes in *Escherichia coli* K-12 using PCR products. Proc Natl Acad Sci USA 97:6640–6645
69. Jarboe LR, Zhang X, Wang X et al (2010) Metabolic engineering for production of biorenewable fuels and chemicals: contributions of synthetic biology. J Biomed Biotechnol 2010:761042
70. Gaj T, Sirk SJ, Barbas CF (2014) Expanding the scope of site-specific recombinases for genetic and metabolic engineering. Biotechnol Bioeng 111:1–15
71. Posfai G, Koob MD, Kirkpatrick HA et al (1997) Versatile insertion plasmids for targeted genome manipulations in bacteria: isolation, deletion, and rescue of the pathogenicity island LEE of the *Escherichia coli* O157:H7 genome. J Bacteriol 179:4426–4428
72. Jantama K, Zhang X, Moore JC et al (2008) Eliminating side products and increasing succinate yields in engineered strains of *Escherichia coli* C. Biotechnol Bioeng 101:881–893
73. Zhang X, Jantama K, Moore JC et al (2007) Production of L -alanine by metabolically engineered *Escherichia coli*. Appl Microbiol Biotechnol 77:355–366
74. Gay P, Le Coq D, Steinmetz M et al (1985) Positive selection procedure for entrapment of insertion sequence elements in gram-negative bacteria. J Bacteriol 164:918–921
75. Hammer K, Mijakovic I, Jensen PR (2006) Synthetic promoter libraries–tuning of gene expression. Trends Biotechnol 24:53–55

76. Nevoigt E, Kohnke J, Fischer CR et al (2006) Engineering of promoter replacement cassettes for fine-tuning of gene expression in *Saccharomyces cerevisiae*. Appl Environ Microbiol 72:5266–5273
77. Alper H, Fischer C, Nevoigt E et al (2005) Tuning genetic control through promoter engineering. Proc Natl Acad Sci USA 102:12678–12683
78. Solem C, Jensen PR (2002) Modulation of gene expression made easy. Appl Environ Microbiol 68:2397–2403
79. Jensen PR, Hammer K (1998) The sequence of spacers between the consensus sequences modulates the strength of prokaryotic promoters. Appl Environ Microbiol 64:82–87
80. Jensen PR, Hammer K (1998) Artificial promoters for metabolic optimization. Biotechnol Bioeng 58:191–195
81. de la Cueva-Mendez G, Pimentel B (2007) Gene and cell survival: lessons from prokaryotic plasmid R1. EMBO Rep 8:458–464
82. Keasling JD (2008) Synthetic biology for synthetic chemistry. ACS Chem Biol 3:64–76
83. Lu J, Tang J, Liu Y et al (2012) Combinatorial modulation of *galP* and *glk* gene expression for improved alternative glucose utilization. Appl Microbiol Biotechnol 93:2455–2462
84. De Mey M, Maertens J, Boogmans S et al (2010) Promoter knock-in: a novel rational method for the fine tuning of genes. BMC Biotechnol 10:26
85. Meynial-Salles I, Cervin MA, Soucaille P (2005) New tool for metabolic pathway engineering in *Escherichia coli*: one-step method to modulate expression of chromosomal genes. Appl Environ Microbiol 71:2140–2144
86. Millard CS, Chao YP, Liao JC et al (1996) Enhanced production of succinic acid by overexpression of phosphoenolpyruvate carboxylase in *Escherichia coli*. Appl Environ Microbiol 62:1808–1810
87. Santos CN, Stephanopoulos G (2008) Combinatorial engineering of microbes for optimizing cellular phenotype. Curr Opin Chem Biol 12:168–176
88. Pfleger BF, Pitera DJ, Smolke CD et al (2006) Combinatorial engineering of intergenic regions in operons tunes expression of multiple genes. Nat Biotechnol 24:1027–1032
89. Dueber JE, Wu GC, Malmirchegini GR et al (2009) Synthetic protein scaffolds provide modular control over metabolic flux. Nat Biotechnol 27:753–759
90. Cobb RE, Sun N, Zhao H (2013) Directed evolution as a powerful synthetic biology tool. Methods 60:81–90
91. Wang HH, Isaacs FJ, Carr PA et al (2009) Programming cells by multiplex genome engineering and accelerated evolution. Nature 460:894–898
92. Yoo SM, Na D, Lee SY (2013) Design and use of synthetic regulatory small RNAs to control gene expression in *Escherichia coli*. Nat Protoc 8:1694–1707
93. Xu P, Wang W, Li L et al (2014) Design and kinetic analysis of a hybrid promoter-regulator system for malonyl-CoA sensing in *Escherichia coli*. ACS Chem Biol 9:451–458
94. Farmer WR, Liao JC (2000) Improving lycopene production in *Escherichia coli* by engineering metabolic control. Nat Biotechnol 18:533–537
95. Zhang F, Carothers JM, Keasling JD (2012) Design of a dynamic sensor-regulator system for production of chemicals and fuels derived from fatty acids. Nat Biotechnol 30:354–359
96. Cobb RE, Si T, Zhao H (2012) Directed evolution: an evolving and enabling synthetic biology tool. Curr Opin Chem Biol 16:285–291
97. Zhou S, Yomano LP, Shanmugam KT et al (2005) Fermentation of 10 % (w/v) sugar to D: (-)-lactate by engineered *Escherichia coli* B. Biotechnol Lett 27:1891–1896
98. Zhou S, Shanmugam KT, Yomano LP et al (2006) Fermentation of 12 % (w/v) glucose to 1.2 M lactate by *Escherichia coli* strain SZ194 using mineral salts medium. Biotechnol Lett 28:663–670
99. Grabar TB, Zhou S, Shanmugam KT et al (2006) Methylglyoxal bypass identified as source of chiral contamination in l(+) and d(-)-lactate fermentations by recombinant *Escherichia coli*. Biotechnol Lett 28:1527–1535
100. Zhang X, Shanmugam KT, Ingram LO (2010) Fermentation of glycerol to succinate by metabolically engineered strains of *Escherichia coli*. Appl Environ Microbiol 76:2397–2401

101. Zhang X, Jantama K, Moore JC et al (2009) Metabolic evolution of energy-conserving pathways for succinate production in *Escherichia coli*. Proc Natl Acad Sci USA 106:20180–20185
102. Jiang M, Wan Q, Liu R et al (2014) Succinic acid production from corn stalk hydrolysate in an *E. coli* mutant generated by atmospheric and room-temperature plasmas and metabolic evolution strategies. J Ind Microbiol Biotechnol 41:115–123
103. Trinh CT, Srienc F (2009) Metabolic engineering of *Escherichia coli* for efficient conversion of glycerol to ethanol. Appl Environ Microbiol 75:6696–6705
104. Fernandez-Sandoval MT, Huerta-Beristain G, Trujillo-Martinez B et al (2012) Laboratory metabolic evolution improves acetate tolerance and growth on acetate of ethanologenic *Escherichia coli* under non-aerated conditions in glucose-mineral medium. Appl Microbiol Biotechnol 96:1291–1300
105. Zheng H, Wang X, Yomano LP et al (2013) Improving *Escherichia coli* FucO for furfural tolerance by saturation mutagenesis of individual amino acid positions. Appl Environ Microbiol 79:3202–3208
106. Liu H, Yan M, Lai C et al (2010) gTME for improved xylose fermentation of *Saccharomyces cerevisiae*. Appl Biochem Biotechnol 160:574–582
107. Tyo KE, Alper HS, Stephanopoulos GN (2007) Expanding the metabolic engineering toolbox: more options to engineer cells. Trends Biotechnol 25:132–137
108. Alper H, Stephanopoulos G (2007) Global transcription machinery engineering: a new approach for improving cellular phenotype. Metab Eng 9:258–267
109. Alper H, Moxley J, Nevoigt E et al (2006) Engineering yeast transcription machinery for improved ethanol tolerance and production. Science 314:1565–1568
110. Shi DJ, Wang CL, Wang KM (2009) Genome shuffling to improve thermotolerance, ethanol tolerance and ethanol productivity of *Saccharomyces cerevisiae*. J Ind Microbiol Biotechnol 36:139–147
111. Otte B, Grunwaldt E, Mahmoud O et al (2009) Genome shuffling in *Clostridium diolis* DSM 15410 for improved 1,3-propanediol production. Appl Environ Microbiol 75:7610–7616
112. Hida H, Yamada T, Yamada Y (2007) Genome shuffling of *Streptomyces sp.* U121 for improved production of hydroxycitric acid. Appl Microbiol Biotechnol 73:1387–1393
113. Zhang YX, Perry K, Vinci VA et al (2002) Genome shuffling leads to rapid phenotypic improvement in bacteria. Nature 415:644–646
114. Warner JR, Reeder PJ, Karimpour-Fard A et al (2010) Rapid profiling of a microbial genome using mixtures of barcoded oligonucleotides. Nat Biotechnol 28:856–862
115. Royce LA, Boggess E, Fu Y et al (2014) Transcriptomic analysis of carboxylic acid challenge in *Escherichia coli*: beyond membrane damage. PLoS One 9:e89580
116. McCloskey D, Gangoiti JA, King ZA et al (2014) A model-driven quantitative metabolomics analysis of aerobic and anaerobic metabolism in *E. coli* K-12 MG1655 that is biochemically and thermodynamically consistent. Biotechnol Bioeng 111:803–815
117. Wiback SJ, Mahadevan R, Palsson BO (2004) Using metabolic flux data to further constrain the metabolic solution space and predict internal flux patterns: the *Escherichia coli* spectrum. Biotechnol Bioeng 86:317–331
118. Oliver DJ, Nikolau B, Wurtele ES (2002) Functional genomics: high-throughput mRNA, protein, and metabolite analyses. Metab Eng 4:98–106
119. Patterson SD, Aebersold RH (2003) Proteomics: the first decade and beyond. Nat Genet 33 (Suppl):311–323
120. Alonso S, Rendueles M, Diaz M (2014) Microbial production of specialty organic acids from renewable and waste materials. Crit Rev Biotechnol . doi:10.3109/07388551.2014.904269
121. Tsao GT, Cao NJ, Du J et al (1999) Production of multifunctional organic acids from renewable resources. In: Tsao GT, Brainard AP, Bungay HR et al (eds) Recent progress in bioconversion of Lignocellulosics. Springer, Berlin, pp 243–280
122. Zhou S, Causey TB, Hasona A et al (2003) Production of optically pure D-lactic acid in mineral salts medium by metabolically engineered *Escherichia coli* W3110. Appl Environ Microbiol 69:399–407

123. Shukla VB, Zhou S, Yomano LP et al (2004) Production of D(-)-lactate from sucrose and molasses. Biotechnol Lett 26:689–693
124. Wang Q, Yang P, Liu C et al (2013) Biosynthesis of poly(3-hydroxypropionate) from glycerol by recombinant *Escherichia coli*. Bioresour Technol 131:548–551
125. Meng DC, Shi ZY, Wu LP et al (2012) Production and characterization of poly(3-hydroxypropionate-co-4-hydroxybutyrate) with fully controllable structures by recombinant *Escherichia coli* containing an engineered pathway. Metab Eng 14:317–324
126. Werpy T GPe (2004) Top value added chemicals from biomass. U.S. Department of Energy, Washington, DC. http://www1.eere.energy.gov/biomass/pdfs/35523.pdf
127. Holo H (1989) Chloroflexus aurantiacus secretes 3-hydroxypropionate, a possible intermediate in the assimilation of CO_2 and acetate. Arch Microbiol 151:252–256
128. Strauss G, Eisenreich W, Bacher A et al (1992) 13C-NMR study of autotrophic CO_2 fixation pathways in the sulfur-reducing Archaebacterium *Thermoproteus neutrophilus* and in the phototrophic *Eubacterium Chloroflexus aurantiacus*. Eur J Biochem 205:853–866
129. Hugler M, Huber H, Stetter KO et al (2003) Autotrophic CO_2 fixation pathways in archaea (Crenarchaeota). Arch Microbiol 179:160–173
130. Berg IA, Kockelkorn D, Buckel W et al (2007) A 3-hydroxypropionate/4-hydroxybutyrate autotrophic carbon dioxide assimilation pathway in Archaea. Science 318:1782–1786
131. Ansede JH, Pellechia PJ, Yoch DC (1999) Metabolism of acrylate to beta-hydroxypropionate and its role in dimethylsulfoniopropionate lyase induction by a salt marsh sediment bacterium, *Alcaligenes faecalis* M3A. Appl Environ Microbiol 65:5075–5081
132. Loh KD, Gyaneshwar P, Papadimitriou EM et al (2006) A previously undescribed pathway for pyrimidine catabolism. Proc Natl Acad Sci U S A 103:5114–5119
133. Kim KS, Pelton JG, Inwood WB et al (2010) The Rut pathway for pyrimidine degradation: novel chemistry and toxicity problems. J Bacteriol 192:4089–4102
134. Andersen G, Bjornberg O, Polakova S et al (2008) A second pathway to degrade pyrimidine nucleic acid precursors in eukaryotes. J Mol Biol 380:656–666
135. Valdehuesa KN, Liu H, Nisola GM et al (2013) Recent advances in the metabolic engineering of microorganisms for the production of 3-hydroxypropionic acid as C_3 platform chemical. Appl Microbiol Biotechnol 97:3309–3321
136. Kumar V, Ashok S, Park S (2013) Recent advances in biological production of 3-hydroxypropionic acid. Biotechnol Adv 31:945–961
137. Jiang X, Meng X, Xian M (2009) Biosynthetic pathways for 3-hydroxypropionic acid production. Appl Microbiol Biotechnol 82:995–1003
138. Raj S, Rathnasingh C, Jo J-E et al (2008) Production of 3-hydroxypropionic acid from glycerol by a novel recombinant *Escherichia coli* BL21 strain. Process Biochem 43:1440–1446
139. Mohan Raj S, Rathnasingh C, Jung WC et al (2009) Effect of process parameters on 3-hydroxypropionic acid production from glycerol using a recombinant *Escherichia coli*. Appl Microbiol Biotechnol 84:649–657
140. Rathnasingh C, Raj SM, Jo JE et al (2009) Development and evaluation of efficient recombinant *Escherichia coli* strains for the production of 3-hydroxypropionic acid from glycerol. Biotechnol Bioeng 104:729–739
141. Tokuyama K, Ohno S, Yoshikawa K et al (2014) Increased 3-hydroxypropionic acid production from glycerol, by modification of central metabolism in *Escherichia coli*. Microb Cell Fact 13:64
142. Bunch PK, Mat-Jan F, Lee N et al (1997) The *ldhA* gene encoding the fermentative lactate dehydrogenase of Escherichia coli. Microbiology 143(Pt 1):187–195
143. Chatterjee R, Millard CS, Champion K et al (2001) Mutation of the *ptsG* gene results in increased production of succinate in fermentation of glucose by *Escherichia coli*. Appl Environ Microbiol 67:148–154
144. Donnelly MI, Millard CS, Clark DP et al (1998) A novel fermentation pathway in an *Escherichia coli* mutant producing succinic acid, acetic acid, and ethanol. Appl Biochem Biotechnol 70–72:187–198

145. Stols L, Donnelly MI (1997) Production of succinic acid through overexpression of NAD(+)-dependent malic enzyme in an *Escherichia coli* mutant. Appl Environ Microbiol 63:2695–2701
146. Escalante A, Cervantes AS, Gosset G et al (2012) Current knowledge of the *Escherichia coli* phosphoenolpyruvate-carbohydrate phosphotransferase system: peculiarities of regulation and impact on growth and product formation. Appl Microbiol Biotechnol 94:1483–1494
147. Gabor E, Gohler AK, Kosfeld A et al (2011) The phosphoenolpyruvate-dependent glucose-phosphotransferase system from *Escherichia coli* K-12 as the center of a network regulating carbohydrate flux in the cell. Eur J Cell Biol 90:711–720
148. Postma PW, Lengeler JW, Jacobson GR (1993) Phosphoenolpyruvate:carbohydrate phosphotransferase systems of bacteria. Microbiol Rev 57:543–594
149. Vemuri GN, Eiteman MA, Altman E (2002) Succinate production in dual-phase *Escherichia coli* fermentations depends on the time of transition from aerobic to anaerobic conditions. J Ind Microbiol Biotechnol 28:325–332
150. Sanchez AM, Bennett GN, San KY (2005) Novel pathway engineering design of the anaerobic central metabolic pathway in *Escherichia coli* to increase succinate yield and productivity. Metab Eng 7:229–239
151. Cheng KK, Wang GY, Zeng J et al (2013) Improved succinate production by metabolic engineering. Biomed Res Int 2013:538790
152. Clark DP (1989) The fermentation pathways of *Escherichia coli*. FEMS Microbiol Rev 5:223–234
153. Shanmugam KT, Ingram LO (2008) Engineering biocatalysts for production of commodity chemicals. J Mol Microbiol Biotechnol 15:8–15
154. Maloy SR, Nunn WD (1982) Genetic regulation of the glyoxylate shunt in *Escherichia coli* K-12. J Bacteriol 149:173–180
155. Lorca GL, Ezersky A, Lunin VV et al (2007) Glyoxylate and pyruvate are antagonistic effectors of the *Escherichia coli* IclR transcriptional regulator. J Biol Chem 282:16476–16491
156. Yamamoto K, Ishihama A (2003) Two different modes of transcription repression of the *Escherichia coli* acetate operon by IclR. Mol Microbiol 47:183–194
157. Cozzone AJ, El-Mansi M (2005) Control of isocitrate dehydrogenase catalytic activity by protein phosphorylation in *Escherichia coli*. J Mol Microbiol Biotechnol 9:132–146
158. Balzer GJ, Thakker C, Bennett GN et al (2013) Metabolic engineering of *Escherichia coli* to minimize byproduct formate and improving succinate productivity through increasing NADH availability by heterologous expression of NAD(+)-dependent formate dehydrogenase. Metab Eng 20:1–8
159. Stols L, Kulkarni G, Harris BG et al (1997) Expression of Ascaris suum malic enzyme in a mutant *Escherichia coli* allows production of succinic acid from glucose. Appl Biochem Biotechnol 63–65:153–158
160. Jantama K, Haupt MJ, Svoronos SA et al (2008) Combining metabolic engineering and metabolic evolution to develop nonrecombinant strains of *Escherichia coli* C that produce succinate and malate. Biotechnol Bioeng 99:1140–1153
161. Gosset G (2005) Improvement of *Escherichia coli* production strains by modification of the phosphoenolpyruvate:sugar phosphotransferase system. Microb Cell Fact 4:14
162. Zhang X, Jantama K, Shanmugam KT et al (2009) Reengineering *Escherichia coli* for Succinate Production in Mineral Salts Medium. Appl Environ Microbiol 75:7807–7813
163. Zhu X, Tan Z, Xu H et al (2014) Metabolic evolution of two reducing equivalent-conserving pathways for high-yield succinate production in *Escherichia coli*. Metab Eng 24:87–96
164. Steiert PS, Stauffer LT, Stauffer GV (1990) The *lpd* gene product functions as the L protein in the *Escherichia coli* glycine cleavage enzyme system. J Bacteriol 172:6142–6144
165. Guest JR, Creaghan IT (1974) Further studies with lipoamide dehydrogenase mutants of *Escherichia coli* K12. J Gen Microbiol 81:237–245
166. Guest JR, Creaghan IT (1972) Lipoamide dehydrogenase mutants of *Escherichia coli* K 12. Biochem J 130:8

167. Kim YM, Cho HS, Jung GY et al (2011) Engineering the pentose phosphate pathway to improve hydrogen yield in recombinant *Escherichia coli*. Biotechnol Bioeng 108:2941–2946
168. Sauer U, Canonaco F, Heri S et al (2004) The soluble and membrane-bound transhydrogenases UdhA and PntAB have divergent functions in NADPH metabolism of *Escherichia coli*. J Biol Chem 279:6613–6619
169. Sauer U, Lasko DR, Fiaux J et al (1999) Metabolic flux ratio analysis of genetic and environmental modulations of *Escherichia coli* central carbon metabolism. J Bacteriol 181:6679–6688
170. Battat E, Peleg Y, Bercovitz A et al (1991) Optimization of L-malic acid production by *Aspergillus flavus* in a stirred fermentor. Biotechnol Bioeng 37:1108–1116
171. Peleg Y, Rokem JS, Goldberg I (1990) A simple plate-assay for the screening of L-malic acid producing microorganisms. FEMS Microbiol Lett 55:233–236
172. Pines O, Even-Ram S, Elnathan N et al (1996) The cytosolic pathway of L-malic acid synthesis in *Saccharomyces cerevisiae*: the role of fumarase. Appl Microbiol Biotechnol 46:393–399
173. Taing O, Taing K (2007) Production of malic and succinic acids by sugar-tolerant yeast *Zygosaccharomyces rouxii*. Eur Food Res Technol 224:343–347
174. Kawagoe M, Hyakumura K, Suye S-I et al (1997) Application of bubble column fermentors to submerged culture of *Schizophyllum commune* for production of L-malic acid. J Ferment Bioeng 84:333–336
175. Zhang X, Wang X, Shanmugam KT et al (2011) L-malate production by metabolically engineered *Escherichia coli*. Appl Environ Microbiol 77:427–434
176. Straathof AJ, van Gulik WM (2012) Production of fumaric Acid by fermentation. Subcell Biochem 64:225–240
177. Ehrlich F (1911) Über die bildung von fumarsäure durch schimmelpilze. Ber Dtsch Chem Ges 44:3737–3742
178. Wang G, Huang D, Qi H et al (2013) Rational medium optimization based on comparative metabolic profiling analysis to improve fumaric acid production. Bioresour Technol 137:1–8
179. Zhou Y, Nie K, Zhang X et al (2014) Production of fumaric acid from biodiesel-derived crude glycerol by *Rhizopus arrhizus*. Bioresour Technol 163:48–53
180. Gu S, Xu Q, Huang H et al (2014) Alternative respiration and fumaric acid production of Rhizopus oryzae. Appl Microbiol Biotechnol 98:5145–5152
181. Goldberg I, Steiglitz B (1985) Improved rate of fumaric acid production by Tweens and vegetable oils in *rhizopus arrhizus*. Biotechnol Bioeng 27:1067–1069
182. Zhou Y, Du J, Tsao GT (2002) Comparison of fumaric acid production by *Rhizopus oryzae* using different neutralizing agents. Bioprocess Biosyst Eng 25:179–181
183. Fu YQ, Li S, Chen Y et al (2010) Enhancement of fumaric acid production by *Rhizopus oryzae* using a two-stage dissolved oxygen control strategy. Appl Biochem Biotechnol 162:1031–1038
184. Xu Q, Li S, Fu Y et al (2010) Two-stage utilization of corn straw by *Rhizopus oryzae* for fumaric acid production. Bioresour Technol 101:6262–6264
185. Ding Y, Li S, Dou C et al (2011) Production of fumaric acid by *Rhizopus oryzae*: role of carbon-nitrogen ratio. Appl Biochem Biotechnol 164:1461–1467
186. Roa Engel CA, van Gulik WM, Marang L et al (2011) Development of a low pH fermentation strategy for fumaric acid production by *Rhizopus oryzae*. Enzyme Microb Technol 48:39–47
187. Zhang B, Skory CD, Yang ST (2012) Metabolic engineering of *Rhizopus oryzae*: effects of overexpressing pyc and pepc genes on fumaric acid biosynthesis from glucose. Metab Eng 14:512–520
188. Gu C, Zhou Y, Liu L et al (2013) Production of fumaric acid by immobilized *Rhizopus arrhizus* on net. Bioresour Technol 131:303–307
189. Ling LB, Ng TK (1989). US Patent 4,877,731. Google Patents
190. Song CW, Kim DI, Choi S et al (2013) Metabolic engineering of *Escherichia coli* for the production of fumaric acid. Biotechnol Bioeng 110:2025–2034

191. Singh J, Gupta KP (2003) Calcium glucarate prevents tumor formation in mouse skin. Biomed Environ Sci 16:9–16
192. Singh J, Gupta KP (2007) Induction of apoptosis by calcium D-glucarate in 7,12-dimethyl benz [a] anthracene-exposed mouse skin. J Environ Pathol Toxicol Oncol 26:63–73
193. Walaszek Z, Szemraj J, Hanausek M et al (1996) d-Glucaric acid content of various fruits and vegetables and cholesterol-lowering effects of dietary d-glucarate in the rat. Nutr Res 16:673–681
194. Moon TS, Yoon SH, Lanza AM et al (2009) Production of glucaric acid from a synthetic pathway in recombinant *Escherichia coli*. Appl Environ Microbiol 75:589–595
195. Moon TS, Dueber JE, Shiue E et al (2010) Use of modular, synthetic scaffolds for improved production of glucaric acid in engineered *E. coli*. Metab Eng 12:298–305
196. Shiue E, Prather KL (2014) Improving D-glucaric acid production from myo-inositol in *E. coli* by increasing MIOX stability and myo-inositol transport. Metab Eng 22:22–31
197. Xie NZ, Liang H, Huang RB et al (2014) Biotechnological production of muconic acid: current status and future prospects. Biotechnol Adv 32:615–622
198. Lin Y, Sun X, Yuan Q et al (2014) Extending shikimate pathway for the production of muconic acid and its precursor salicylic acid in *Escherichia coli*. Metab Eng 23:62–69
199. Sun X, Lin Y, Huang Q et al (2013) A novel muconic acid biosynthesis approach by shunting tryptophan biosynthesis via anthranilate. Appl Environ Microbiol 79:4024–4030
200. Draths KM, Frost JW (1994) Environmentally Compatible Synthesis of Adipic Acid from D-Glucose. J Am Chem Soc 116:399–400
201. Niu W, Draths KM, Frost JW (2002) Benzene-free synthesis of adipic acid. Biotechnol Prog 18:201–211
202. Bui V, Lau MK, MacRae D et al (2013) Methods for producing isomers of muconic acid and muconate salts, United States Patent application. US 2013/0030215 A1, Google Patents
203. Polen T, Spelberg M, Bott M (2013) Toward biotechnological production of adipic acid and precursors from biorenewables. J Biotechnol 167:75–84
204. Musser MT (2000) Adipic Acid, in *Ullmann's Encyclopedia of Industrial Chemistry*. Wiley-VCH Verlag GmbH & Co, KGaA
205. Yu JL, Xia XX, Zhong JJ et al (2014) Direct biosynthesis of adipic acid from a synthetic pathway in recombinant *Escherichia coli*. Biotechnol Bioeng. doi:10.1002/bit.25293
206. Han L, Chen W, Yuan F et al (2013) Biosynthesis of adipic acid. Sheng Wu Gong Cheng Xue Bao 29:1374–1385
207. Picataggio S, Beardslee T (2012) Biological methods for preparing adipic acid. Google Patents US 8,241,879 B2
208. Parthasarathy A, Pierik AJ, Kahnt J et al (2011) Substrate specificity of 2-hydroxyglutaryl-CoA dehydratase from *Clostridium symbiosum*: toward a bio-based production of adipic acid. Biochemistry 50:3540–3550
209. Noack H, Georgiev V, Blomberg MR et al (2011) Theoretical insights into heme-catalyzed oxidation of cyclohexane to adipic acid. Inorg Chem 50:1194–1202
210. Burgard AP, Pharkya P, Osterhout RE (2010) Microorganisms for the production of adipic acid and other compounds. Google Patents US 2010/0330626 A1
211. Dugal M, Sankar G, Raja R et al (2000) Designing a Heterogeneous Catalyst for the Production of Adipic Acid by Aerial Oxidation of Cyclohexane. Angew Chem Int Ed Engl 39:2310–2313
212. Nakamura CE, Whited GM (2003) Metabolic engineering for the microbial production of 1,3-propanediol. Curr Opin Biotechnol 14:454–459
213. Jain R, Yan Y (2011) Dehydratase mediated 1-propanol production in metabolically engineered *Escherichia coli*. Microb Cell Fact 10:97
214. Altaras NE, Cameron DC (1999) Metabolic engineering of a 1,2-propanediol pathway in *Escherichia coli*. Appl Environ Microbiol 65:1180–1185
215. Altaras NE, Cameron DC (2000) Enhanced production of (R)-1,2-propanediol by metabolically engineered *Escherichia coli*. Biotechnol Prog 16:940–946

216. Soma Y, Inokuma K, Tanaka T et al (2012) Direct isopropanol production from cellobiose by engineered *Escherichia coli* using a synthetic pathway and a cell surface display system. J Biosci Bioeng 114:80–85
217. Lan EI, Liao JC (2013) Microbial synthesis of n-butanol, isobutanol, and other higher alcohols from diverse resources. Bioresour Technol 135:339–349
218. Zhang K, Sawaya MR, Eisenberg DS et al (2008) Expanding metabolism for biosynthesis of nonnatural alcohols. Proc Natl Acad Sci USA 105:20653–20658
219. Dhande YK, Xiong M, Zhang K (2012) Production of C5 carboxylic acids in engineered *Escherichia coli*. Process Biochem 47:1965–1971
220. Bhan N, Xu P, Koffas MA (2013) Pathway and protein engineering approaches to produce novel and commodity small molecules. Curr Opin Biotechnol 24:1137–1143
221. Baez A, Cho KM, Liao JC (2011) High-flux isobutanol production using engineered *Escherichia coli*: a bioreactor study with in situ product removal. Appl Microbiol Biotechnol 90:1681–1690
222. Bastian S, Liu X, Meyerowitz JT et al (2011) Engineered ketol-acid reductoisomerase and alcohol dehydrogenase enable anaerobic 2-methylpropan-1-ol production at theoretical yield in *Escherichia coli*. Metab Eng 13:345–352
223. Machado HB, Dekishima Y, Luo H et al (2012) A selection platform for carbon chain elongation using the CoA-dependent pathway to produce linear higher alcohols. Metab Eng 14:504–511
224. Tseng HC, Prather KL (2012) Controlled biosynthesis of odd-chain fuels and chemicals via engineered modular metabolic pathways. Proc Natl Acad Sci USA 109:17925–17930
225. Yu P, Tai Y-S, Woodruff AP et al (2012) Engineering artificial metabolic pathways for biosynthesis. Curr Opin Chem Eng 1:373–379
226. Pitera DJ, Paddon CJ, Newman JD et al (2007) Balancing a heterologous mevalonate pathway for improved isoprenoid production in *Escherichia coli*. Metab Eng 9:193–207
227. Chowdhury R, Sahu G, Das J (1996) Stress response in pathogenic bacteria. J Bioscience 21:149–160

Adv Biochem Eng Biotechnol (2016) 155: 141–163
DOI 10.1007/10_2015_306

Published Online: 8 February 2015

Engineering *Escherichia coli* Cell Factories for *n*-Butanol Production

Hongjun Dong, Chunhua Zhao, Tianrui Zhang, Zhao Lin, Yin Li and Yanping Zhang

Abstract The production of *n*-butanol, as a widely applied solvent and potential fuel, is attracting much attention. The fermentative production of butanol coupled with the production of acetone and ethanol by *Clostridium* (ABE fermentation) was once one of the oldest biotechnological processes, ranking second in scale behind ethanol fermentation. However, there remain problems with butanol production by *Clostridium*, especially the difficulty in genetically manipulating clostridial strains. In recent years, many efforts have been made to produce butanol using non-native strains. Until now, the most advanced effort was the engineering of the user-friendly and widely studied *Escherichia coli* for butanol production. This paper reviews the current progress and problems relating to butanol production by engineered *E. coli* in terms of prediction using mathematical models, pathway construction, novel enzyme replacement, butanol toxicity, and tolerance engineering strategies.

Keywords: *n*-butanol · Metabolic engineering · *Escherichia coli* · Pathway reconstruction · Butanol tolerance

Contents

H. Dong · C. Zhao · T. Zhang · Y. Li · Y. Zhang (✉)
CAS Key Laboratory of Microbial Physiological and Metabolic Engineering, Institute of Microbiology, Chinese Academy of Sciences, Beijing 100101, China
e-mail: zhangyp@im.ac.cn

C. Zhao · Z. Lin
University of Chinese Academy of Sciences, Beijing 100049, China

T. Zhang
Tianjin Institute of Industrial Biotechnology, Chinese Academy of Sciences, Tianjin 300308, China

1 Introduction

While the infrastructure of the current economy is established on the petrochemical industry, the oil crisis of the 1970s was a warning that humanity's dependence on oil is not sustainable [23]. It is generally accepted that crude oil will be depleted in the twenty-first century at the speed of current consumption. The shift from a petroleum-based economy to a biomass-based economy has become a global objective. In the drive to find alternatives to fossil products, the production of butanol from renewable resources attracts much attention nowadays [39].

Butanol (butyl alcohol or 1-butanol or *n*-butanol, C_4H_9OH, MW 74.12) is a colorless liquid with a distinct odor. It is mainly used to synthesize butyl acrylate and methacrylate esters for latex surface coatings and the production of enamels and lacquers, butyl glycol ether, butyl acetate, and plasticizers. Additionally, butanol can be used directly as the diluent for formulations of brake fluid and as solvent in the production of hormones, vitamins, and antibiotics [23]. Although ethanol has been extensively recognized as a typical biofuel, butanol, as an alternative biofuel, has several important advantages over ethanol, such as higher energy content, lower water absorption, better blending ability with gasoline, and direct use in conventional combustion engines without modification [12].

Butanol is naturally produced via the anaerobic fermentation of biomass substrates by some clostridia species; this is referred to as ABE fermentation because it is coupled with the production of acetone and ethanol. Much progress has been achieved over a century of study on ABE fermentation, such as the development of genetic manipulation tools and omic analyses of the physiology of solventogenic bacteria. However, there are still problems with butanol production by *Clostridium*: (1) It remains time consuming and difficult to genetically manipulate *Clostridium* strains although new tools have been developed; (2) it is difficult to improve the butanol yield because of the naturally coupled production of acetone and ethanol; (3) the relatively slow growth and spore-forming life cycle are problems for industrial fermentation; and (4) the relatively unknown genetic system and complex physiology of the microorganism present difficulties in engineering the metabolism for optimal production of butanol. Therefore, construction of the next generation of butanol producers from user-friendly organisms would be an alternative way for producing butanol with lower cost than clostridial strains [20].

However, heterologous production of butanol in non-clostridial microbes is not as simple as simply transferring several known genes. First, the host needs to be genetically manipulated easily to support multiple steps of engineering and many trial-and-error experiments. Second, the butanol pathway needs to be carefully designed and new genes from other organisms need to be tested in establishing an efficient pathway in a new host. Third, the native pathway and carbon flux need to be readjusted through genome engineering. Fourth, the butanol tolerance and use of cheap substrates need to be improved. In this regard, *Escherichia coli* seems to be an optimal microorganism with well-studied genetic background and rich genetic tools. More importantly, *E. coli* has been proved to be the successful horses for the microbial cell factories of some products [36]. In recent years, many groups have reported successful butanol production in *E. coli* [13, 45]. Here, we reviewed the current progress and problems relating to butanol production in non-native microbes, especially in *E. coli*.

2 Theoretical Prediction to Improve Butanol Production in *E. coli* Using Computational Models

The first butanol-producing microbe was found by Louis Pasteur in 1861 [23]. The process of natural butanol production is well-known ABE fermentation in which butanol production is coupled with the production of acetone and ethanol. Owing to the demand for large amounts of acetone in the manufacture of cordite in World Wars I and II, ABE fermentation peaked in the 1950s [23]. In the past more than 100 years of ABE fermentation, scientists have learned the butanol synthetic pathway and can now transfer it to many other microbes for the heterologous production of butanol with the help of molecular biology tools [10]. However, the genetic modification of butanol production is not always effective, because engineering of a single gene may lead to unanticipated dramatic changes in the metabolic network. Comprehensive in silico models and highly accurate prediction methods are thus desired to reduce the trial-and-error risk and to improve our understanding of microbial physiology. In recent years, efforts have been made to construct genome-scale metabolic models related to butanol production based on genome annotation and metabolome analysis.

2.1 *Flux Balance Analysis (FBA)*

Flux balance analysis (FBA) is a mathematical modeling approach often used by metabolic engineers to quantitatively simulate microbial metabolism. FBA assumes that metabolic networks will reach a steady state constrained by the stoichiometry [25]. By performing FBA while maximizing the cell growth and butanol production rate in *Clostridium*, the relationship between acetate accumulation and butanol production was investigated. It was revealed that the rate of butanol production decreased with a decreasing rate of acetate production [21]. Additionally, by adding reactions involved in butanol production catalyzed by butyryl-CoA dehydrogenase (BCD), butanal dehydrogenase, and butanol dehydrogenase to the metabolic model of *E. coli*, a genome-scale FBA model was constructed to simulate triple reaction knockouts that contribute to improving butanol production. The model indicated that the knockout of *adhE* and *pta* was essential for the high production of butanol. It was confirmed that, by disrupting ethanol and acetate production pathways, 27 % of glucose was converted into butanol. Additionally, it has been evaluated experimentally that the disruption resulted in 1.4-fold butanol yield of the control strain [40].

2.2 *Kinetic Simulation Model*

Besides FBA analysis, a kinetic model was constructed to simulate the dynamic profiles of microbial metabolism. Shinto et al. designed three kinetic simulation models that describe the dynamic behaviors of metabolites in ABE fermentation by *Clostridium saccharoperbutylacetonicum* N1-4. The simulation results showed that an increase in kinetic parameters (V_{max1}, K_{m1}) at R_1 (glucose to fructose-6-P) had the greatest negative impact on butanol production. However, a decrease in acetone production was responsible for butanol production [46]. These results provide targets for further genetic modification of butanol-producing strains.

3 Engineering *E. coli* for Butanol Production

The paper that James Liao group from University of California, Los Angeles, submitted to the journal *Metabolic Engineering* on May 18, 2007, is the first work on the production of butanol in a non-native microbe [3]. In the following years, scientists from different countries reported works on the hetero-production of butanol in different hosts and made much progress in strain improvement (Table 1). The best heterologous butanol-producing strains are presently derived from *E. coli*, which can produce 14–15 g/L butanol with a yield of 31–33 % [13, 45] and thus have industrial advantages over clostridial strains. Here, we mainly summarize the progress made in butanol production by *E. coli*.

Table 1 Butanol production by engineered *E. coli* strains

E. coli strain	Overexpressed genes	Disrupted genes	Media and conditions	Titer (g/L)	Yield (w/w) (%)	Productivity (g/L/h)	References
BW25113	*atoB*, *hbd*, *crt*, *bcd*, *etfA*, *etfB*, *adhE2*	*adhE*, *ldhA*, *frdBC*, *pta*, *fnr*	**TB medium with glycerol**, shaken in a sealed 12-ml glass tube for 24 h	0.552	–	–	Metab Eng, 2008, 10:305–311
JM109	*thl*, *hbd*, *crt*, *bcd*, *etfA*, *etfB*, *adhE2*	no	**M9 medium**, initial OD660 = 20, stirred in a 100-ml bottle in an anaerobic chamber for 60 h	1.2	6.1	–	Appl Microbiol Biotechnol, 2008, 77:1305–1316
BW25113	*thrA*^{fbr}*BC*, *ilvA*, *leuABCD*, *kilvd*, *adhE2*	*metA*, *tdh*, *ilvI*, *ilvB*, *adhE*	**M9 medium** with 5 g/L YE, shaken in a 250-ml screw-capped conical flask for 4 days	0.8	–	–	Metab Eng, 2008, 10:312–320
BL21Star (DE3)	*atoB*, *hbd*, *crt*, *bcd*, *etfA*, *etfB*, *adhE2*, *gapA*	no	**TB medium**, shaken in a 250-ml screw-capped flask for 48 h	0.580	–	–	Metab Eng, 2009, 11:262–273
MG1655	*atoB*, *hbd*, *crt*, *bcd*, *etfA*, *etfB*, *adhE2*, *atoC(c)*	*adhE*, *fadR*::*IS5*	**Minimal medium with palmitic acid as carbon source**, initial OD550 = 10, shaken in a sealed 50-ml baffled shake flask for 72 h	2.05		–	Appl Environ Microbiol, 2010, 76(15):5067
DH1	*phaA*, *hbd*, *crt*, *ter*, *adhE2*, *aceEF.lpd*	no	**TB medium**, shaken in a sealed 250-ml baffled flask for 3 days	4.650	28	–	Nat Chem Biol, 2011, 7:222–227
BW25113	*atoB*,*hbd*, *crt*, *ter*, *hbd*, *adhE2*, *fdh*$_{CB}$	*adhE*, *ldhA*, *frdBC*, *pta*	**TB medium**, shaken in a 10-ml BD Vacutainer sealed tubes for 75 h	15	30.8[a]	0.2	Appl Environ Microbiol, 2011, 77(9):2905

(continued)

Table 1 (continued)

E. coli strain	Overexpressed genes	Disrupted genes	Media and conditions	Titer (g/L)	Yield (w/w) (%)	Productivity (g/L/h)	References
MG1655	*atoC*©, *crp**, *yqeF*, *fucO*	*fadR::IS5*, *yqhD*, *eutE*, *arcA*, *adhE*, *frdA*, *pta*	**Minimal medium**, conducted in a SixFors multi-fermentation system (Infors HT) with control of oxygen at 5 % of saturation and pH at 7	14	33	0.39	Nature, 2011, 476:355–359
DH1	*atoB*, *hbd*, *crt*, *bcd*, *etfA*, *etfB*, *adhE2*, *cel3A*, *osmY-cel*, *gly43F*, *osmY-xyn10B*	*adhE*	**EZ-rich medium (Teknova) with 3.3 % w/v ionic liquid-treated switchgrass**	~0.028	–	–	Proc Natl Acad Sci USA, 2011, 108 (50):19949–19954
MG1655 $lacI^Q$	*atoB*, *fadB*, *fadE*, *adhE (G568A)*	no	**M9 medium**, shaken in 20 * 200-mm test tube with ventilation plugs for 24 h	0.614	–	–	Biotechnol Lett, 2012, 34:463–469
ATCC11303	*thl*, *hbd*, *crt*, *bcd*, *etfA*, *etfB*,*adhE2*, *aceEF-lpd*	*adhE*, *ackA*, *frdABCD*, *pflB*, *ldhA*	**LB with 50 g/l glucose**, initial OD550=32, stirred in 9-ml screw-capped tubes in an anaerobic jar for 60 h	1.254	15	–	J Ind Microbiol Biotechnol, 2012, 39(8):1101–1107
MG1655 (DE3)	*thl*, *hbd*, *crt*, *bcd*, *etfA*, *etfB*, *adhE2*, *GlpF*, fdh_{SC}	*adhE*, *ldhA*, *frdBC*	**TB medium**, initial OD600=1.5, shaken in a 50-ml sealed flasks for 48 h	0.154	5.16		Ann Microbiol, 2014, 64(1):219–227

[a] The value is obtained from 36.18 % (g/g, equal to 88 % mol/mol) × 85 % (15 % of butanol is synthesized from TB medium rather than glucose, as indicated in the reference)

3.1 Establishing a Butanol Synthetic Pathway in E. coli and Selection of Efficient Enzymes

In the initial stage of engineering *E. coli* for butanol production, it is natural to transfer the whole butanol pathway from *Clostridium* to *E. coli*, which includes seven genes *thl*, *hbd*, *crt*, *bcd*, *etfA*, *etfB*, and *adhE2*, catalyzing two molecules of acetyl-CoA to one molecule of butanol in six steps (Fig. 1). However, when scientists from the USA and Japan firstly transferred the clostridial butanol pathway to *E. coli* through plasmids in 2007, the engineered strain produced less than 1 g/L butanol (*vs.* clostridial butanol titer 10–20 g/L) [3, 19], although by-product pathways were disrupted. The results indicate that engineering an efficient butanol-producing *E. coli* is not as simple as simply expressing several clostridial genes. Determining the rate-limiting step and selecting alternative genes to fit the *E. coli* host are the key to the heterologous production of butanol (Fig. 1).

3.1.1 Thiolase

The first step in butanol synthesis is the condensing of two acetyl-CoA moles to one acetoacetyl-CoA mole by a thiolase (encoded by the *thl* gene). Thiolase is a ubiquitous enzyme that plays key roles in many vital biochemical pathways, including beta oxidation in the degradation of fatty acids and various biosynthetic pathways. *E. coli* synthesizes two distinct 3-ketoacyl-CoA thiolase enzymes. One is a protein product of the *fadA* gene; the second is a product of the *atoB* gene. To date, FadA has not been tested for butanol production in published work. The *atoB* gene is known to be induced by growth on acetoacetate and exhibits strict substrate specificity for acetoacetyl-CoA. More importantly, AtoB has higher specific activity (1,078 U/mg) than clostridial Thl enzyme (216 U/mg). Hence, when the *thl* gene was replaced with *atoB* gene, the titer of butanol increased more than 3-fold [3]. Additionally, *E. coli* has a *yqeF* gene that encodes a predicted acetyl-CoA acetyltransferase. Overexpression of the *yqeF* gene supports a functional reversal of the beta-oxidation cycle in the synthesis of butanol, which has a better effect than the overexpression of *atoB* [13]. The Chang group at the University of California, Berkeley, constructed a butanol synthetic pathway inspired by the efficient production of polyhydroxyalkanoates in *E. coli*, which transplanted a three-gene pathway from *Ralstonia eutrophus* for monomer biosynthesis (*phaAB*) and polymerization (*phaC*) to yield a biodegradable plastic that can be produced at 50 % dry cell weight at near-theoretical yields. Overexpression of the *phaA* gene can support butanol synthesis at 4.65 g/L in laboratory-scale shake-flask experiments [7]. Additionally, the *ERG10* gene from *Saccharomyces cerevisiae* has been shown to be functional in a butanol synthetic pathway in *S. cerevisiae* [48], but has not been tested in an *E. coli* host. It should be noted that although several gene candidates encoding acetyl-CoA acetyltransferase for butanol synthesis have been improved, the best effect of one gene should depend on the host context and expression mode.

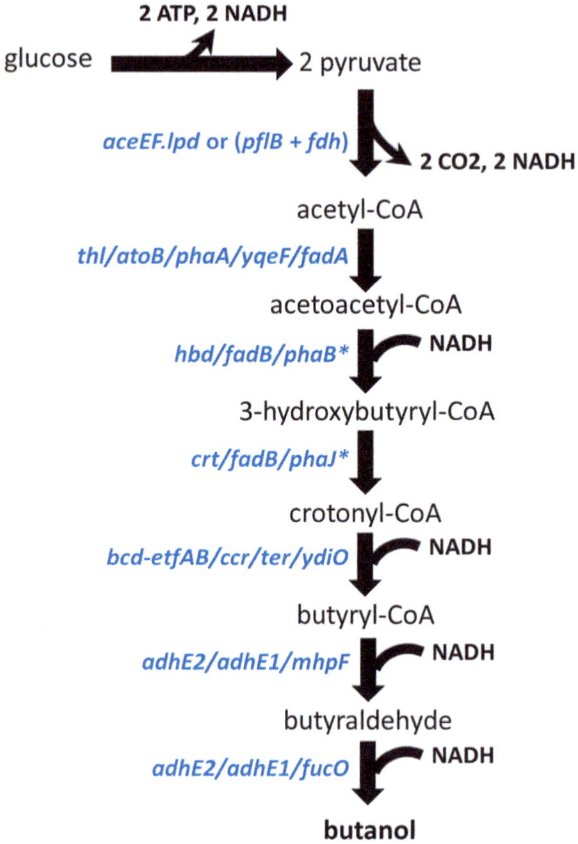

Fig. 1 Pathway and genes for a heterologous butanol pathway in *E. coli*. *aceEF.lpd*: pyruvate dehydrogenase complex from *E. coli*, *pflB*: pyruvate–formate lyase, *fdh*: NAD+-dependent formate dehydrogenase from *Candida boidinii*, *thl*: thiolase from *Clostridium acetobutylicum*, *atoB*: acetyltransferase from *E. coli*, *phaA*: polyhydroxyalkanoate synthase from *Ralstonia eutrophus*, *yqeF*: acetyl-CoA C-acetyltransferase from *E. coli*, *fadA*: acetyl-CoA acyltransferase from *E. coli*, *hbd*: 3-hydroxybutyryl-CoA dehydrogenase from *C. acetobutylicum*, *fadB*: fused 3-hydroxybutyryl-CoA epimerase from *E. coli*, *phaB*: acetyoacetyl CoA reductase from *R. eutrophus*,*crt*: 3-hydroxybutyryl-CoA dehydratase from *C. acetobutylicum*, *phaJ*: (R)-specific enoyl-CoA hydratase from *R. eutrophus*, *bcd-etfAB*: butyryl-CoA dehydrogenase complex from *C. acetobutylicum*, *ccr*: crotonyl-CoA reductase from *Streptomyces collinus*, *ter*: NADH-dependent crotonyl-CoA from *Treponema denticola*, *ydiO*: acyl-CoA dehydrogenase from *E. coli*, *adhE2*: bifunctional acetaldehyde-CoA/alcohol dehydrogenase (CAP0162) from *C. acetobutylicum*, *adhE1*: bifunctional acetaldehyde-CoA/alcohol dehydrogenase (CAP0035) from *C. acetobutylicum*, *mhpF*: acetaldehyde-CoA dehydrogenase II from *E. coli*, *fucO*: L-1,2-propanediol oxidoreductase from *E. coli*. * *phaB* and *phaJ* should be used together

3.1.2 3-Hydroxybutyryl-CoA Dehydrogenase and 3-Hydroxybutyryl-CoA Dehydratase

Most published work directly uses the *hbd* gene (3-hydroxybutyryl-CoA dehydrogenase) and *crt* gene (3-hydroxybutyryl-CoA dehydratase) from *C. acetobutylicum* for the two reactions of acetoacetyl-CoA to 3-hydroxybutyryl-CoA and 3-hydroxybutyryl-CoA to crotonyl-CoA, respectively, in *E. coli* [55]. The native bifunctional *fadB* gene (fused 3-hydroxybutyryl-CoA epimerase) in *E. coli* was also improved to be able to catalyzing the two reactions [18]. Besides, the *phaB* (acetyoacetyl CoA reductase) and *phaJ* ((R)-specific enoyl-CoA hydratase) for the production of polyhydroxyalkanoates from *R. eutrophus* also could be used for above two reactions [7]. It is worthy to note that *phaB* and *phaJ* should be used together, because of the stereoisomerism specificity of these enzymes [7]. However, no evidence indicates which enzyme is the best for the synthetic butanol pathway in *E. coli.*

3.1.3 Butyryl-CoA Dehydrogenase

The fourth step of butanol synthesis is the reduction of crotonyl-CoA to butyryl-CoA by BCD, which needs EtfAB as an electron carrier. Studies have indicated that BCD catalysis is involved in clostridial ferredoxins, which may not fit the cellular context of *E. coli.* In practice, all artificial butanol pathways containing BCD in *E. coli* produced a limited titer of butanol. In the first case of constructing butanol-producing *E. coli* by the Liao group, BCD-EtfAB was replaced with Ccr (encoding a crotonyl-CoA reductase) from *Streptomyces coelicolor.* However, the resulting *E. coli* strain produced less butanol [3]. The Chang group also tested the effects of the *ccr* gene (from *S. collinus*) on butanol production. They found that the butanol titer is related to the expression strength of the *ccr* gene [7], which indicates that this step is rate limiting in the butanol pathway in a non-native *E. coli* host. Studies have also indicated that the Ccr-catalyzed reduction of crotonyl-CoA to butyryl-CoA is a side reaction of the native reductive carboxylation reaction to form ethylmalonyl-CoA, confirming the Ccr activity is low for butanol production [7]. The biological reduction reaction of enoyl-CoA is ubiquitous in nature, such as in fatty acid synthesis and the beta-oxidation pathway of fatty acid; the reaction requires flavin as factors and is reversible. According to the principles of thermodynamics, the direct hydride transfer from NAD(P)H to the enoyl-CoA that increases the barrier for the reverse oxidation reaction and thus potentially kinetically traps crotonyl-CoA in the synthetic butanol pathway can be achieved by eliminating the less-downhill intermediate state produced in the substrate reduction by the flavin cofactor [7]. Fortunately, a crotonyl-CoA-specific trans-enoyl-CoA reductase (Ter) from *Euglena gracilis* was improved to catalyze the irreversible oxidation of crotonyl-CoA to butyryl-CoA in the presence of NAD^+ or $NADP^+$. The Liao group investigated *ter* genes from *Treponema denticola, Treponema vincentii, Flavobacterium johnsoniae,* and *Fibrobacter succinogenes* and found that the *ter* gene from *T. denticola* was the best [45]. The Chang group also selected the *ter* gene from *T. denticola* according to enzymatic mechanism analysis for their butanol pathway, which resulted in a butanol titer of 4.65 g/L without removing any by-product pathways [7]. The engineered *E. coli* containing this *ter* gene constructed by the Liao group could produce 15 g/L butanol [45].

3.1.4 Aldehyde/Alcohol Dehydrogenase

The final two steps of butanol synthesis are the reduction of butyryl-CoA to butyraldehyde by aldehyde dehydrogenase and the subsequent reduction to butanol by alcohol dehydrogenase, consuming two NADH molecules. In the native butanol-producing model of the bacterium *C. acetobutylicum*, the two steps can be catalyzed by one enzyme, bifunctional aldehyde/alcohol dehydrogenase, which is encoded by the *adhE1* gene (active in the solvent production phase) or *adhE2* gene (active in the alcohol production phase). Using the same promoter for the expression of the two genes in *E. coli*, compared with the *adhE1* strain, the *adhE2* strain has 8-fold activity for butyrate dehydrogenase but no increase for butanol dehydrogenase activity, leading to 4-fold butanol production [19]. The *adhE2* gene was also compared with the *adhE* gene from *E. coli*. The results showed that versus *adhE*, the *adhE2* showed 1.5-fold activity when using butyryl-CoA as substrate, and 6-fold selectivity of butyryl-CoA: acetyl-CoA [3]. Although *adhE2* was successfully used for butanol production in *E. coli*, ethanol is still one of the main products (ethanol:butanol ratio exceeding 1:10), thus limiting the butanol yield. Hence, more gene candidates of aldehyde/alcohol dehydrogenase may be screened to reduce the ethanol titer and decrease the ethanol:butanol ratio in future work to improve the strain.

3.2 Optimization of the Gene Expression in the Butanol Pathway

In the initial configuration of a heterologous pathway, the gene expression profile is usually not optimal for maximal carbon flux. Hence, fine tuning of the gene is an essential step in the construction of an efficient microbial cell factory. Different methods of controlling gene expression have been developed, such as the use of a promoter library, the use of an RBS strength prediction algorithm, and the MAGE fine-tuning method. The Yang group from the Institute of Plant Physiology and Ecology at Chinese Academy of Sciences used the strong promoter Alper PLTetO1 or the weak promoter Alper BB to express the *thl* gene and used the strong promoter Braatsch20 or the weak promoter Braatsch10 to express other genes (one operon) of the butanol pathway in *E. coli*. The results showed that the combination of Alper PLTetO1-thl and Braatsch10-operon is best and provided a butanol titer that was 3- to 5-fold higher than that of other combinations [49]. There have been few other published works on the systematic fine tuning of genes for butanol production, which should be considered one of the main directions of constructing high-carbon-flux butanol pathways.

3.3 Engineering Reducing Power Balance for Efficient Butanol Production

In the biosynthetic pathway from glucose to butanol, a precise redox balance can be achieved with a maximal theoretical butanol yield of 41.1 % (w/w). However, such balance is difficult to achieve in the practical engineering of *E. coli*. The main problem relates to the conversion of pyruvate to acetyl-CoA. If the reaction is catalyzed by the PDH complex (encoded by *aceEF.lpd* genes), two NADH molecules are generated that can provide the redox balance of butanol synthesis. However, the PDH complex is inactive in the anaerobic condition owing to the anaerobic sensitivity of E3 component Lpd (dihydrolipoamide dehydrogenase); an active PDH complex is essential for butanol production. To solve this problem, the Chang group overexpressed *aceEF.lpd* genes in a plasmid, resulting in a 3-fold increase in PDH activity, a 53 % increase in the NADH concentration, and a 1.6-fold increase in the butanol titer [7]. Another research group from Northern Illinois University employed the same strategy with the aim to engineer a homobutanol fermentation pathway in *E. coli*; the resulting strain only produced a measurable amount of butanol under anaerobic conditions [17], indicating that other factors should be optimized to couple this strategy. In previous studies, the Ingram group from the University of Florida found an anaerobic active *lpd* mutant *lpd101* (E354K) in the process of the laboratory evolution of *E. coli* [53], and the Zhang group from the Tianjin Institute of Industrial Biotechnology at the Chinese Academy of Sciences found another anaerobic active *lpd* mutant *lpdA** (C242T, C823T, and C1073T) in an adapted succinate-producing *E. coli* [59]. However, these *lpd* mutants have not been used for butanol production in *E. coli* to date, which should be an efficient strategy for obtaining NADH for butanol production. For the anaerobic growth of *E. coli*, the cell mainly uses the pyruvate formate-lyase (encoded by the *pflB* gene) to catalyze pyruvate into acetyl-CoA and formate. The formate is secreted or converted to carbon dioxide and hydrogen by native formate-hydrogen lyase complex. Hence, the reducing power from pyruvate is wasted in the form of formate or hydrogen. It is known that formate can be converted into carbon dioxide and NADH by the specific formate dehydrogenase (encoded by the *fdh* gene) from yeast. The Liao group successfully used NADH obtained from formate by overexpression of the *fdh* gene from *Candida boidinii* as the driving force, to improve the butanol titer and yield, with the reduced formation of by-products [45]. It is worth noting that the Gonzalez group from Rice University obtained a butanol yield of 33 % (vs. max. 41 %) in *E. coli* with active reversal of the beta-oxidation cycle, without manipulating the reaction of pyruvate to acetyl-CoA [13]. The mechanism of reducing the power supply in the above strain may provide new insights into improving butanol production by *E. coli*. Although the above cases and strategies improve the capability of butanol production, the best yield of butanol produced by engineered *E. coli* was only 80 % of the maximal value, indicating that barriers remain to be solved.

3.4 Removing by-product Pathways to Supply Sufficient Precursors for Butanol Production

Butanol production by clostridial strains is naturally coupled with the production of acetone, ethanol, and small amounts of acetate and butyrate, resulting in a low yield of butanol and high feedstock cost. The main purpose of engineering *E. coli* for butanol production is to improve the butanol yield from sugars, reducing the feedstock cost. According to the well-studied metabolic pathway of *E. coli*, the key genes for the production of by-products are known, namely *frdABCD* for succinate, *ldhA* for lactate, *pta-ack* for acetate, and *adhE* for ethanol. In most butanol-producing *E. coli* strains, these genes were disrupted to provide adequate precursors for butanol production. It is notable that the *ldhA* gene was not disrupted in the engineered *E. coli* with active beta-oxidation cycle of the Gonzalez group, which could still produce butanol with 33 % yield [13]. Although the typical by-product pathways were disrupted, known and unknown by-products were still produced by the engineered strains more or less. To solve this problem, more genes of the corresponding by-products need to be disrupted, and the butanol pathway needs to be further optimized to trap more carbon flux from other pathways.

3.5 Using Cheap Substrates for the Low Cost of Butanol Production

Butanol produced from biomass as a bulk chemical or biofuel must have a low production cost to compete with products of crude oil. It is thus important to select cheap feedstocks for butanol production. In constructing butanol-producing *E. coli*, scientists tested different cheap substrates for butanol production, which included palmitic acid, ionic liquid-treated switchgrass, glycerol, and xylose (Table 1). However, butanol titers from these substrates are lower than 2 g/L. The low titers can be explained that the tested strains were not the best strains, and insufficient effort was made in engineering the substrate utilization. The use of cheap feedstock for butanol production by *E. coli* should be the key to an economical industrial process and thus needs to be strongly promoted.

4 Butanol Toxicity and Engineering Butanol Tolerance in *E. coli*

Although *E. coli* can convert sugars (glucose and xylose) to butanol at a relatively high level, it cannot tolerate 2 % (*v*/*v*) butanol [27]. *E. coli* is unable to produce butanol at a very high level as a result. Considering the relationship between butanol tolerance and butanol production by *Clostridial* strains [16, 29, 30, 34], butanol toxicity to *E. coli* is considered a bottleneck for butanol production. It is thus important to develop a butanol-tolerant strain in *E. coli* for the production of high-titer butanol at levels needed for economic efficiency.

4.1 Butanol Toxicity to Microbes

The toxicity of butanol, as a solvent, to cells begins with the butanol impact on the cell membrane. Cell membranes are composed of a phospholipid bilayer interspersed with proteins. In addition to providing structural integrity and maintaining a barrier to the extracellular environment, they facilitate transport in and out of the cell and are responsible for signal transduction, communication, and energy production [37]. When cells are exposed to butanol, the butanol accumulates in the phospholipid bilayer, the hydroxyl moiety accumulates near the phospholipid polar headgroup, and the aliphatic chains are intercalated between the fatty acyl chains of the phospholipids [54]. The hydroxyl group of the butanol spends more time hydrogen bonded to the phosphate group of the lipid than the more hydrophobic longer-chain n-alkanols, which are more deeply embedded in the bilayer. As a result, butanol generates larger disordering in the phospholipid bilayer than the other n-alkanols [56]. Hereafter, the membrane loses its integrity, and the structural and functional properties of the membranes are affected. An increase in permeability to protons and ions has been observed. Consequently, dissipation of the proton motive force and impairment of intracellular pH homeostasis occur. In addition to the effects of lipophilic compounds on the lipid part of the membrane, proteins embedded in the membrane are affected. The effects on the membrane-embedded proteins probably result to a large extent from changes in the lipid environment [47]. In addition, it has been shown that butanol can affect cells by damaging and denaturing biological molecules, including damage to DNA and lipid damage by oxidative and related mechanisms [37]. These results provide insights into butanol toxicity to *E. coli*, from which promising strategies for improving the tolerance to butanol can be obtained.

4.2 Mechanisms of Butanol Tolerance

4.2.1 Omic Analyses Revealing Molecular Mechanisms of Butanol Tolerance

Although butanol is toxic to microbes, some species or strains can tolerate butanol to some degree. As shown in Table 2, *Pseudomonas putida* strains possess a high tolerance to butanol and can grow in 6 % (vol/vol) butanol [43]. Some *Lactobacillus* and *Pediococcus* species can tolerate butanol of up to 3 % or more. The tolerance mechanisms are useful in engineering butanol-tolerant strains. In recent

Table 2 Butanol tolerance of some species

Genus	Species	Strain	Butanol tolerance (%)	Reference
Bacillus sp.		SB-1	3 (v/v)	Curr Sci India, 2002, 82: 622–623
Enterococcus	*casseliflavus*	IMAU10148	3.5 (v/v)	Lett Appl Microbiol, 2010, 50: 373–379
Enterococcus	*faecium*	IB1	2.5–3 (w/v)	Appl Biochem Biotech, 2012, 168: 1672–1680
Lactobacillus	*brevis*		3 (w/v)	Appl Microbiol Biotechnol, 2010, 87: 635–646
Lactobacillus	*delbrueckii*		2.5 (v/v)	Appl Biochem Biotech, 2009, 153: 13–20
Lactobacillus	*plantarum*	E4	3 (v/v)	Lett Appl Microbiol, 2010, 50: 373–379
Pediococcus	*acidilactici*	IMAU20068	3.5 (v/v)	Lett Appl Microbiol, 2010, 50: 373–379
Pediococcus	*pentosaceus*	IMAU20032	3.5 (v/v)	Lett Appl Microbiol, 2010, 50: 373–379
Pseudomonas	*putida*	DOT-T1E, S12, VLB120	6 (v/v)	Appl Environ Microbiol, 2009, 75: 4653–4656

years, system biotechnological approaches have been widely used to investigate the molecular mechanism of butanol tolerance.

Comparative proteomic analyses revealed that glycerol metabolism genes (*glpA* and *glpF*), numerous stress genes (*dnaK*, *groES*, *groEL*, *hsp90*, *hsp18*, *clpC*, and *htrA*), the solventogenic operon *aad–ctfA–ctfB*, and other solventogenic genes were up-regulated in response to butanol stress [1] in the native butanol producer *Clostridium acetobutylicum*. Most were up-regulated in advance (acidogenic phase) [34]. This suggests that the strain Rh8 may have developed a mechanism to prepare itself for coping with butanol challenges before butanol was produced, leading to increased butanol production [34]. Additionally, the butanol-tolerant mutant strain was shown to have evolved a more stabilized membrane structure and to have developed a cost-efficient energy metabolism strategy, to cope with the butanol challenge [33]. Further, comparative genomic analysis indicated a surprisingly high ratio of rRNA mutations that might contribute to improved butanol tolerance [5]. This suggests that strain Rh8 might mutate some rRNA genes to change the structure and function of the whole ribosome. Engineering the factor involved in the translation process can therefore be considered a new strategy of improving microbial stress tolerance worthy of testing [5]. In addition, it was found that in response to butanol on the membrane, *C. acetobutylicum* synthesized increased levels of saturated acyl chains [52]. The growth of cells in the stationary phase coincides with a gradual increase in the percentage ratio of saturated to unsaturated fatty acids. An increased synthesis of saturated fatty acids may provide a more stable membrane environment under butanol stress [4].

Besides clostridia, species that tolerate a high concentration of butanol were used to investigate the mechanism of butanol tolerance. The most interesting findings were solvent efflux pumps and the ability to shift from *cis* isomers to *trans* isomers. For example, *P. putida* strains contain mainly palmitoleic acid and vaccenic acid as *trans* isomers and are directly synthesized from the *cis* isomer within 1 min of exposure to the solvent with no shift in the position of the double bond. Because organic solvents increase membrane fluidity, *P. putida* strains shifting their *cis*-to-*trans* ratio could counteract this alteration [41]. Efflux pumps are membrane transporters and play an important role in cell survival by exporting a wide range of substrates, including bile salts, antimicrobial drugs, and solvents. The efflux pump *srpABC* from *P. putida* S12 has been shown to export hexane, octanol, and several other hydrocarbons. Three efflux pumps (TtgABC, TtgDEF, and TtgGHI) are found in *P. putida* DOT-T1E and are collectively known as the toluene tolerance genes [14].

These results suggest that the molecular mechanism of butanol tolerance is complex; however, the results suggest candidates to be engineered to improve microbial tolerance to butanol. Some candidates have been confirmed by genetic modification, as summarized below.

4.2.2 Investigation of Candidate Targets Contributing to Butanol Tolerance

1. Glycerol metabolism genes
   The expression of the *gldA* gene that encodes glycerol dehydrogenase can be reduced by antisense ribonucleic acid (RNA). It has been shown that the butanol tolerance of *C. beijerinckii* is increased by the reduced activity of glycerol dehydrogenase [31].
2. Heat-shock proteins (HSPs)
   According to the above studies, many stress-responding proteins, including HSPs, are induced by butanol. The HSP system is a cellular stress response system that works during the folding and degradation of proteins. Overexpression of HSP *groESL* in *C. acetobutylicum* ATCC824 resulted in prolonged metabolism and increased butanol production and tolerance [50, 51]. Overexpression of HSPs *grpE* and *htpG* improved the butanol tolerance of *C. acetobutylicum* but did not increase butanol production [32]. Expression of HSP33 from solvent-tolerant *Bacillus psychrosaccharolyticus* in *C. acetobutylicum* ATCC824 did not confer increased solvent tolerance during growth, but increased the total solvent titer by 22 % [9]. This suggests that most HSPs contribute to butanol tolerance, which might be applied in engineering a butanol-tolerant *E. coli* strain.

3. Transcriptional regulator related to solvent production
   Spo0A is a multivalent transcription factor regulator. Expression of *spo0A* in *C. acetobutylicum* promoted expression of the solvent formation genes in the stationary phase, induced the conversion of acid into solvent, and provided increased tolerance and solvent production under butanol stress [2]. By genomic-library enrichment and DNA microarray analysis, CAC1869 categorized as a singleton transcriptional regulator was found. Overexpression of CAC1869 in *C. acetobutylicum* ATCC824 increased butanol tolerance by 81 % and prolonged the metabolic activity [8].
4. Other targets contributing to butanol tolerance
   Glutathione (GSH) is also involved in protein stabilization, antioxidation, and detoxification; so, a study was conducted by introducing GSH synthetic genes *gshAB* into *C. acetobutylicum* DSM1731. The engineered strain DSM1731(pI-TAB) produced GSH and exhibited improved butanol tolerance and increased butanol production capability [58]. Furthermore, the gene SMB_G1518 in *C. acetobutylicum* DSM1731 that codes the cysteine-rich zinc-finger domain putatively interacting with alcohol and the close gene SMB_G1519 were shown to be possible negative regulators involved in butanol tolerance [22].

4.3 Engineering E. coli to Improve Butanol Tolerance

On the basis of molecular mechanisms of butanol tolerance and confirmed strategies for *Clostridium*, efforts were made to improve the butanol tolerance of *E. coli* (Table 3).

4.3.1 Overexpression or Deletion of Genes to Improve the Butanol Tolerance of *E. coli*

Butanol is known to affect the membrane by increasing the membrane fluidity. For *E. coli*, several transcriptional analyses have been performed to clarify the stress caused by butanol. The results indicate an increase in reactive oxygen species during butanol stress. The free radicals directly attack the membrane by lipid peroxidation [44].

To relieve the oxidative stress in the host cell, metallothioneins (MTs), which are known as scavengers of reactive oxygen species (ROS), were engineered in *E. coli* hosts for both cytosolic and outer-membrane-targeted (osmoregulatory membrane protein OmpC fused) expressions. Cytosolic expression was conducted for the alcohol tolerance measurements of the engineered *E. coli* strains of MTs from human (HMT), mouse (MMT), and tilapia fish (TMT), while the OmpC-fused MT strains (OmpC-HMT, OmpC-MMT, and OmpC-TMT) were expressed for

Table 3 Engineering strategies to improve butanol tolerance in *E. coli*

Strategy	Results on butanol tolerance	Reference
Expression of cytosolic and outer-membrane-targeted metallothioneins (MTs)	From 0.5–1 % (v/v) to 1.5–2 %	Biotechnol Biofuels, 2013, 6: 130
Overexpression of *groESL*	GroESL overexpression strain demonstrated a 2.8-fold increase in integrated growth under curve (IGUC) over the control strain with a challenge of 0.75 % (v/v)	Metab Eng, 2013, 15: 196–205
Overexpression of *entC*	From 0.5 % (v/v) to 0.66 %	PloS One, 2011, 6 (3): e17678
Overexpression of *feoA*	From 0.5 % (v/v) to 0.75 %	PloS One, 2011, 6 (3): e17678
Deletion of *astE*	From 0.5 % (v/v) to 0.74 %	PloS One, 2011, 6 (3): e17678
Evolution	Only 6 g/L (0.74 %, v/v) was tested, both *E. coli* SA481 and TW306 (mutant strains) showed an increased tolerance to butanol relative to JCL260 (parent strain)	Mol Syst Biol, 2010, 6(1): 1–11
Mutation of RNAP alpha subunit	Mutant strain grew faster and exhibited a higher accumulated cell mass than the control in the presence of 0.9 % (v/v) butanol	Appl Environ Microbiol, 2009, 75(9): 2705–2711
Utilization of artificial transcription factors (ATFs)	Among 10^6 ATF transformants screened, 75 ATF transformants survived in LB medium containing 1.5 % (v/v) butanol, when cultured in a range of 1–2 % butanol	Biotechnol Bioeng, 2011, 108(4): 742–749
Mutation of cyclic AMP receptor protein (CRP)	When butanol concentration increased from 0.8 % (v/v) to 1.2 %, the growth rate of a mutant MT5 (0.18 h^{-1}) became twice that of the wild type (0.09 h^{-1})	Appl Microbiol Biotechnol, 2012, 94(4): 1107–1117

membrane-targeted MTs. The abilities of these engineered *E. coli* to scavenge intracellular or extracellular ROS were examined, and TMT was found to perform best among the three MTs, growing in a medium with 1 % (v/v) butanol. Additionally, the membrane-targeted fusion protein, OmpC-TMT, improved host tolerance to 1.5 % butanol, above the tolerance of 1 % for TMT [11].

Efflux pumps play an important role in solvent tolerance. In *E. coli*, the AcrAB-TolC system acts as an efflux pump, with AcrB being the inner membrane transporter, AcrA being the membrane fusion protein, and TolC being the outer membrane protein. A library of heterologously expressed efflux pumps was examined and none of the pumps were able to increase *E. coli* tolerance to butanol [15]. Many studies have suggested that efflux pumps are ineffective at exporting short-chain alcohols.

A molecular chaperone is a cellular stress response molecule that works during the folding and degradation of proteins, with HSPs being well-known examples. Overexpression of *groESL* (a heat-shock gene) in *E. coli* provided an effective outcome. Cultures of 0.75 % butanol were the only challenged samples in which the strain 10-β(pACYC184) showed a net increase in cell density above the starting point, doubling across the entirety of the experiment, while 10-β(pAC-groESL) doubled more than twice in the same time frame. In 0.75 % butanol, the overexpressed *groESL* demonstrated a 2.8-fold increase in integrated growth under the curve over the control [60]. In addition, the Hsp33 of *B. psychrosaccharolyticus* overexpressed in *E. coli* increased the *E. coli*'s tolerance to isopropyl alcohol, demonstrating that a psychrophilic protein is functional at higher temperatures and confers a tolerant phenotype [24]. This protein might be functional for improving butanol tolerance *in E. coli* as well.

An enrichment strategy involving the serial transfer of batch cultures in increasing butanol concentrations (0, 0.9, 1.3, and 1.7 % butanol) along with respective controls was performed recently. The overexpressed genes that conferred the largest increase in butanol tolerance, *entC* and *feoA*, were related to iron transport and metabolism and increased the butanol tolerance by 32.86 ± 4.0 % and 49.16 ± 3.3 %, respectively (compared with the initial butanol tolerance of 0.5 %). The gene whose deletion resulted in the largest increase in resistance to butanol was *astE*, with butanol tolerance being enhanced by 48.76 ± 6.3 % [42].

4.3.2 Transcriptional Engineering of *E. coli* to Improve Butanol Tolerance

To select a butanol-tolerant *E. coli* strain, transcriptional engineering of the bacterial RNA polymerase alpha subunit was studied. Results showed a mutant strain with a mutant RNA polymerase alpha subunit grew well in LB medium containing 0.9 % (v/v) butanol [26].

Lee et al. developed a new method of increasing the butanol tolerance of *E. coli* with artificial transcription factor (ATF) libraries that consist of zinc-finger DNA-binding proteins and an *E. coli* cyclic AMP receptor protein. Using these ATFs, they selected a butanol-tolerant *E. coli* that can tolerate butanol up to 1.5 % (v/v), with a concomitant increase in heat resistance [28].

Zhang et al. demonstrated that the butanol tolerance of *E. coli* can be greatly enhanced through random mutagenesis of global transcription factor cyclic AMP receptor protein. Four mutants (MT1–MT4) with elevated butanol tolerance were isolated from error-prone PCR libraries through enrichment screening. A DNA shuffling library was then constructed using MT1–MT4 as templates, and one mutant (MT5) that exhibited the best tolerance ability among all variants was selected. In the presence of 0.8 % (v/v, 6.5 g/l) butanol, the growth rate of MT5 was found to be 0.28 h^{-1} while that of wild type was 0.20 h^{-1}. When the butanol concentration increased to 1.2 % (9.7 g/l), the growth rate of MT5 (0.18 h^{-1}) became twice that of the wild type (0.09 h^{-1}) [57].

4.3.3 Evolution Engineering of *E. coli* Strains to Improve Butanol Tolerance

For *E. coli*, the ethanol-caused stress is well studied; these results were used for the construction of ethanol-producing strains. Nevertheless, butanol-resistant mutant strains are not so well understood owing to a series of unclear mechanisms. Experimental evolution is an effective method used for chemical tolerance while fermentation is limited by chemical products. However, the phenotype cannot be clearly explained sometimes because of the complex mechanisms.

Researchers isolated three *E. coli* clones capable of growth in 2 % (w/v) isobutanol in glucose media and two clones capable of growth in 1.75 % isobutanol in xylose media, representing 60 and 40 % improvements in tolerance, respectively, compared with the wild-type strain [35]. On the basis of the similarity of isobutanol and butanol, we suppose this strategy also works for butanol tolerance.

Atsumi et al. employed a method of sequential transfer to the isobutanol production host strain, *E. coli* JCL260. JCL260 was initially inoculated into LB broth containing 4 g/L isobutanol. After 15 sequential transfers, the isobutanol concentration in the medium had increased to 6 g/L. The isobutanol concentration then reached 8 g/L after the next 15 transfers. After a total of 45 transfers, we isolated the largest single colony, denoted SA481, on an LB agar plate with 8 g/L isobutanol. SA481 showed increased growth compared with JCL260 in the presence of 6 and 8 g/L isobutanol, while maintaining similar growth in the absence of isobutanol. The study demonstrated the isobutanol-tolerant mutants also had increased tolerance to butanol (6 g/L) and 2-methyl-1-butanol (3 g/L).

5 Discussion and Perspectives

E. coli has been improved to be an excellent butanol producer through metabolic engineering of a new synthetic pathway. The butanol yield of 33 % by *E. coli* is a great advantage over the use of clostridial strains. The maximal butanol titer was 15 g/L, which is lower than the maximal titer of 20 g/L produced by some clostridial strains, and the butanol productivity is lower than that of clostridial strains. Therefore, more effort should be made to improve the performance of *E. coli*.

Besides *E. coli*, other species, such as *B. subtilis* [38], *S. cerevisiae* [48], *P. putida* [38], and *L. brevis* [6], have been used as the host to produce butanol. However, none of these species produce more than 3 g/L of butanol. It is suggested that both the enzymes involved in the butanol synthetic pathway and the matching of the pathway with the host are important in engineering an efficient butanol producer.

Besides the metabolic pathway, the butanol tolerance of host strains is a critical factor affecting butanol production performance. Butanol tolerance is a complex mechanism related to mutagenic changes. Although much progress on the mechanism of butanol toxicity has been achieved and new strategies for improving butanol tolerance developed, such work has not been performed on a butanol-producing strain. The further improvement of the butanol titer may depend on butanol tolerance engineering. Researchers are now using genomics, transcriptomics, proteomics, and metabonomics as tools to analyze the global changes in response to butanol challenge. They hope to understand the tolerance mechanisms clearly and connect the butanol tolerance with yield in *E. coli*. We suppose the system approach will improve butanol production through metabolic engineering in *E. coli*.

Finally, cheaper feedstocks such as glycerol and cellulose hydrolysates should be considered, and this will require additional genetic engineering or metabolic evolution of a butanol-producing strain.

Acknowledgments This work was supported by the National High Technology Research and Development Program of China (Grant No. 2011AA02A208 and 2012AA022100) and the National Program on Key Basic Research Project of China (Grant No. 2011CBA00807).

References

1. Alsaker KV, Paredes C, Papoutsakis ET (2010) Metabolite stress and tolerance in the production of biofuels and chemicals: gene-expression-based systems analysis of butanol, butyrate, and acetate stresses in the anaerobe *Clostridium acetobutylicum*. Biotechnol Bioeng 105(6):1131–1147
2. Alsaker KV, Spitzer TR, Papoutsakis ET (2004) Transcriptional analysis of *spo0A* overexpression in *Clostridium acetobutylicum* and its effect on the cell's response to butanol stress. J Bacteriol 186(7):1959–1971
3. Atsumi S, Cann AF, Connor MR, Shen CR, Smith KM, Brynildsen MP, Chou KJ, Hanai T, Liao JC (2008) Metabolic engineering of *Escherichia coli* for 1-butanol production. Metab Eng 10(6):305–311
4. Baer SH, Blaschek HP, Smith TL (1987) Effect of butanol challenge and temperature on lipid composition and membrane fluidity of butanol-tolerant *Clostridium acetobutylicum*. Appl Environ Microbiol 53(12):2854–2861
5. Bao G, Dong H, Zhu Y, Mao S, Zhang T, Zhang Y, Chen Z, Li Y (2014) Comparative genomic and proteomic analyses of *Clostridium acetobutylicum* Rh8 and its parent strain DSM 1731 revealed new understandings on butanol tolerance. Biochem Biophys Res Commun 450 (4):1612–1618
6. Berezina OV, Zakharova NV, Brandt A, Yarotsky SV, Schwarz WH, Zverlov VV (2010) Reconstructing the clostridial *n*-butanol metabolic pathway in *Lactobacillus brevis*. Appl Microbiol Biotechnol 87(2):635–646
7. Bond-Watts BB, Bellerose RJ, Chang MC (2011) Enzyme mechanism as a kinetic control element for designing synthetic biofuel pathways. Nat Chem Biol 7(4):222–227
8. Borden JR, Papoutsakis ET (2007) Dynamics of genomic-library enrichment and identification of solvent tolerance genes for *Clostridium acetobutylicum*. Appl Environ Microbiol 73 (9):3061–3068
9. Bormann S, Baer ZC, Sreekumar S, Kuchenreuther JM, Toste FD, Blanch HW, Clark DS (2014) Engineering *Clostridium acetobutylicum* for production of kerosene and diesel blendstock precursors. Metab Eng 25:124–130
10. Branduardi P, de Ferra F, Longo V, Porro D (2014) Microbial *n*-butanol production from Clostridia to non-Clostridial hosts. Eng Life Sci 14(1):16–26

11. Chin WC, Lin KH, Chang JJ, Huang CC (2013) Improvement of *n*-butanol tolerance in *Escherichia coli* by membrane-targeted tilapia metallothionein. Biotechnol Biofuels 6:130
12. Dürre P (2007) Biobutanol: an attractive biofuel. Biotech J 2(12):1525–1534
13. Dellomonaco C, Clomburg JM, Miller EN, Gonzalez R (2011) Engineered reversal of the beta-oxidation cycle for the synthesis of fuels and chemicals. Nature 476(7360):355–359
14. Dunlop MJ (2011) Engineering microbes for tolerance to next-generation biofuels. Biotechnol Biofuels 4:32
15. Dunlop MJ, Dossani ZY, Szmidt HL, Chu HC, Lee TS, Keasling JD, Hadi MZ, Mukhopadhyay A (2011) Engineering microbial biofuel tolerance and export using efflux pumps. Mol Syst Biol 7:487
16. Formanek J, Mackie R, Blaschek HP (1997) Enhanced butanol production by *Clostridium beijerinckii* BA101 grown in semidefined P2 medium containing 6 percent maltodextrin or glucose. Appl Environ Microbiol 63(6):2306–2310
17. Garza E, Zhao J, Wang Y, Wang J, Iverson A, Manow R, Finan C, Zhou S (2012) Engineering a homobutanol fermentation pathway in *Escherichia coli* EG03. J Ind Microbiol Biotechnol 39 (8):1101–1107
18. Gulevich AY, Skorokhodova AY, Sukhozhenko AV, Shakulov RS, Debabov VG (2012) Metabolic engineering of *Escherichia coli* for 1-butanol biosynthesis through the inverted aerobic fatty acid beta-oxidation pathway. Biotechnol Lett 34(3):463–469
19. Inui M, Suda M, Kimura S, Yasuda K, Suzuki H, Toda H, Yamamoto S, Okino S, Suzuki N, Yukawa H (2008) Expression of *Clostridium acetobutylicum* butanol synthetic genes in *Escherichia coli*. Appl Microbiol Biotechnol 77(6):1305–1316
20. Jang YS, Lee JM, Malaviya A, Seung DY, Cho JH, Lee SY (2012) Butanol production from renewable biomass: rediscovery of metabolic pathways and metabolic engineering. Biotechnol J 7(2):186–198
21. Jang YS, Lee JY, Lee JM, Lee SY (2009) Metabolic engineering of *Clostridium acetobutylicum* M5 for butanol-ethanol production by complementation of *adhE1* and *ctfAB* genes. New Biotechnol 25:S273–S274
22. Jia K, Zhang Y, Li Y (2012) Identification and characterization of two functionally unknown genes involved in butanol tolerance of *Clostridium acetobutylicum*. PloS One 7(6):e38815
23. Jones DT, Woods DR (1986) Acetone-butanol fermentation revisited. Microbiol Rev 50 (4):484–524
24. Kang HJ, Heo DH, Choi SW, Kim KN, Shim J, Kim CW, Sung HC, Yun CW (2007) Functional characterization of Hsp33 protein from *Bacillus psychrosaccharolyticus*; additional function of HSP33 on resistance to solvent stress. Biochem Biophys Res Commun 358 (3):743–750
25. Kauffman KJ, Prakash P, Edwards JS (2003) Advances in flux balance analysis. Curr Opin Biotechnol 14(5):491–496
26. Klein-Marcuschamer D, Santos CN, Yu H, Stephanopoulos G (2009) Mutagenesis of the bacterial RNA polymerase alpha subunit for improvement of complex phenotypes. Appl Environ Microbiol 75(9):2705–2711
27. Knoshaug EP, Zhang M (2009) Butanol tolerance in a selection of microorganisms. Appl Biochem Biotechnol 153(1–3):13–20
28. Lee JY, Yang KS, Jang SA, Sung BH, Kim SC (2011) Engineering butanol-tolerance in *Escherichia coli* with artificial transcription factor libraries. Biotechnol Bioeng 108(4): 742–749
29. Lin YL, Blaschek HP (1983) Butanol production by a butanol-tolerant strain of *Clostridium acetobutylicum* in extruded corn broth. Appl Environ Microbiol 45(3):966–973
30. Liu XB, Gu QY, Yu XB (2013) Repetitive domestication to enhance butanol tolerance and production in *Clostridium acetobutylicum* through artificial simulation of bio-evolution. Bioresour Technol 130:638–643
31. Liyanage H, Young M, Kashket ER (2000) Butanol tolerance of *Clostridium beijerinckii* NCIMB 8052 associated with down-regulation of *gldA* by antisense RNA. J Mol Microbiol Biotechnol 2(1):87–93

32. Mann MS, Dragovic Z, Schirrmacher G, Lutke-Eversloh T (2012) Over-expression of stress protein-encoding genes helps *Clostridium acetobutylicum* to rapidly adapt to butanol stress. Biotechnol Lett 34(9):1643–1649
33. Mao S, Luo Y, Bao G, Zhang Y, Li Y, Ma Y (2011) Comparative analysis on the membrane proteome of *Clostridium acetobutylicum* wild type strain and its butanol-tolerant mutant. Mol Biosyst 7(5):1660–1677
34. Mao S, Luo Y, Zhang T, Li J, Bao G, Zhu Y, Chen Z, Zhang Y, Li Y, Ma Y (2010) Proteome reference map and comparative proteomic analysis between a wild type *Clostridium acetobutylicum* DSM 1731 and its mutant with enhanced butanol tolerance and butanol yield. J Proteome Res 9(6):3046–3061
35. Minty JJ, Lesnefsky AA, Lin F, Chen Y, Zaroff TA, Veloso AB, Xie B, McConnell CA, Ward RJ, Schwartz DR, Rouillard JM, Gao Y, Gulari E, Lin XN (2011) Evolution combined with genomic study elucidates genetic bases of isobutanol tolerance in *Escherichia coli*. Microb Cell Fact 10:18
36. Murphy CD (2012) The microbial cell factory. Org Biomol Chem 10(10):1949–1957
37. Nicolaou SA, Gaida SM, Papoutsakis ET (2010) A comparative view of metabolite and substrate stress and tolerance in microbial bioprocessing: from biofuels and chemicals, to biocatalysis and bioremediation. Metab Eng 12(4):307–331
38. Nielsen DR, Leonard E, Yoon SH, Tseng HC, Yuan C, Prather KLJ (2009) Engineering alternative butanol production platforms in heterologous bacteria. Metab Eng 11(4–5): 262–273
39. Nigam PS, Singh A (2011) Production of liquid biofuels from renewable resources. Prog Energy Combust Sci 37(1):52–68
40. Ohno S, Furusawa C, Shimizu H (2013) *In silico* screening of triple reaction knockout *Escherichia coli* strains for overproduction of useful metabolites. J Biosci Bioeng 115(2): 221–228
41. Ramos JL, Duque E, Gallegos MT, Godoy P, Ramos-Gonzalez MI, Rojas A, Teran W, Segura A (2002) Mechanisms of solvent tolerance in gram-negative bacteria. Annu Rev Microbiol 56:743–768
42. Reyes LH, Almario MP, Kao KC (2011) Genomic library screens for genes involved in *n*-butanol tolerance in *Escherichia coli*. PloS One 6(3):e17678
43. Ruhl J, Schmid A, Blank LM (2009) Selected *Pseudomonas putida* strains able to grow in the presence of high butanol concentrations. Appl Environ Microbiol 75(13):4653–4656
44. Rutherford BJ, Dahl RH, Price RE, Szmidt HL, Benke PI, Mukhopadhyay A, Keasling JD (2010) Functional genomic study of exogenous *n*-butanol stress in *Escherichia coli*. Appl Environ Microbiol 76(6):1935–1945
45. Shen CR, Lan EI, Dekishima Y, Baez A, Cho KM, Liao JC (2011) Driving forces enable high-titer anaerobic 1-butanol synthesis in *Escherichia coli*. Appl Environ Microbiol 77(9): 2905–2915
46. Shinto H, Tashiro Y, Yamashita M, Kobayashi G, Sekiguchi T, Hanai T, Kuriya Y, Okamoto M, Sonomoto K (2007) Kinetic modeling and sensitivity analysis of acetone-butanol-ethanol production. J Biotechnol 131(1):45–56
47. Sikkema J, de Bont JA, Poolman B (1995) Mechanisms of membrane toxicity of hydrocarbons. Microbiol Rev 59(2):201–222
48. Steen EJ, Chan R, Prasad N, Myers S, Petzold CJ, Redding A, Ouellet M, Keasling JD (2008) Metabolic engineering of *Saccharomyces cerevisiae* for the production of *n*-butanol. Microb Cell Fact 7(1):36
49. Tang W, Li J, Chen J, Yang S (2012) Butanol pathway construction and promoter optimization in *Escherichia coli*. Chinese J Biotechnol (in Chinese) 28(11):1328–1336
50. Tomas CA, Beamish J, Papoutsakis ET (2004) Transcriptional analysis of butanol stress and tolerance in *Clostridium acetobutylicum*. J Bacteriol 186(7):2006–2018
51. Tomas CA, Welker NE, Papoutsakis ET (2003) Overexpression of *groESL* in *Clostridium acetobutylicum* results in increased solvent production and tolerance, prolonged metabolism, and changes in the cell's transcriptional program. Appl Environ Microbiol 69(8):4951–4965

52. Vollherbst-Schneck K, Sands JA, Montenecourt BS (1984) Effect of butanol on lipid composition and fluidity of *Clostridium acetobutylicum* ATCC 824. Appl Environ Microbiol 47(1):193–194
53. Wang Q, Ou MS, Kim Y, Ingram LO, Shanmugam KT (2010) Metabolic flux control at the pyruvate node in an anaerobic *Escherichia coli* strain with an active pyruvate dehydrogenase. Appl Environ Microbiol 76(7):2107–2114
54. Weber FJ, de Bont JA (1996) Adaptation mechanisms of microorganisms to the toxic effects of organic solvents on membranes. Biochim Biophys Acta 1286(3):225–245
55. Wen M, Bond-Watts BB, Chang MC (2013) Production of advanced biofuels in engineered *E. coli*. Curr Opin Chem Biol 17(3):472–479
56. Westerman PW, Pope JM, Phonphok N, Doane JW, Dubro DW (1988) The interaction of *n*-alkanols with lipid bilayer membranes: a ^{2}H-NMR study. Biochim Biophys Acta 939(1): 64–78
57. Zhang H, Chong H, Ching CB, Song H, Jiang R (2012) Engineering global transcription factor cyclic AMP receptor protein of *Escherichia coli* for improved 1-butanol tolerance. Appl Microbiol Biotechnol 94(4):1107–1117
58. Zhu L, Dong H, Zhang Y, Li Y (2011) Engineering the robustness of *Clostridium acetobutylicum* by introducing glutathione biosynthetic capability. Metab Eng 13(4):426–434
59. Zhu X, Tan Z, Xu H, Chen J, Tang J, Zhang X (2014) Metabolic evolution of two reducing equivalent-conserving pathways for high-yield succinate production in *Escherichia coli*. Metab Eng 24:87–96
60. Zingaro KA, Terry Papoutsakis E (2013) GroESL overexpression imparts *Escherichia coli* tolerance to *i*-, *n*-, and 2-butanol, 1,2,4-butanetriol and ethanol with complex and unpredictable patterns. Metab Eng 15:196–205

Adv Biochem Eng Biotechnol (2016) 155: 165–197
DOI 10.1007/10_2015_330

Published Online: 17 October 2015

Microbial Cell Factories for Diol Production

W. Sabra, C. Groeger and An-Ping Zeng

Abstract Diols are compounds with two hydroxyl groups and have a wide range of appealing applications as chemicals and fuels. In particular, five low molecular diol compounds, namely 1,3-propanediol (1,3-PDO), 1,2-propanediol (1,2-PDO), 2,3-butanediol (2,3-BDO), 1,3-butanediol (1,3-BDO), and 1,4-butanediol (1,4-BDO), can be biotechnologically produced by direct microbial bioconversion of renewable materials. In this review, we summarize recent developments in the microbial production of diols, especially regarding the engineering of typical microbial strains as cell factory and the development of corresponding bioconversion processes.

Graphical Abstract

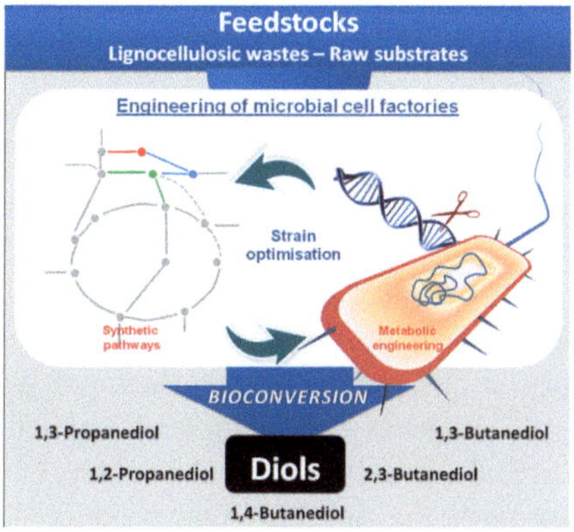

W. Sabra · C. Groeger · A.-P. Zeng (✉)
Institute of Bioprocess and Biosystems Engineering, Hamburg University of Technology, Denickestr. 15, 21073 Hamburg, Germany
e-mail: AZE@tuhh.de

W. Sabra
e-mail: wael.sabra@tuhh.de

C. Groeger
e-mail: Christin.Groeger@tuhh.de

Keywords Propanediol · Butanediol · Diols · Bioconversion · Recovery of Diols

Contents

1 Introduction

At the beginning of the twentieth century, before petroleum was introduced as raw material, the chemical industry had to rely on coal and renewable resources. Until 1930, the most important bulk products of that time, such as fuels (ethanol, butanol), organic acids (acetic acid, citric acid, lactic acid) and other basic chemicals, were biotechnologically produced from biomass. With the development of the petroleum industry, many of these biotechnological processes were replaced by chemical synthesis routes based on petroleum or natural gas. Nowadays, over 80 million tons of industrial chemicals are manufactured globally each year from fossil-based feedstocks [1]. These petrochemicals, which encompass building blocks, intermediate chemicals, and derived final products like polymers, are valued at over $2 trillion and provide the materials and products that impact and enable virtually every aspect of our daily existence [1]. However, these great benefits historically have come at great cost. While the chemicals themselves play a positive role in society, the petroleum-based processes used to manufacture chemicals engender challenges that can jeopardize the economy, the environment, and overall global security. Nowadays, the rapid advances in plant biotechnology, molecular biology, and new tools and concepts such as systems and synthetic biology, and biorefinery of renewable biomaterials have created new opportunities and markets for many biotechnologically produced (bio)chemicals. Many chemicals, which could only be produced by chemical processes in the past, could potentially be generated biologically from renewable resources. The microbial production of diols is a prominent example of success of the so-called white or industrial biotechnology

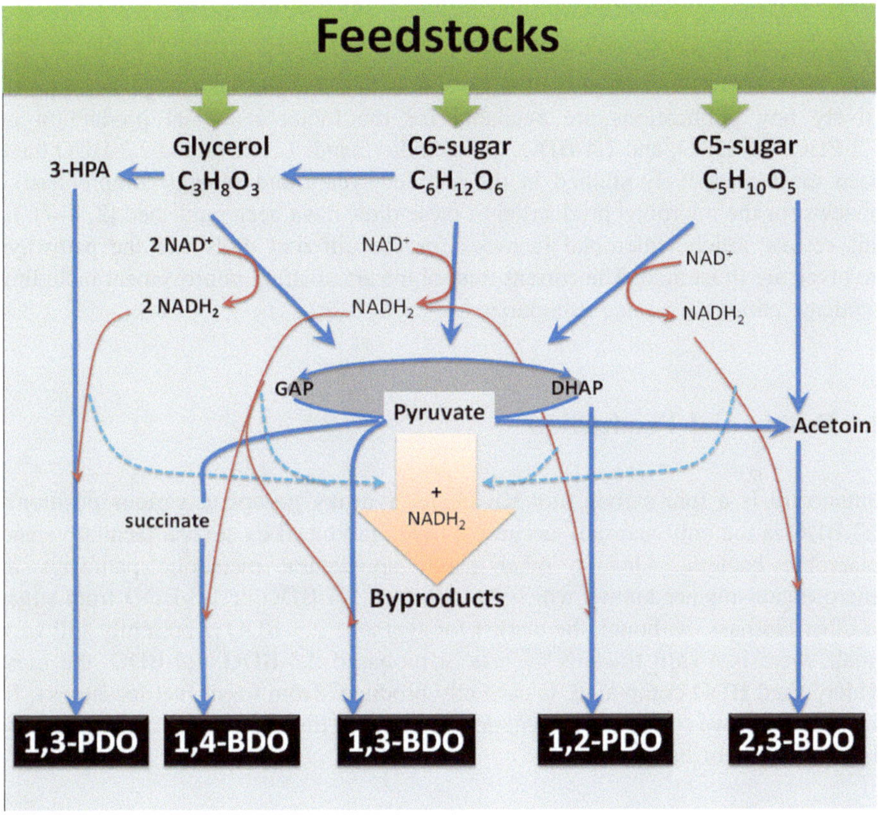

Fig. 1 Major routes for bioproduction of diols from different feedstocks (modified after [2]). *3-HPA* 3-hydroxypropionaldehyde, *GAP* glyceraldehyde 3-phosphate, *DHAP* dihydroxyacetone phosphate

in recent years. Diols are compounds with two hydroxyl groups which have a wide range of important applications as chemicals and fuels. They are considered as platform green chemicals for many industries. In particularly, the microbial production of 1,3-propanediol (1,3-PDO), 1,2-propanediol (1,2-PDO), 2,3-butanediol (2,3-BDO) and, more recently 1,4-butanediol (1,4-BDO) and 1,3-butanediol (1,3-BDO) has received much interest in industrial biotechnology. These diols can be produced from different renewable feedstocks and even waste materials from biofuel production (Fig. 1) [2].

The production of 1,3-PDO and 1,4-BDO has reached commercial scales. They are especially useful as biomonomers for the polyesters polypropylene terephthalate (PPT) and polybutylene terephthalate (PBT). Both PPT and PBT have the potential to steal market share from the classic polyester polyethylene terephthalate (PET) [3]. Pilot plant-scale production of 1,2-PDO and 2,3-BDO has also been reported. All these diols are of immense industrial interests because they are either established chemicals presently produced from fossil resources in large production volumes (e.g. more than 1.5 Mio. t per year for 1,2-PDO and about 1.3 Mio. t per year for 1,4-BDO), or large market potentials [2]. Despite their large impact, relatively few publications are available for the biotechnological production of 1,2-PDO, 1,3-BDO, and 1,4-BDO. On the other hand, 1,3-PDO and 2,3-BDO have been most intensively studied in the last few years and several comprehensive reviews for the microbial production of these diols have been published [2, 4–7]. In this review article, microbial factories for the different diols and the pathways involved are illustrated. The current state of the art of strain improvement including synthetic pathways is also summarized.

2 Butanediol Production

Butanediol is a four-carbon diol having its hydroxy groups at various positions. 2,3-BDO is the only naturally occurring BDO, produced by several facultative and anaerobic bacteria. On the other hand, no natural metabolic pathways or micro-organisms are known which can produce 1,4-BDO or 1,3-BDO from sugar or other biomass. Although the market for synthetic 2,3-BDO is presently still very small, there is a shift towards the use of biobased 2,3-BDO. 1,4-BDO, the most widely used BDO compound, is currently produced from fossil fuel feedstocks. In the following, we summarize recent development in the microbial production of the different types of butanediol.

2.1 2,3-Butanediol

2,3-Butanediol (2,3-BDO) is one of the promising bulk chemicals which exhibits a wide range of potential applications [8–10]. It is used as the starting material for bulk chemicals such as methyl ethyl ketone, gamma-butyrolactone, and 1,3-butadiene [11]. Nowadays, the manufacture of 2,3-BDO is growing by an annual rate of 4–7 % due to the increased demand for many of its derivatives [12, 13]. 2,3-BDO is widely used in chemical, food, fuel, aeronautical, and other fields. Due to the presence of two chiral centres, 2,3-BDO has three isomers: *levo* (2*R*, 3*R*) and *dextro* (2*S*, 3*S*) forms with optical activity and the *meso*-form with no optical activity.

The optically active forms of 2,3-BDOs are very valuable chemicals in the directed asymmetric synthesis of chiral chemicals using boronic esters. Moreover, chiral compounds are especially important to provide chiral groups in drugs, in high-value pharmaceutical or for liquid crystals manufacture [14, 15]. The various applications of this polymer are summarized elsewhere [5, 16].

Although the first commercial production of 2,3-BDO was biotechnological one operated in Germany in the middle of the last century, currently the commercialized process for its synthesis is based entirely on a chemical route. However, the synthetic (petroleum-based) 2,3-BDO does not have a very large market due to its unique structure and costly chemical synthesis. Also, there is no efficient method to convert the intermediate into downstream derivatives such as butadiene, methyl ethyl ketone, and butenes. Therefore, 2,3-BDO has not been produced on a large scale and is currently available as a laboratory chemical and is being sold as a small-volume intermediate for some niche applications like food flavouring. Moreover, its high price led also to inadequate development of its application [17]. Therefore, biobased 2,3-BDO is considered to be a highly attractive market and is expected to provide immense opportunities to the main players involved in the market.

2.1.1 Micro-organisms of Potential Significance for 2,3-BDO Production

Bacteria effectively producing 2,3-BDO belong mainly to the Enterobacteriaceae family. Their representative species are *Klebsiella pneumoniae, K. oxytoca, and E. aerogenes. Pseudomonas chlororaphis and Paenibacillus polymyxa* belonging to the families *Pseudomonadaceae* and *Paenibacillaceae*, respectively, have received attention due to the formation of a pure optically active stereoisomer (*L-form*) in plant rhizospheres. In general, the highest 2,3-BDO concentrations were obtained with pathogen (risk group 2) micro-organisms (Table 1) and thus not desirable for industrial-scale production. Interestingly, Jurchescu et al. [18] reported recently the production of 2,3-BDO by *Bacillus licheniformis* DSMZ 8785 grown on glucose in fed-batch cultivation. The maximum 2,3-BDO concentration obtained was 144.7 g/L, which was comparable to that achieved by the risk group 2 strains. Moreover, by using thermophilic *B. licheniformis* strains, high concentrations (103–115 g/L) of 2,3-BDO could be produced either from glucose [19, 20] or from plant polysaccharide inulin in a simultaneous saccharification and fermentation process [20]. Advantages of the thermophilic process include less contamination risk at high temperature and more efficient utilization of the plant substrate by simultaneous saccharification [19]. Indeed, species of *Bacillus* or *Paenibacillus* appear to be more suitable for commercial 2,3-BDO production. While a mixture of *levo* and *meso* (1:1 ratio) was formed by *B. licheniformis, P. polymyxa* has the ability to form almost exclusively the *levo*-isomer (over 98 %) when grown under anaerobic conditions [13, 21–25]. Recently, Fu et al. [26] showed that NADH played a vital role for chirally pure D-2,3-BDO production in *Bacillus subtilis* grown under limited oxygen conditions. Although the final concentrations in the 2,3-BDO fermentation are lower than those of *B. licheniformis,* the optical purity of the produced diol could be of interest for the fine chemical industry and specific synthesis. Under microaerobic conditions, the 2,3-BDO productivity of this bacterium is higher, but the optical purity decreases, since the *meso*-form is increasingly formed [23].

Table 1 Comparison of 2,3-BDO production by different species, substrates, and fermentation modes

Organism	2,3-BDO isomers	Substrate	Fermentation mode	Concentration (g/L)		Yield	Productivity (g/L h)	References
				2,3-BDO	Acetoin	(g/g)		
Risk group 2								
Klebsiella pneumoniae	*Meso*	Glucose	Fed batch	150	10	–	4.21	[8, 10]
		Glycerol	Fed batch	49.2		0.356	–	[9]
		Xylose + glucose	Fed batch	88	113	0.212	–	[21]
		Jerusalem artichoke powder	Fed-batch SSF	84	7.6	0.294	2.1	[21]
Enterobacter aerogenes	*Meso* *L*	Glucose	Fed batch with cell recycle	110		0.49	5.4	[27]
Enterobacter cloacae		Cassava powder	Fed batch	78		0.42	3.3	[28]
Klebsiella oxytoca	*Meso*	Glucose	Fed batch	95.5	1.9	0.478	1.71	[29]
	L	Xylose + glucose	Batch	23.2	2.5	0.387	0.52	[29]
Serratia marcescens	*Meso*	Sucrose	Batch	152	–	0.41	2.67	[30]
Risk group 1								
Bacillus polymyxa	*L*	Glucose	Fed batch	40.5	6	–	0.81	[21]
		Xylose	Shaking flask	7	–	0.248	–	
Bacillus licheniformis	*L, m*	Glucose Inulin	Fed-batch culture Fed batch	144 103	–	0.4	1.14 3.4	[18, 20]
		Corn stover	Fed batch	74			2.1	[31]
Paenibacillus polymyxa	*L*	Glucose	Fed-batch culture	50	–		0.8	[24]
	L	Glucose		20	–			[13]

(continued)

Table 1 (continued)

Organism	2,3-BDO isomers	Substrate	Fermentation mode	Concentration (g/L)		Yield	Productivity (g/L h)	References
				2,3-BDO	Acetoin	(g/g)		
Genetically modified micro-organisms								
E. coli *E. coli*	*D*	Glucose diacetyl	*Fed batch* *Fed batch*	74 31.7	–	0.4	1.2 2.3	[17, 32]
Saccharomyces cerevisiae		Glucose and galactose		100				[33]
Enterobacter cloacae	*L*	Glucose xylose	*Fed batch*	152			3.5	[11]
	L	Biomass hydrolysate	*Fed batch*	119.4			2.3	[11]

2.1.2 Metabolic Pathways of 2,3-BDO Biosynthesis

Several bacteria, yeasts, or even algae have the capability to produce 2,3-BDO, but the observed yields are often quite different [5, 34, 35]. The 2,3-BDO biosynthetic pathway has been intensively studied in bacteria (Fig. 2). 2,3-BDO synthesis is typically a part of a mixed-acid fermentation pathway observed under anaerobic or microaerobic growth conditions of different micro-organisms (Fig. 2). In addition to 2,3-BDO and depending on the micro-organism and cultivation conditions, other end products are formed, such as ethanol, acetate, lactate, formate, and succinate. In order to enhance the 2,3-BDO yield (theoretical maximum yield 0.5 g/g on glucose), most of the work done was concentrated on an efficient channelling of pyruvate to 2,3-BDO and not to the different by-products. The formation and selectivity of 2,3-BDO stereoisomers and in particular the control of their purity have not been completely understood. Consequently, various metabolic pathways have been proposed (Fig. 3).

Acetoin is the precursor of 2,3-BDO and is formed in bacteria from pyruvate through several enzymatic reactions. Under anaerobic conditions, α-acetolactate synthase catalyses the condensation of two pyruvate molecules with a single decarboxylation to form α-acetolactate that is converted to acetoin by α-acetolactate decarboxylase. Under low oxygen concentration, α-acetolactate can undergo a spontaneous decarboxylation, leading to the formation of diacetyl. Subsequently, a

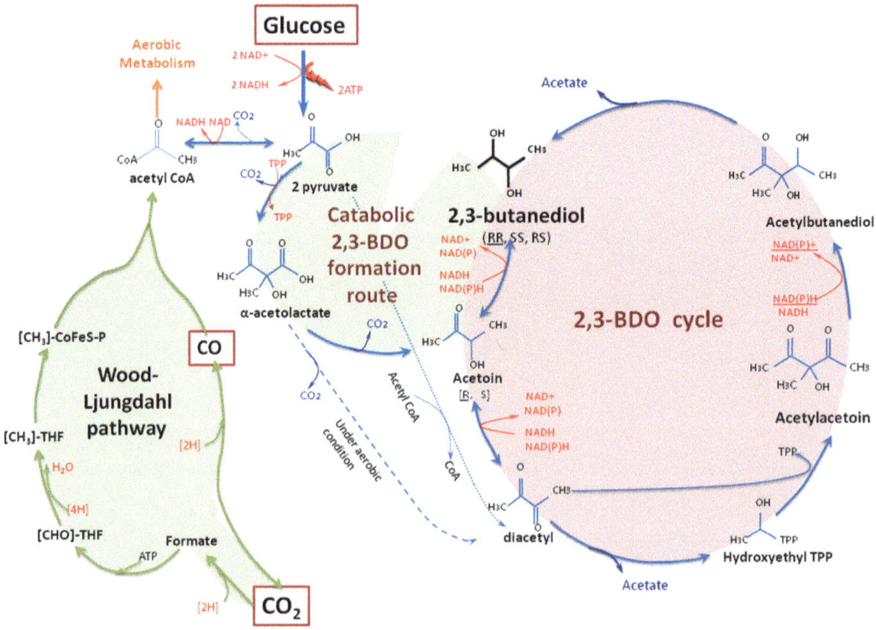

Fig. 2 Different pathways involved in the formation of 2,3-BDO in bacteria (after [5, 35]). *THF* tetrahydrofolate

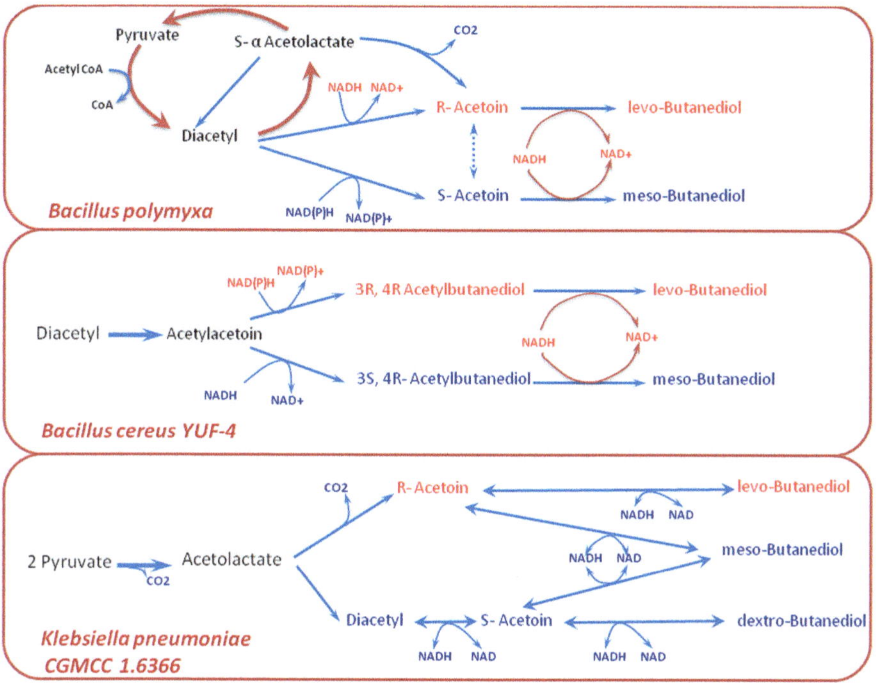

Fig. 3 Mechanisms of the formation of 2,3-BDO stereoisomers (modified after [5])

NADH-linked diacetyl reductase converts the latter to acetoin. Finally, 2,3-BDO of different isomeric forms is formed from acetoin by the action of different acetoin reductase enzymes with different stereospecificities, or by a cyclic pathway (the so-called butanediol cycle), the existence of which has been reported in different bacteria as shown in Fig. 3 [14, 36, 37]. Recently, autotrophic 2,3-BDO synthesis from CO_2 and/or CO plus H_2 was shown to exist in different acetogenic *Clostridium* species [35, 38]. Wood–Ljungdahl pathway was shown to be involved in which CO and/or CO_2 feeds the methyl and carbonyl branches of the pathway. In the methyl branch, CO or CO_2 is fixed in a sequence of tetrahydrofolate (THF)- and cobalamin-dependent reactions into a methyl group, which is then combined with CO (used either directly or after enzymatic reduction of CO_2) to form acetyl-CoA, in which the latter is catalysed by the CODH/ACS (carbon monoxide dehydrogenase/acetyl-CoA synthase) complex. Acetyl-CoA serves as a precursor for growth and 2,3-BDO production [35] (Fig. 2).

In the BDO cycle (Fig. 2), acetoin is oxidized to diacetyl by acetoin dehydrogenase, and then, 2 diacetyl molecules are converted to acetylacetoin and acetate by the enzyme acetylacetoin synthase. Acetylacetoin is further reduced to acetylbutanediol with different stereospecificities by either NAD(P)H- or NADH-linked acetylacetoin reductase. Different 2,3-BDO stereoisomers are then formed by the action of acetylbutanediol reductase. Through this butanediol cycle, 2 forms of stereoisomers are formed in *B. cereus* as reported by Ui et al. [39]. Interestingly, in *P. polymyxa* grown under microaerobic conditions, diacetyl is converted to *S*-acetoin by a NAD(P)H-linked diacetyl reductase. Anaerobically, this bacterium produces 98 % of the *levo*-form through the catabolic 2,3-BDO formation route (Figs. 2, 3). Moreover, an acetoin racemase catalysing the conversion between the different forms of acetoin was proposed for the same bacterium [39]. Recently, Chen et al. [14] elaborated the mechanism of the different stereoisomer formation in *K. pneumonia.* They reported that glycerol dehydrogenase exhibited *2R,3R*-butanediol dehydrogenase activity and was responsible for *levo*-butanediol synthesis from *R*-acetoin. This enzyme also contributed to *meso*-2,3-butanediol synthesis from *S*-acetoin. Butanediol dehydrogenase was the only enzyme that catalyses the conversion of diacetyl to *S*-acetoin and further to *dextro*-butanediol (Fig. 3).

2.1.3 Pathway Engineering and Synthetic Pathway for 2,3-BDO Formation

Despite the intensive research done on enhancing 2,3-BDO production by its native risk group 2 bacteria (e.g. see [11, 40, 41]), the concerns associated with the utilization of potential pathogenic bacteria and/or the inefficient utilization of cellulosic sugars have led many scientists to engineer more safer strains. Oliver et al. [42] have developed a 2,3-BDO biosynthetic pathway in the photosynthetic cyanobacterium *Synechococcus elongatus*. The strain still has a limited productivity (2.38 g/L 2,3-BDO), and more research is needed to reach a desirable titre suitable for industrial application. Efforts were also done to enhance the production of optically active 2,3-BDO in native strains. A mutant of *P. polymyxa* with constitutive synthesis of the α-acetolactate synthase was constructed [21]. The mutant obtained grew more slowly than the wild type but produced fourfold more 2,3-BDO. By knocking out some by-product-producing genes in *Enterobacter cloacae*, Li et al. [11] were able to produce 119 g/L of enantiomerically pure 2,3-BDO using lignocellulosic hydrolysates.

Moreover, *E. coli* was extensively used as a host for many metabolic engineering studies for the production of 2,3-BDO, especially for the production of optically active one. Until recently, the synthetic pathways constructed in *E. coli* for enantiomerically pure 2,3-BDO using different stereospecific dehydrogenases from diverse species gave relatively low concentration of 2,3-BDO [5, 43]. Recently, applying a systematic metabolic engineering approach, Xu et al. [17] optimized the production of 2,3-BDO in recombinant *E. coli* strains. 2,3-BDO biosynthesis gene clusters were cloned from several native 2,3-BDO producers, including *B. subtilis*, *B. licheniformis*, *K. pneumoniae*, *Serratia marcescens*, and *E. cloacae*, inserted into the expression vector pET28a, and compared for 2,3-BDO synthesis. The best strain was then studied in fed-batch fermentation and was found to produce 74 g/L within 62 h [17].

Since no natural producers for the dextro-2,3-BDO (*2S,3S*) have been found, biosynthesis of this diol enantiomer has been achieved using engineered *E. coli* [32,

44]. Li et al. [32, 44] obtained 26.8 g/L of highly pure (>99 %) (*2S,3S*)-2,3-BDO in a fed-batch culture from diacetyl. Moreover, through introducing NADH regeneration enzymes into *E. coli*, a higher product titre (31.7 g/L) of (2*S*, 3*S*)-2,3-BDO was obtained [32].

Many industrial biotechnological processes are moving towards the use of yeast as a platform. Engineered yeast strains were also reported that are capable of producing 100 g/L of enantiomerically pure *levo*-2-3-BDO from a mixture of glucose and galactose with a yield over 70 % of the theoretical value [33, 45]. The high titre and yield of the optically active 2,3-BDO produced make the engineered yeast strain promising hosts for a cost-effective production of biobased 2,3-BDO.

2.2 *1,4-Butanediol*

1,4-Butanediol (1,4-BDO) is an important commodity chemical used to manufacture over 2.5 million tons of valuable products annually. The major use of 1,4-butanediol is in the production of tetrahydrofuran (THF) and PBT [46]. THF is used to produce spandex fibres and other performance polymers, resins, solvents, and printing inks for plastics. PBT is an engineering-grade thermoplastic that combines excellent mechanical and electrical properties with robust chemical resistance. The automotive and electronics industries heavily rely on PBT to produce connectors, insulators, wheel covers, gearshift knobs, and reinforcing beams. There is also growing demand in the apparel industry for renewable, biobased spandex. 1,4-BDO is also used as a plasticizer (e.g. in polyesters and cellulosics), as a carrier solvent in printing ink, a cleaning agent, an adhesive (in leather, plastics, polyester laminates, and polyurethane footwear), in agricultural and veterinary chemicals, and in coatings (in paints, varnishes, and films). 1,4-butanediol is also reportedly used as a solvent in cosmetic formulations and as a humectant in pharmaceuticals [47]. Recently, Diaz et al. [48] reviewed the various biodegradable polymers that can be synthesized from 1,4-BDO and dicarboxylic acids. Application of a series of polymers that cover a wide range of properties, namely materials from elastomeric to rigid characteristics that are suitable for applications such as hydrogels, soft tissue engineering, drug delivery systems, and liquid crystals, is reported.

In nature, no metabolic pathway and no micro-organisms are found so far that can produce 1,4-BDO from sugar or other biomass. Therefore, fossil fuel-based feedstocks such as acetylene, butane, propylene, and butadiene are the current sources for its production. Recently, using genome-scale metabolic model of *E. coli* and biopathway prediction algorithms, the company Genomatica has established unnatural synthetic pathways and correspondingly engineered *E. coli* strains for 1,4-BDO bioproduction from sugars such as glucose, xylose, sucrose, and biomass-derived mixed-sugar streams [46, 49]. In one pathway, sugar is first

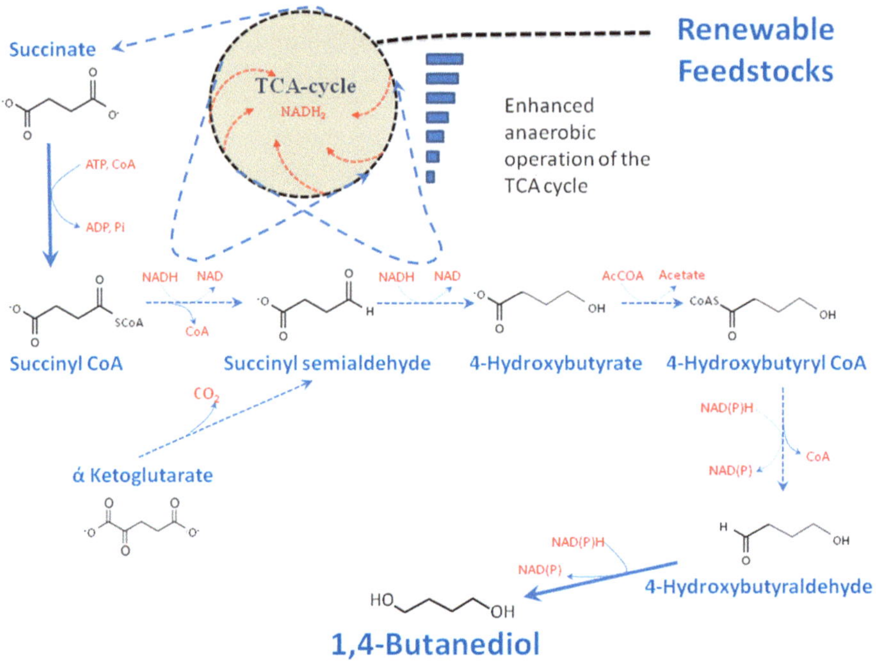

Fig. 4 1,4-BDO biosynthetic pathways introduced in *E. coli* (modified after Yim et al. [46]). *Solid lines* show reactions occurring naturally in *E. coli*, whereas *dotted lines* represent introduced synthetic reaction steps

converted into succinyl-CoA which is then further converted into 1,4-BDO over 4-hydroxybutyrate and other intermediates (Fig. 4), and a strain capable of producing 18 g/L 1,4-butanediol was engineered. The engineered *E. coli* has an enhanced anaerobic operation of the oxidative tricarboxylic acid cycle, thereby generating more reducing power to drive the synthetic 1,4-BDO pathway. According to Genomatica, they have done extensive work to optimize the yield and the rate of 1,4-BDO production, to minimize the by-products, and to enhance the 1,4-BDO tolerance of the engineered strain. Yim et al. [46] proposed that by rising the rates of key steps in the pathway, removing metabolic inefficiencies and substantially reducing by-products may increase the titre further. Burk [1] stated that the commercial production of 1,4-BDO from sugar will require much less energy and release significantly less carbon dioxide and is expected to have a substantial cost advantage relative to the current petrochemical process. Indeed, systems biology and fermentation process engineering approaches can identify and address bottlenecks that are obstacles to commercialization like achieving higher cell densities with improved specific productivity [46]. Recently, strains able to produce 30–40 g/L of 1,4-BDO in a continuous bioreactor were developed and patented by Genomatica [50].

2.3 1,3-Butanediol

1,3-Butanediol (1,3-BDO) is used as a chemical intermediate in the manufacture of polyester plasticizers, as a solvent for flavouring, and as a humectant in pet foods and tobacco. Its uses in cosmetics have been reviewed by the Cosmetic Ingredient Review which concluded that 1,3-butanediol is safe as normally used in cosmetics [51]. (*R*)-1,3-BDO, a non-natural alcohol, is a valuable building block for the synthesis of various optically active compounds such as pheromones, fragrances, and insecticides by direct incorporation into the target molecules, or is used as chiral template in the Lewis acid-mediated reactions of acetals with nucleophiles [52]. (*R*)-1,3-BDO is especially interesting as a starting material of chiral azetidinone derivatives and key intermediate of penems and carbapenems for industrial synthesis of ß-lactam antibiotics. Because these antibiotics are the mostly used antibacterial agents in clinical practice worldwide, the demand for *R*-1,3-BDO has been drastically increased, and as a consequence, the production method of *R*-1,3-BDO has been intensively studied [53–55]. So far, 1,3-BDO has been synthesized as a racemic mixture of *R*- and *S*-forms, mainly from petroleum-based chemicals such as a prochiral precursor, 4-hydroxy-2-butanone. Moreover, Eguchi and Mochida [56] attempted a kinetic resolution of 1,3-BDO by lipase-catalysed diacylations in organic solvent, resulting in (*R*)-1,3-diacetoxybutane with 23.4 % yield and 98.6 % enantiomeric purity. Using whole cells of recombinant *E. coli* expressing exogenous dehydrogenase from *Candida parapsilosis*, Daicel Chemical Industries Ltd. produce *R*-1,3-BDO with 48.4 % yield and 95 % enantiomeric purity [57]. Recently, Kataoka et al. [53] constructed an effective synthetic production route of 1,3-BDO from glucose in *E. coli* (Fig. 5). The high demand on reducing equivalents and cofactors for the production of 1,3-BDO (Fig. 5) reflects the importance of the aerobic catabolism of glucose for reducing equivalent regeneration. Hence, Kataoka et al. [54] optimized 1,3-BDO in an engineered *E. coli* by strict regulation of the overall oxygen transfer coefficient (k_La) during the cultivation. With optimized fermentation conditions, this recombinant *E. coli* strain was able to produce up to 9 g/L of 98.5 % enantiomeric purity of *R*-1,3-BDO. Although the titre reported by Kataoka et al. [53] was more than 8-fold higher than that reported in the patent published earlier in 2009 [58], still much work has to be done to reach an acceptable concentration suitable for commercialization.

3 Propanediol Production

Propanediol is a three-carbon diol having its hydroxy groups, at the first and the last carbon atom, in case of 1,3-PDO, or at the first and the second carbon atom in 1,2-PDO. 1,2-PDO is a chiral molecule and mostly available as a racematic mixture. Both 1,3-PDO and 1,2-PDO offer broad application spectra, either directly as solvents or as platform chemicals for a broad product spectrum. Even though a chemical synthesis is possible, the interest in biological production of propanediols

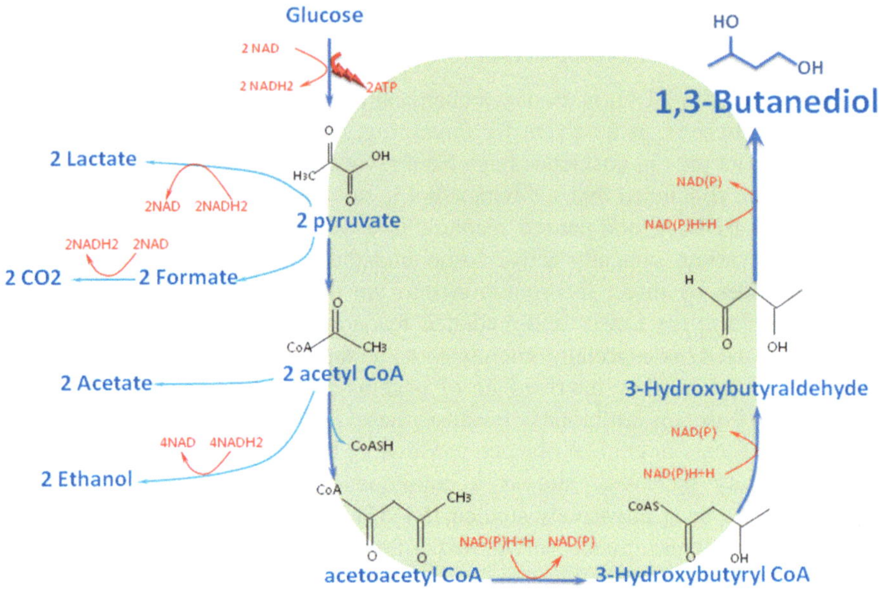

Fig. 5 Schematic diagram of 1,3-BDO biosynthetic pathways from glucose in an engineered *E. coli* (modified after [53, 54])

increases. Fermentation processes need less pressure, ambient temperature, and no expensive catalysts. Furthermore, they allow a sustainable process by transferring waste streams of biodiesel production or lignocellulosic residues into valuable side products.

3.1 1,3-Propanediol

Because of the attractive physical and chemical properties, and hence the various applications of 1,3-PDO, the interest in such polymer increased significantly in the last few years. On the one hand, it is used directly as solvent and antifreeze component in varnish, adhesives, or resins [6], as polyglycol-type lubricant, and in cosmetic products [59]. On the other hand, it is a very suitable monomer for synthetic reactions like polycondensation. 1,3-PDO is well known for the production of polytrimethylene terephthalate (PTT), biodegradable polyester which is utilized fibre not only in textiles and carpets but also in coatings. Furthermore, it can be used for the production of other biodegradable plastics polyesters, polyethers, and polyurethanes [60].

Till recently, biotechnology could not economically compete with the chemical synthesis of 1,3-PDO. In 2004, however, DuPont constructed a biochemical plant in Loudon for manufacturing 1,3-propanediol using *E. coli* with a synthetic pathway from glucose. The plant was commissioned in November 2006. Very recently, the two companies METabolic Explorer (France) and SK Chemicals (South Korea) recently announced a joint agreement to manufacture 1,3-propanediol from crude glycerol. Together, they will market it in Europe and Asia to fulfil the expanding global demand for 1,3-PDO (www.metabolic-explorer.com, www.skchemicals.com, 2014). Biotechnological plants for 1,3-PDO from glycerol were also built in China. It is not clear if any of these plants are in operation.

3.1.1 Micro-organism of Potential Significance for 1,3-PDO Production

1,3-PDO is one of the natural products of the anaerobic degradation of glycerol in many bacteria. Therefore, and as by-products of biodiesel industry, crude glycerol was intensively used for the production of 1,3-PDO. However, the productive strains should be used that can tolerate impurities normally found in crude glycerol (salts, free fatty acids, and methanol [61]). The production of 1,3-PDO from glycerol is mainly performed by micro-organisms of the families *Clostridiaceae* and *Enterobacteriaceae, and* several species of *Klebsiella, Clostridia, Citrobacte*r, and *Enterobacter* are known to convert glycerol to 1,3-PDO under anaerobic conditions. The most-studied and well-known species are *K. pneumoniae* and *Clostridium butyricum*, because of their high substrate tolerance as well as high yield and productivity. Although *C. butyricum* is strictly anaerobic and *K. pneumonia* is facultative anaerobic (easier to handle), species of *Clostridia* are more interesting for industrial application. *K. pneumoniae* is classified as an opportunistic pathogen, and hence, special safety precautions are needed to use *K. pneumonia* for fermentation. Recently, in a cocultivation of cyanobacteria with *K. pneumoniae,* Wang et al. reported the production of 1,3-PDO from CO_2 [62]. Moreover, it was shown that 1,3-PDO can be produced in an unsterile process from raw glycerol using either mixed culture [63] or pure culture of *C. butyricum* [64] and *C. pasteurianum* [65]. This new development makes it economically very competitive. Moreover, the incorporation of the 1,3-PDO production into a biorefinery concept can further increase the ecological advantage and the commercial chance of the glycerol-based process. Friedmann and Zeng [66] proposed to use a mixed culture to produce 1,3-PDO and methane from glycerol. This concept was successfully demonstrated within a European 7th Framework research project (www.propnergy.eu) in laboratory and pilot scale. The basic idea was to use acidogenic and methanogenic bacteria for converting the by-products simultaneously into methane. Alternatively, the by-products can be degraded in a following biogas bioreactor. Formerly, a theoretical and metabolic flux study of syntrophic-like growth of *C. butyricum* and *Methanosarcina mazei*, a methanogenic archeon, under anaerobic conditions was carried out to analyse the several possible scenarios, especially to examine the preference of *M. mazei* in scavenging acetate and formate under conditions of different substrate availability, including methanol as a cosubstrate in biodiesel-derived raw glycerol [67]. Zhou et al. [68] studied the bioconversion of glycerol to 1,3-PDO with a mixed population in a microbial bioelectrochemical system (BES). Though the mixed population used in this study was less effective, the use of BES system for delivering the necessary reducing power for 1,3-PDO production represents an interesting development. More recently, Choi et al. [69] showed that *C. pasteurianum*, a promising 1,3-PDO producer as mentioned above, can directly use electrons from cathode for the regeneration of reducing power in glycerol fermentation. However, the electron flow from the cathode was relatively low and the effect on the glycerol fermentation was not significant. In fact, microbial electrochemical processes for biosynthesis are still poorly understood [70, 71]. The use of a mixed culture in BSE is even more complicated. In general, it is essential to better understand the regulation and metabolic interactions and to control the dynamics of microbial consortia suitable for such processes and to inhibit the 1,3-PDO degradation.

Mixtures of glucose and glycerol have also been used for the production of 1,3-PDO by using members of *Lactobacillaceae*. *Lactobacilli* have only the reductive conversion and need an additional substrate for the growth and generation of the reducing equivalents. *L. reuteri*, *L. brevis*, *L. buchneri*, *L. collonoides,* and *L. panis* were reported to produce 1,3-PDO in mixed substrate fermentation [72]. Pflügl et al. [73] reported the production of 42 g/L 1,3-PDO from glycerol by *L. diolivorans.* However, after the addition of glucose, the 1,3-PDO production increased up to 74 g/L [73]. Recently, Sabra et al. [74] reported the simultaneous production of 1,3-PDO and *n*-butanol in mixtures of glucose and glycerol in different ratios using *C. pasteurianum.* On the other hand, with glucose as monosubstrate, several approaches with genetically modified organisms have been reported (see Sect. 4.1.2). An overview of the potential 1,3-PDO productive strains is given in Table 2.

3.1.2 Biosynthetic Pathways and Pathway Engineering of 1,3-PDO

The natural pathway for the production of 1,3-PDO in different micro-organisms is shown in Fig. 6. Generally, the pathway is divided into two parallel routes, a reductive route for the production of 1,3-PDO (A) and an oxidative route (B) where glycerol is metabolized via glycolysis into pyruvate and energy is produced. Only about 5 % of the glycerol is used for biomass production, when it is the sole carbon source [59].

In the reductive route, glycerol is dehydrated into 3-hydroxypropionaldehyde (3-HPA) by glycerol dehydratase. 3-HPA is subsequently reduced to 1,3-PDO by 1,3-propanediol oxidoreductase (PDOR) under consumption of nicotinamide adenine dinucleotide ($NADH_2$). This reducing equivalent is generated in the oxidative route through the synthesis of pyruvate and the transformation of pyruvate into acetyl-CoA. Different micro-organisms convert pyruvate into different by-products (Fig. 6). Indeed, the yield of 1,3-PDO per glycerol depends on the availability of $NADH_2$. The availability is not only determined by the micro-organism itself but also dependent on the process conditions of the fermentation [89]. Hence, the yield

Table 2 Comparison of 1,3-PDO production by different species, substrates, and fermentation modes

Organism	Substrate	Fermentation mode	Media Supplements	1,3-PDO (g/L)	Yield (g/g)	Productivity (g/L h)	References
C. butyricum DSM 5431	Glycerol	Fed batch	1 g/L yeast extract	58	0.56	2.7	[75]
C. butyricum VPI 3266			0.04 mg/L biotin 8 mg/L *p*-aminobenzoic acid	65	0.57	1.2	[76]
C. butyricum DSM 5431 mutant 2/2			1 g/L yeast extract	70.6	0.54	0.8	[77]
C. butyricum E5			2 g/L yeast extract	65.6	0.54	1.4	[78]
C. butyricum DSM 5431			1 g/L yeast extract	47.5	0.51	2.4	[79]
C. butyricum mutant 2/2			1 g/L yeast extract	70.4	0.56	1.4	[79]
C. butyricum			–	86.6	–	–	[80]
C. butyricum VPI 1718			1 g/L yeast extract crude glycerol	68	0.5	–	[64]
Clostridia mixed culture			0.024 mg/L biotin 0.015 mg/L pantothenate	70	0.46	2.6	[63]
C. butyricum DSP1		Repeated batch	Crude glycerol	62	0.53	0.76	[81]
C. butyricum CNCM 1211		batch	0.004 mg/L biotin	67	0.52	–	[82]
C. acetobutylicum DG1 (pSPD5)		Fed batch	0.04 mg/L biotin 8 mg/L *p*-aminobenzoic acid Crude glycerol	84	0.54	1.8	[83]
		Continuous		60	0.53	3	[83]

(continued)

Table 2 (continued)

Organism	Substrate	Fermentation mode	Media	1,3-PDO	Yield	Productivity (g/L h)	References
			Supplements	(g/L)	(g/g)		
Clostridium IK124		Fed batch	5 g/L yeast extract	87.7	0.54	1.9	[84]
C. pasteurianum mutant			1 g/L yeast extract in situ *gas stripping*	27	0.21	1.2	[85]
Citrobacter freundii [FMCC-B 294 (VK-19)			5 g/L peptone 5 g/L meat extract 2.5 g/L yeast extract	68	0.4	0.79	[86]
K. pneumoniae				83.5	–	–	[80]
Klebsiella oxytoca M5alΔ ldhA	Sucrose + glycerol	Fed batch	1 g/L yeast extract	84	0.62 mol/mol glycerol	1.6	[87]
Lactobacillus reuteri	Glucose + glycerol		(Repeated fed batch) 10 g/L bact. peptone 10 g/L meat extract 5 g/L yeast extract	65	0.97 mol/mol glycerol	1.2	[88]
L. diolivorans	Glucose + glycerol		5 mg/L vitamin B_{12} 10 g/L casein peptone 10 g/L meat extract 5 g/L yeast extract	84.5	0.55 mol/mol substrate	0.45	[73]

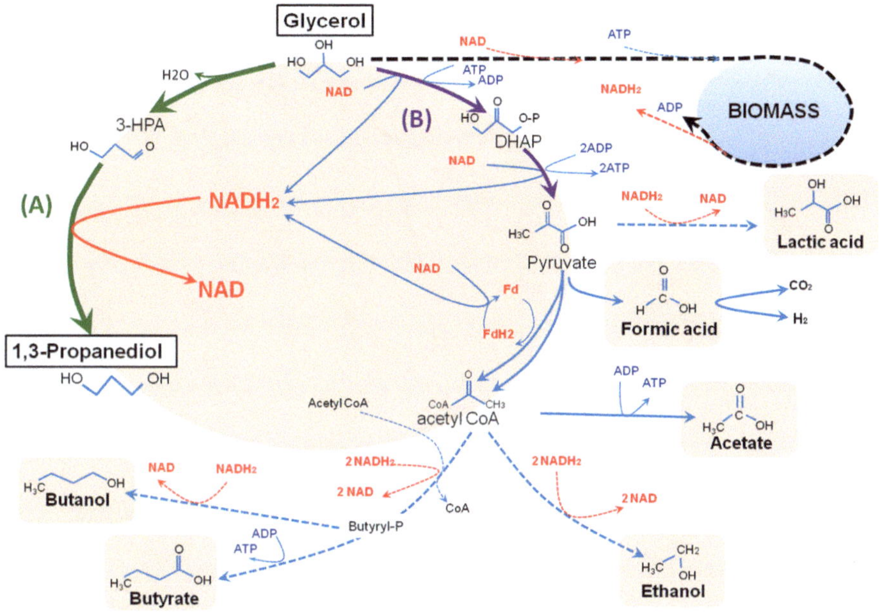

Fig. 6 Metabolic pathway of glycerol fermentation

of 1,3-PDO depends on the combination and stoichiometry of the reductive and oxidative pathways. Consequently, the maximum yield of 1,3-PDO formation from glycerol in clostridia represents 0.67 mol/mol and is achieved under conditions where acetic acid is the main by-product and not butyric acid, ethanol, or butanol [89–91]. If no hydrogen and butyric acid are produced at all during the fermentation, the theoretical yield can be further increased to 0.72 mol/mol [89, 90, 92]. The 1,3-PDO yield from glycerol can be additionally enhanced with an in vitro approach using crude enzymes from different organisms [93]. These systems feature several biomanufacturing advantages, such as fast reaction rate, easy product separation, broad reaction condition and tolerance to toxic substrates or products [94]. Nevertheless, the cost and stability of enzyme and coenzymes restrict the use of such systems in industrial scale.

Intensive work has been done to genetically modify micro-organisms to convert glucose to 1,3-PDO in one micro-organism. In the DuPont PDO process, a synthetic pathway was successfully developed to produce PDO from glucose, in which the glycerol synthesis pathway from *S. cerevisiae* (catalysed by glycerol 3-phosphate dehydrogenase (DAR1) and glycerol 3-phosphate phosphatase (GPP1/2) and the metabolic pathway of converting glycerol to PDO from *K. pneumonia* (glycerol dehydratase, encoded by the genes *dhaB1*, *dhaB2*, and *dhaB3*) were integrated into *E. coli* (Fig. 7, [95]). The last step, the formation of 1,3-PDO is realized by a 1,3-propanediol oxidoreductase isoenzyme from *E. coli* (YqhD). Continuous strain development was made by DuPont/Genencor, and the most fundamental changes

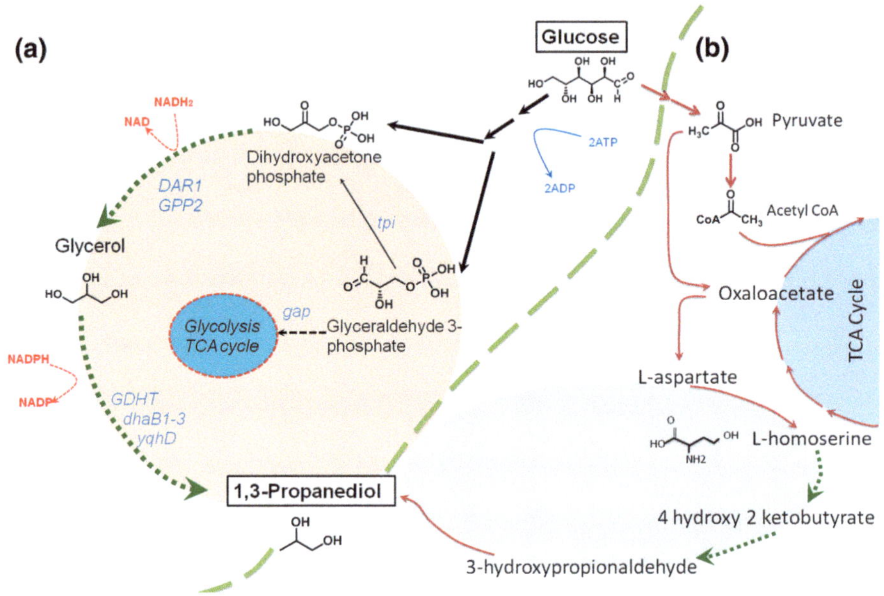

Fig. 7 Engineered *E. coli* strains for the production of 1,3-PDO from glucose. **a** Glycerol-dependent synthetic pathway [95] and **b** non-glycerol-dependent pathway [96]. *Dotted arrows* indicate introduced synthetic pathway steps

done were probably the elimination of D-glucose transport by the phosphotransferase system (PTS) and the downregulation of glyceraldehyde 3-phosphate dehydrogenase (gap) together with reactivation of tpi. Finally, the yield could be increased to 135 g/L with a productivity of 3.5 g/L h [95].

Still, in such a production system, the substrate suicide of glycerol dehydratase (GDHt) that could limit the productivity has to be overcome [97]. Recently, Chen et al. [96] constructed a new non-glycerol-derived synthetic 1,3-PDO synthesis in *E. coli* (Fig. 7). With protein engineering of glutamate dehydrogenase, they extended the pathway of homoserine, a natural intermediate of cellular amino acid metabolism. At first, homoserine is converted by deamination into 4-hydroxy-2-ketobutyrate, followed by decarboxylation into 3-hydroxypropionaldehyde (3HPA). Like in the conventional pathway, 3HPA is subsequently transformed by alcohol dehydrogenase into 1,3-PDO. The theoretical maximum yield (1.5 mol 1,3-PDO/mol glucose) of the new 1,3-PDO pathway is the same as that of the DuPont route. Since homoserine synthesis is a common pathway in most of the bacteria, the proposed route can be engineered into selected hosts with the more favourable ability to utilize different and cheap sugars. Moreover, the proposed pathway does not utilize GDHt and thus can avoid the serious problems associated with vitamin B12 and substrate suicide. This non-natural pathway is thus very appealing for 1,3-PDO production.

3.2 1,2-Propanediol

1,2-propanediol (1,2-PDO), generally called propylene glycol, is a major commodity chemical with a global demand estimated to be around 1.36 Mio. t/a for several industries [98]. It appears as a colourless hygroscopic liquid with low volatility and an oily consistency. This industrial important compound is mainly utilized as solvent, antifreeze, de-icer, and heat transfer fluids [99]. Furthermore, it could be applied as colour compound and flavour and fragrance carrier in foods, beverages, cosmetics, and pharmaceuticals, or even as tobacco humectants [100]. The interest in 1,2-PDO increases since it is less toxic than products based on ethylene glycol for humans and animals. The US Food and Drug Administration (FDA) has determined 1,2-PDO to be "generally recognized as safe" for use in food, cosmetics, and medicines [98].

3.2.1 Microbial Cell Factories for the Production of 1,2-PDO

The biological route for producing 1,2-PDO from sugars is known since many years. Early studies on *Thermoanaerobacterium thermosaccharolyticum* [101, 102], *Bacteroides ruminicola* [103], *C. sphenoides* [104], *L. Buchneri* [105]), and *E. coli* [99] have demonstrated 1,2-PDO formation. In comparison with other diols, the 1,2-PDO yields are much lower, either from sugars or from glycerol [92]. The biosynthesis of 1,2-PDO requires the conversion of the main carbon source into DHAP with the glycolytic pathways (Fig. 8). Therefore, due to higher reduction degree of glycerol, the yield of 1,2-PDO is higher than that from glucose (theoretical maximum yield of 0.63 and 0.72 g/g from glucose and glycerol, respectively). In either way, the biosynthesis consumes redox equivalent and ATP [99].

There are two possible pathways for the biosynthesis of 1,2-PDO. The first one metabolizes deoxy sugars (methyl pentoses (Fig. 8a)), whereas the second one converts DHAP into methylglyoxal (Fig. 8b) and further to 1,2-PDO. The deoxy pathway is well studied in *E. coli* and is reviewed by Bennett and San [106]. At first, *L*-rhamnose is converted into *L*-rhamnulose-1-phosphate, which is subsequently split into dihydroxyacetone phosphate and S-lactaldehyde by the enzyme RhaD (*L*-rhamnose dehydrogenase). Fucose, on the other hand, is first isomerized into *L*-fuculose and transformed into *L*-fuculose-1-phosphate by the enzyme *L*-fuculose kinase. Another enzyme, fucA (*L*-fuculose-1-phosphate aldolase), cleaves it into dihydroxyacetone phosphate and *L*-lactaldehyde. Depending on the redox conditions, the lactaldehyde can be either reduced to 1,2-PDO or oxidized to lactic acid. Anaerobic conditions lead to conversion into *S*-1,2-PDO, catalysed by a NAD-oxidoreductase fucO (*S*)-1,2-propanediol oxidoreductase [106]. Since deoxy sugars are quite expensive as substrate, the deoxy pathway is considered to be uneconomical as an industrial process.

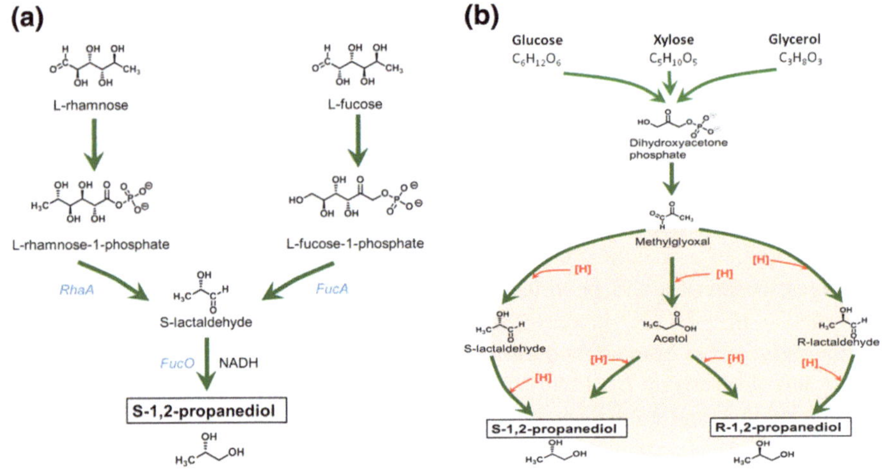

Fig. 8 **a** Metabolic pathways of deoxy sugar fermentation into 1,2-propanediol by *E. coli* and **b** the various routes of the conversion of methylglyoxal to 1,2-PDO in different micro-organisms (modified after [106])

In *T. thermosaccharolyticum*, the second pathway is found. At first, DHAP is produced from glucose, xylose, mannose, or cellobiose. DHAP is then converted with methylglyoxal synthase into methylglyoxal (MG). Subsequently, MG is reduced to *R*-1,2-PDO with aldose reductase or glycerol dehydrogenase. If glucose is fermented only into *R*-1,2-PDO, acetate, and CO_2, a theoretical yield of 0.42 g *R*-1,2-PDO/g glucose is possible. In *E. coli*, MG is converted into acetol with the NADPH- or NADH-dependent lactaldehyde oxidoreductase, and alcohol or aldehyde dehydrogenases. *E. coli* also converts MG into *R*-lactaldehyde with NADH-dependent glycerol dehydrogenase. On the contrary, the yeast *S. cerevisiae* produces S-lactaldehyde from MG, which is subsequently converted into *S*-1,2-PDO by a NADPH-dependent aldose reductase. However, the MG production in *S. cerevisiae* is non-enzymatic and spontaneous, and the final 1,2-PDO titre is quite low [101].

Recently, Clomburg and Gonzales [99] developed a new strain of *E. coli* with increased production of 1,2-PDO. The functional pathway was engineered by combining different strategies (Fig. 9a): (I) to ensure DHAP availability, they changed the PEP-dependent DHAK (dihydroxyacetone kinase) with the ATP-dependent DHAK from *Citrobacter freundii*; (II) they overexpressed the genes for 1,2-PDO synthesis from DHAP; and (III) competitive pathways for acetate and lactate were deleted. Other side products were maintained to ensure the necessary redox balance and ATP generation. The recombinant *E. coli* strain produced 5.6 g/L 1,2-PDO with a yield of 0.21 g/g glycerol [99]. More recently, Koch et al. [107] established a recombinant *E. coli* to enhance 1,2-PDO production from several carbon sources with three newly integrated and highly expressed enzymes.

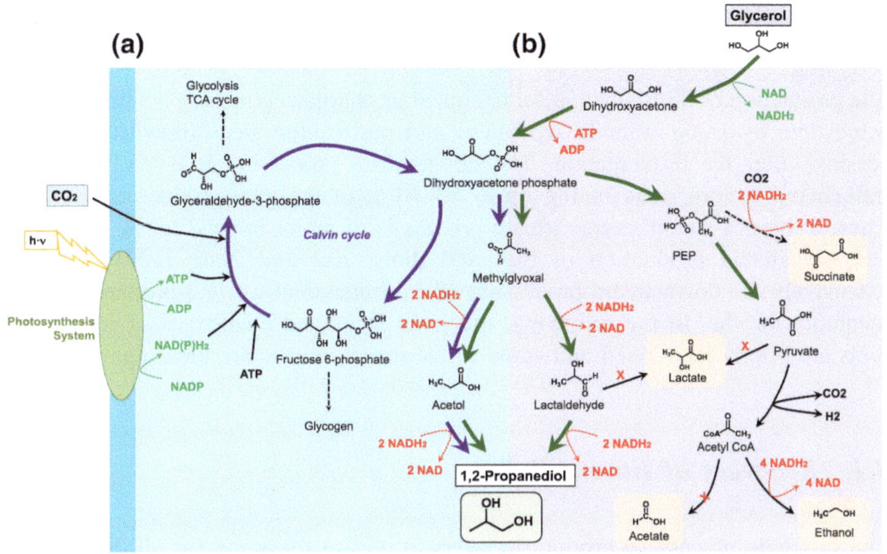

Fig. 9 Synthetic pathways for the production of 1,2-PDO: **a** from CO_2 by *Synechococcus elongates* (according to [108]) and **b** from glycerol by *E. coli*. *Red crosses* indicate deleted pathways, and *dashed lines* represent intermediate steps. (modified after [99])

This new pathway avoids the toxic intermediate methylglyoxal and use the natural formation of lactate. The latter one is transferred into lactyl-CoA by lactate-CoA transferase and then into lactaldehyde by lactyl-CoA reductase. In the last step 1,2-PDO is formed with the help of lactaldehyde reductase (also 1,2-propanediol oxidoreductase). Due to this, a maximum yield of 0.55 g PDO/glycerol and additional ATP can be achieved.

Furthermore, in a more sustainable approach, Li and Liao [108] described a photosynthetic conversion of carbon dioxide with a newly engineered cyanobacterium *S. elongatus* PCC 7942 (Fig. 9b). For the production of 1,2-PDO, genes for methylglyoxal synthase (mgsA), glycerol dehydrogenase (gldA), and aldehyde reductase (yqhD) from *E. coli* have been inserted. Additionally, the alcohol dehydrogenase (sADHs) from *C. beijerinkii* is induced into the cyanobacterium. The NADPH pool of *S. elongates* itself was taken into account for the 1,2-PDO production that requires many reducing equivalents. Therefore, the NADPH-specific secondary alcohol dehydrogenase was newly implemented in the pathway, resulting in the production of 150 mg/L 1,2-PDO from environmental CO_2 and light. Despite the progresses made in the implementation of metabolic engineering strategies and developing different new strains, the low reaction rate and product concentration are the most important barriers in its industrial production.

4 Recovery of Diols

The production of diols suitable for chemical or pharmaceutical applications is only achievable by using suitable separation and purification steps (downstream processing) after the fermentation. The downstream processing is one of the most influencing factors, contributing up to 50–70 % of the total product costs [109]. Thus, to have a suitable downstream process is of major interest for an economic and sustainable production of biobased diols. Xiu and Zeng [109] reviewed extensively the downstream processing of 2,3-butanediol and 1,3-propanediol fermentation broths. In the following, the main steps and challenges of recovering diols are briefly mentioned and some of the recent studies are then highlighted.

4.1 Recovery of Butanediol

The principle process of product recovery is almost the same for all diols. After cultivation, the final fermentation broth is a multicomponent mixture not only consisting of water, residual substrate and salts, and side products (e.g. alcohols, organic acids), but also consisting of cells and cell debris in addition to the target product. The initial step is the separation of biomass, which can be performed via centrifugation, filtration, or flocculation [110]. Residual salts may cause fouling on heating devices or inactivation of catalysts. Therefore, they may need to be removed, e.g., by electrodialysis, salting out, or ion exchange chromatography. The excess amount of water in the broth can be reduced by evaporation. The last step to obtain high purification grades is mainly conducted via distillation.

Difficulties in the recovery of butanediol-like 2,3-BDO are mainly caused by the high boiling point (180 °C for 2,3-BDO) and high hydrophilicity. Extractive separation is hampered by its low selectivity and a relatively low distribution coefficient towards extracting solvents. Promising solvents studied include ethyl acetate, tributyl phosphate, diethyl ether, *n*-butanol, dodecanol, and oleyl alcohol. For example, Anvari and Kayati used the non-toxic oleyl alcohol for an in situ extraction, but separated only 68 % of the total 2,3-BDO produced by *K. pneumonia* [111]. Improvement of extraction methods represents the combination of solvent extraction and salting-out techniques. The salting-out technique is based on a system of two aqueous phases: one with a hydrophilic solvent and one with highly concentrated salts. The increased ionic strength in the salt phase forces more diols to dissolve in the solvent phase, which is in this case an extractant. Li et al. [112] use a mixture of 32 % (w/w) ethanol and 16 % (w/w) ammonium sulphate to recover 91.7 % of 2,3-BDO next to 99.7 % of cells and 91.2 % of proteins. With 34 % (w/w) 2-propanol and 20 % (w/w) ammonium sulphate, Sun et al. [113] separated 93.7 % of 2,3-BDO. Also here, 99 % of the cells could be removed and reused for a new inoculation, which has a positive effect on the process economics. With butanol and potassium, phosphate salts up to 99 % of the 2,3-BDO can be separated, as revealed by an Aspen Plus simulation performed by Birajdar et al. [114]. However, the 2,3-BDO has to be separated again from the extractants by evaporation, which means additional downstream units with additional costs. The purification of 2,3-BDO directly by distillation is hampered due to the high boiling, and it might be only used to enhance the concentration. Qureshi et al. [115] described a vacuum membrane distillation process, where the membrane retains the 2,3-BDO and let the more volatile compounds (water, ethanol) pass through. The concentration could be increased from 40 to 430 g/L. However, medium components caused membrane fouling, and the water flux decreased at higher 2,3-BDO concentrations [115]. A newly developed process combines reactive extraction and reactive distillation. Li et al. [116] used *n*-butyraldehyde (BA) as reactant and extractant at the same time. It reacts with 2,3-BDO to 2-propyl-4,5-dimethyl-1,3-dioxolane (PDD), which is extracted by BA itself. Both BA and PDD are transferred into a reactive distillation column, where the catalysts sulphuric acid and hydrochloric acid cleave PDD again into BA and 2,3-BDO. Li et al. were able to recover 90 % of the 2,3-BDO with purity higher than 99 %.

4.2 Recovery of 1,3-Propanediol and 1,2-Propanediol

The downstream processing of 1,2-PDO and 1,3-PDO from fermentation broth is similar, but could be even more challenging than the recovery of butanediol because of their higher boiling points (188 and 233 °C, respectively).

Liquid–liquid extraction could be more advantageous for PDO because it is selective and more energy efficient than distillation. Malinowsky tested different solvents, such as the series of pentanol until nonanol and hexanal until decanal, and other organic solvent. The best results were achieved with aliphatic alcohols and aldehydes, but the distribution of PDO in the solvents has been very low. Thus, large amounts of solvents would be required [117]. Li et al. [118] described the extraction and salting-out method using ethanol and sodium carbonate. They could separate 97.9 % of the 1,3-PDO and were able to separate 99.1 % of cells and other fermentation products, such as organic acids, in one step with this method. A combination of methanol and dipotassium hydrogen phosphate leads to a slightly higher 1,3-PDO recovery of 98.1 % [119]. In addition, the main side product 2,3-BDO, as well as organic acids, could also be recovered. Müller et al. [120] used ionic liquids as extractants in combination with phosphate salts. Despite the fact that high distribution coefficients for 1,3-PDO could be achieved, extraction with ionic liquids is too expensive and not available for in situ processes due to their high toxicity for the bacteria. Another possibility is the reactive extraction of 1,3-PDO with formaldehyde or acetaldehyde into 2-methyl-1,3-dioxane. The extraction of the product is enabled by the organic solvent extractants *o*-xylene, toluene, or ethylbenzene [121]. In a recent approach, Matsumoto et al. [122] use 1-butanal as reactant and toluene as diluent together with a hydrophobic acidic ionic liquid as a catalyst for the acetalization of 1,3-PDO into a dioxan. With this method, 96 % of the

1,3-PDO could be converted and extracted [122]. Possible drawbacks of this method are undesired reactions of reactant and fermentation by-products, forming further undesired components and causing loss of reactant. The reactive extraction of 1,2-PDO from aqueous environment was described by Broekhuis et al. [123]. 1,2-PDO reacts with acetaldehyde to form 2,4-dimethyl-1,3-dioxolane. In the next step, dioxalan is cleaved via hydrolysis into 1,2-PDO and acetaldehyde. Again, the last step in purification comprises a distillation column. Separation and purification combined in one operation unit can be realized in adsorption processes. With a sulphonate exchange resin, Hilaly and Binder were able to separate 95 % of the 1,3-PDO with a purity of 87 %. This process, however, had a high water demand, resulting in high energy cost [124]. For further cost reduction, Wang et al. [125] used a low-cost cation exchange resin based on polystyrene with high adsorption capacity to recover 1,3-PDO from fermentation broth. Other possibilities are adsorption on silica resin [126] or on beta zeolites [127]. In general, they are very selective, exhibit a simple design, are easy to operate, and are environmentally friendly because the absorbance material can be recovered [125]. However, they are difficult to be implemented in large-scale processes, due to high exchange surfaces and subsequent large pressure loss, together with high tendencies for fouling [109]. In addition, every adsorption process also requires a desorption step with additional costs.

5 Concluding Remarks

As summarized above, significant progresses have been made in the biosynthesis of different diols from various substrates. Quite clearly, for some of the diols, the microbial inherent weaknesses, such as the low product yield, slow reaction rate, high separation cost, and intolerance to toxic products, are the largest obstacles to the cost-competitive biotechnological production (e.g. 1,2-PDO and 1,3-BDO). A more profound comprehension of cell factories' physiology and stress responses would necessarily offer improved tools (at either genetic, metabolic, or system levels) to favour high diol yield and high-quality production. In the past few years, steps taken towards these goals enhanced the bioproduction process economics of some diols significantly. Biobased 1,3-PDO, 1,4-BDO, and 2,3-BDO are successful examples. Still, providing low-cost production process limits the competitiveness of some processes, and hence, much R&D efforts are further needed which may include:

1. Production of high-quality diols suitable for high-value products. This requires system-level understanding of the synthetic pathways to target the formation of desired isomers of diols within cell factories. Pure compounds of optically active 2,3-BDO, 1,3-BDO, 1,4-BDO, or 1,2-PDO are considered as high-value products. It is worth mentioning that chiral synthesis or separation remains a costly step in chemical synthesis, and hence, using enzymes or cells to synthesize compounds with high enantiomeric purity represents an alternative and effective approach.

2. Formation of multiproduct in a biorefinery approach will reduce the process costs significantly. Hence, several conversion technologies (thermochemical, biochemical, etc.) are combined together to reduce the overall cost, as well as to have a better flexibility in product generation and to provide its own power. Examples are the simultaneous production of 1,3-PDO and biogas in unsterile process or the coproduction of 1,3-PDO and butanol.
3. Development of robust microbial cell factories with wide substrate utilization specificities that can dominate in wide number of niches. Lignocellulosic residues that are plentiful and cheap have been widely investigated but their recalcitrance to degradation challenge the production of diols biotechnologically. Hence, adapted cell factories to inhibitors and environmental stresses in such raw substrates are then crucial for forthcoming diol production.
4. Exploring new derivatives or uses of diols that will open new markets. Whereas bioprocesses for 2,3-BDO are well established in terms of productivity, yield, and titre, the market size for 2,3-BDO itself is still relatively small.
5. Downstream processing of diols is technically feasible, and a relatively high purification grade can be achieved, though the costs could be rather high. As a cost-effective method, in situ product recovery integrated with the fermentation process should gain more attention in the future.

Designing new-generation bioprocesses increasingly depend on engineering process-compatible cell factories. The latter, whether through genetic or physiological manipulations, can be greatly assisted by metabolic engineering. To achieve these goals, more fundamental knowledge is needed about metabolic pathways, control mechanisms, and process dynamics to optimally design integrated systems. Chemical engineers, metabolic engineers, and microbial physiologist will have to work for such integrated process. We argue that only by developing cost-efficient processes through integration of fermentation and downstream processing, the microbial production of diols can fulfil their potentials as platform chemicals.

References

1. Burk MJ (2010) Sustainable production of industrial chemicals from sugars. Int Sugar J 112:30–35
2. Zeng AP, Sabra W (2011) Microbial production of diols as platform chemicals: recent progresses. Curr Opin Biotechnol 22:749–757
3. Adkins J, Pugh S, McKenna R, Nielsen DR (2012) Engineering microbial chemical factories to produce renewable "biomonomers". Front Microbiol 3:313
4. Celinska E (2010) Debottlenecking the 1,3-propanediol pathway by metabolic engineering. Biotechnol Adv 28:519–530
5. Sabra W, Dai JY, Quitmann H, Zen, Zeng AP, Xiu ZL (2011) Microbial production of 2, 3-butanediol. Industrial Biotechnology and Commodity Products. Compr Biotechnol 3:87–97. doi:10.1016/B978-0-08-088504-9.00161-6
6. Saxena RK, Anand P, Saran S, Isar J (2009) Microbial production of 1,3-propanediol: recent developments and emerging opportunities. Biotechnol Adv 27:895–913

7. Tjahjasari D, Kaeding T, Zeng AP (2011) 1,3-Propanediol and Polytrimethyleneterephthalate. Compr Biotechnol 3:229
8. Ma C, Wang A, Qin J, Li L, Ai X, Jiang T, Tang H, Xu P (2009) Enhanced 2,3-butanediol production by *Klebsiella pneumoniae* SDM. Appl Microbiol Biotechnol 82:49–57
9. Petrov K, Petrova P (2009) High production of 2,3-butanediol from glycerol by *Klebsiella pneumoniae* G31. Appl Microbiol Biotechnol 84:659–665
10. Qin J, Xiao Z, Ma C, Xie N, Liu P, Xu P (2006) Production of 2,3-butanediol by *Klebsiella pneumoniae* using glucose and ammonium phosphate. Chinese J Chem Eng 14:132–136
11. Li L, Li K, Wang Y, Chen C, Xu Y, Zhang L, Han B, Gao C, Tao F, Ma C (2015) Metabolic engineering of *Enterobacter cloacae* for high-yield production of enantiopure (2R,3R)-2,3-butanediol from lignocellulose-derived sugars. Metab Eng 2015(28):19–27
12. Li L, Su F, Wang Y, Zhang L, Liu C, Li J, Ma C, Xu P (2012) Genome sequences of two thermophilic *Bacillus licheniformis* strains, efficient producers of platform chemical 2,3-butanediol. J Bacteriol 194:4133–4134
13. Li J, Wang W, Ma Y, Zeng AP (2013) Medium optimization and proteome analysis of (R, R)-2,3-butanediol production by Paenibacillus polymyxa ATCC 12321. Appl Microbiol Biotechnol 97:585–597
14. Chen C, Wei D, Shi J, Wang M, Hao J (2014) Mechanism of 2,3-butanediol stereoisomer formation in *Klebsiella pneumoniae*. Appl Microbiol Biotechnol 98:4603–4613
15. Xiao Z, Lv C, Gao C, Qin J, Ma C, Liu Z, Liu P, Li L, Xu P (2010) A novel whole-cell biocatalyst with NAD+ regeneration for production of chiral chemicals. PLoS ONE 5:e8860
16. Ji XJ, Huang H, Ouyang PK (2011a) Microbial 2,3-butanediol production: a state-of-the-art review. Biotechnol Adv 3:351
17. Xu Y, Chu H, Gao C, Tao F, Zhou Z, Li K, Li L, Ma C, Xu P (2014) Systematic metabolic engineering of *Escherichia coli* for high-yield production of fuel bio-chemical 2,3-butanediol. Metab Eng 23:22–33
18. Jurchescu IM, Hamann J, Zhou X, Ortmann T, Kuenz A, Prusse U, Lang S (2013) Enhanced 2,3-butanediol production in fed-batch cultures of free and immobilized *Bacillus licheniformis* DSM 8785. Appl Microbiol Biotechnol 97:6715–6723
19. Li L, Zhang L, Li K, Wang Y, Gao C, Han B, Ma C, Xu P (2013) A newly isolated *Bacillus licheniformis* strain thermophilically produces 2,3-butanediol, a platform and fuel bio-chemical. Biotechnol Biofuels 6:123
20. Li L, Chen C, Li K, Wang Y, Gao C, Ma C, Xu P (2014) Efficient simultaneous saccharification and fermentation of inulin to 2,3-butanediol by thermophilic *Bacillus licheniformis* ATCC 14580. Appl Environ Microbiol 80:6458–6464
21. Celinska E, Grajek W (2009) Biotechnological production of 2,3-butanediol—current state and prospects. Biotechnol Adv 27:715–725
22. Magee RJ, Kosaric N (1987) The microbial production of 2,3-butanediol. Adv Appl Microbiol 32:89–161
23. Nakashimada Y, Kanai K, Nishio N (1998) optimization of dilution rate, pH and oxygen supply on optical purity of 2,3 butanediol produced by *Paenibacillus polymyxa* in chemostat culture. Biotechnol Lett 20:1133–1138
24. Nakashimada Y, Marwoto B, Kashiwamura T, Kakizono T, Nishio N (2000) Enhanced 2,3-butanediol production by addition of acetic acid in *Paenibacillus polymyxa*. J Biosci Bioeng 90:661–664
25. Yu B, Sun J-B, Bommareddy R, Song LF, Zeng AP (2011) A novel (2R,3R)-2,3-Butanediol Dehydrogenase from an industrially potential strain *Paenibacillus polymyxa* ATCC12321. Appl Environ Microbiol 77:4230 (accepted)
26. Fu J, Wang Z, Chen T, Liu W, Shi T, Wang G, Tang YJ, Zhao X (2014) NADH plays the vital role for chiral pure D-(-)-2,3-butanediol production in *Bacillus subtilis* under limited oxygen conditions. Biotechnol Bioeng 111:2126
27. Zeng AP, Biebl H, Deckwer WD (1991) Production of 2,3-butanediol in a membrane bioreactor with cell recycle. Appl Microbiol Biotechnol 34:463–468

28. Wang A, Xu Y, Ma C, Gao C, Li L, Wang Y, Tao F, Xu P (2012) Efficient 2,3-butanediol production from cassava powder by a crop-biomass-utilizer, *Enterobacter cloacae* subsp. dissolvens SDM. PLoS ONE 7:e40442
29. Ji XJ, Huang H, Du J, Zhu JG, Ren LJ, Li S, Nie ZK (2009) Development of an industrial medium for economical 2,3-butanediol production through co-fermentation of glucose and xylose by *Klebsiella oxytoca*. Bioresour Technol 100:5214–5218
30. Zhang L, Sun J, Hao Y, Zhu J, Chu J, Wei D, Shen Y (2010) Microbial production of 2,3-butanediol by a surfactant (serrawettin)-deficient mutant of *Serratia marcescens* H30. J Ind Microbiol Biotechnol 37:857–862
31. Li L, Li K, Wang K, Chen C, Gao C, Ma C, Xu P (2014) Efficient production of 2,3-butanediol from corn stover hydrolysate by using a thermophilic *Bacillus licheniformis* strain. Bioresour Technol 170:256–261
32. Wang Y, Li L, Ma C, Gao C, Tao F, Xu P (2013) Engineering of cofactor regeneration enhances (2S,3S)-2,3-butanediol production from diacetyl. Sci Rep 2013(3):2643
33. Lian J, Chao R, Zhao H (2014) Metabolic engineering of a *Saccharomyces cerevisiae* strain capable of simultaneously utilizing glucose and galactose to produce enantiopure (2R,3R)-butanediol. Metab Eng 23:92–99
34. Xiao Z, Xu P (2007) Acetoin metabolism in bacteria. Crit Rev Microbiol 33:127–140
35. Köpke M, Mihalcea C, Liew F, Tizard JH, Ali MS, Conolly JJ, Al-Sinawi B, Simpson SD (2011) 2,3-butanediol production by acetogenic bacteria, an alternative route to chemical synthesis, using industrial waste gas. Appl Environ Microbiol 77:5467–5475
36. Yu EK, Deschatelets L, Louis-Seize G, Saddler JN (1985) Butanediol production from cellulose and hemicellulose by *Klebsiella pneumoniae* grown in sequential coculture with Trichoderma harzianum. Appl Environ Microbiol 50:924–929
37. Yu EK, Saddler JN (1982) Enhanced production of 2,3-butanediol by *Klebsiella pneumoniae* grown on high sugar concentrations in the presence of acetic acid. Appl Environ Microbiol 44:777–784
38. Köpke M, Gerth ML, Maddock DJ, Mueller AP, Liew F, Simpson SD, Patrick WM (2014) Reconstruction of an acetogenic 2,3-butanediol pathway involving a novel NADPH-dependent primary-secondary alcohol dehydrogenase. Appl Environ Microbiol 80:3394–3403
39. Ui S, Masuda T, Masuda H, Muraki H (1986) Mechanism for the formation of 2,3-butanediol stereoisomers in *Bacillus polymyxa*. J Ferment Technol 64:481–486
40. Ji XJ, Nie ZK, Huang H, Ren LJ, Peng C, Ouyang PK (2011) Elimination of carbon catabolite repression in *Klebsiella oxytoca* for efficient 2,3-butanediol production from glucose-xylose mixtures. Appl Microbiol Biotechnol 89:1119–1125
41. Zhang CY, Peng XP, Li W, Guo XW, Xiao DG (2014) Optimization of 2,3-butanediol production by *Enterobacter cloacae* in simultaneous saccharification and fermentation of corncob residue. Biotechnol Appl Biochem 61:501
42. Oliver JW, Machado IM, Yoneda H, Atsumi S (2013) Cyanobacterial conversion of carbon dioxide to 2,3-butanediol. Proc Natl Acad Sci USA 110:1249–1254
43. Yan Y, Lee CC, Liao JC (2009) Enantioselective synthesis of pure (R, R)-2,3-butanediol in *Escherichia coli* with stereospecific secondary alcohol dehydrogenases. Org Biomol Chem 7:3914–3917
44. Li L, Wang Y, Zhang L, Ma C, Wang A, Tao F, Xu P (2012) Biocatalytic production of (2S,3S)-2,3-butanediol from diacetyl using whole cells of engineered *Escherichia coli*. Bioresour Technol 115:111–116
45. Nan H, Seo SO, Oh EJ, Seo JH, Cate JH, Jin YS (2014) 2,3-Butanediol production from cellobiose by engineered *Saccharomyces cerevisiae*. Appl Microbiol Biotechnol 98:5757–5764
46. Yim H, Haselbeck R, Niu W, Pujol-Baxley C, Burgard A, Boldt J, Khandurina J, Trawick JD, Osterhout RE, Stephen R, Estadilla J, Teisan S, Schreyer HB, Andrae S, Yang TH, Lee SY, Burk MJ, Van DS (2011) Metabolic engineering of *Escherichia coli* for direct production of 1,4-butanediol. Nat Chem Biol 7:445–452

47. Werawattanachai N, Towiwat P, Unchern S, Maher TJ (2007) Neuropharmacological profile of tetrahydrofuran in mice. Life Sci 80:1656–1663
48. Diaz A, Katsarava R, Puiggali J (2014) Synthesis, properties and applications of biodegradable polymers derived from diols and dicarboxylic acids: from polyesters to poly (ester amide)s. Int J Mol Sci 15:7064–7123
49. Burk MJ, Burgard AP, Osterhout RE, Sun J (2012) Microorganisms for the production of 1,4-Butanediol. US Patent 8, 178, 327, B2. (13/009,813)
50. Pharkya P, Burgard AP, Van Dien SJ, Osterhout RE, Burk MJ, Trawick JD, Kuchinskas MP, Steer B (2015) Microorganisms and methods for production of 4-Hydroxybutyrate, 1,4-Butanediol and related compounds. US patent 20150148513
51. Cosmetic Ingredient Review (1985) Final report on the safety assessment of butylene glycol, hexylene glycol, ethoxydiglycol, and dipropylene glycol. J Am Coll Toxicol 5:223–248
52. Matsuyama A, Yamamoto H, Kawada N, Kobayashi Y (2001) Industrial production of (R)-1,3-butanediol by new biocatalysts. J Mol Catal B Enzym 11:513–521
53. Kataoka N, Vangnai AS, Tajima T, Nakashimada Y, Kato J (2013) Improvement of (R)-1,3-butanediol production by engineered *Escherichia coli*. J Biosci Bioeng 115:475–480
54. Kataoka N, Vangnai AS, Ueda H, Tajima T, Nakashimada Y, Kato J (2014) Enhancement of (R)-1,3-butanediol production by engineered *Escherichia coli* using a bioreactor system with strict regulation of overall oxygen transfer coefficient and pH. Biosci Biotechnol Biochem 78:695–700
55. Zheng RC, Ge Z, Qiu ZK, Wang YS, Zheng YG (2012) Asymmetric synthesis of (R)-1,3-butanediol from 4-hydroxy-2-butanone by a newly isolated strain *Candida krusei* ZJB-09162. Appl Microbiol Biotechnol 94:969–976
56. Eguchi T, Mochida K (1992) Lipase-catalyzed diacylation of 1,3-butanediol. Biotechnol Lett 15:955–960
57. Yamamoto H, Matsuyama A, Kobayashi Y (2002) Synthesis of (R)-1,3-butanediol by enantioselective oxidation using whole recombinant *Escherichia coli* cells expressing (S)-specific secondary alcohol dehydrogenase. Biosci Biotechnol Biochem 66:925–927
58. Okabayashi T, Nakajima T, Yamamoto H (2012) Recombinant microorganisms with 1,3-butanediol producing function and uses thereof. Daicel Corporation, editor, Osaka, Japan. US patent 13/504,391(US20120276606 A1)
59. Drozdzynska A, Leja K, Czaczyk K (2011) Biotechnological production of 1,3-Propanediol from crude glycerol. J Biotechnol Comput Biol Bionanotechnol 92:92–100
60. Biebl H, Menzel K, Zeng AP, Deckwer WD (1999) Microbial production of 1,3-propanediol. Appl Microbiol Biotechnol 52:289–297
61. Chatzifragkou A, Dietz D, Komaitis M, Zeng AP, Papanikolaou S (2010) Effect of biodiesel-derived waste glycerol impurities on biomass and 1,3-propanediol production of *Clostridium butyricum* VPI 1718. Biotechnol Bioeng 107:76–84
62. Wang Y, Tao F, Ni J, Li C, Xu P (2015) Production of C3 platform chemicals from CO_2 by genetically engineered cyanobacteria. Green Chem 17:3100–3110
63. Dietz D, Zeng AP (2014) Efficient production of 1,3-propanediol from fermentation of crude glycerol with mixed cultures in a simple medium. Bioprocess Biosyst Eng 37:225–233
64. Chatzifragkou A, Papanikolaou S, Dietz D, Doulgeraki AI, Nychas GJE, Zeng AP (2011) Production of 1,3-propanediol by *Clostridium butyricum* growing on biodiesel-derived crude glycerol through a non-sterilized fermentation process. Appl Microbiol Biotechnol 91:101
65. Kaeding T, DaLuz J, Kube J, Zeng AP (2014) Integrated study of fermentation and downstream processing in a miniplant significantly improved the microbial 1,3-propanediol production from raw glycerol. Bioprocess Biosyst Eng 38:3
66. Friedmann H, Zeng A-P (2008) Process and apparatus for the microbial production of a specific product and methane. EU patent: DE 102007001614 WO 2006/021087 A1 3/2006
67. Bizukojc M, Dietz D, Sun J, Zeng AP (2010) Metabolic modelling of syntrophic-like growth of a 1,3-propanediol producer, *Clostridium butyricum*, and a methanogenic archeon, *Methanosarcina mazei*, under anaerobic conditions. Bioprocess Biosyst Eng 33:507–523

68. Zhou M, Chen J, Freguia S, Rabaey K, Keller J (2013) Carbon and electron fluxes during the electricity driven 1,3-propanediol biosynthesis from glycerol. Environ Sci Technol 47:11199–11205
69. Choi O, Kim T, Woo HM, Um Y (2014) Electricity-driven metabolic shift through direct electron uptake by electroactive heterotroph *Clostridium pasteurianum*. Sci Rep 4:6961
70. Lovley DR (2012) Electromicrobiology. Annu Rev Microbiol 66:391–409
71. Rosenbaum MA, Franks AE (2014) Microbial catalysis in bioelectrochemical technologies: status quo, challenges and perspectives. Appl Microbiol Biotechnol 98:509–518
72. da Cunha MV, Foster MA (1992) Sugar-glycerol cofermentations in lactobacilli: the fate of lactate. J Bacteriol 174:1013–1019
73. Pflügl S, Marx H, Mattanovich D, Sauer M (2012) 1,3-Propanediol production from glycerol with *Lactobacillus diolivorans*. Bioresour Technol 119:133–140
74. Sabra W, Groeger C, Sharma PN, Zeng AP (2014) Improved n-butanol production by a non-acetone producing *Clostridium pasteurianum* DSMZ 525 in mixed substrate fermentation. Appl Microbiol Biotechnol 98:4267–4276
75. Günzel B, Yonsel S, Deckwer WD (1991) Fermentative production of 1, 3-propanediol from glycerol by *Clostridium butyricum* up to a scale of $2m^3$. Appl Microbiol Biotechnol 36:289–294
76. Saint-Amans S, Perlot P, Goma G, Soucaille P (1994) High production of 1,3-propanediol from glycerol by *Clostridium butyricum* VPI 3266 in a simply controlled fed-batch system. Biotechnol Lett 16:831–836
77. Abbad-Andaloussi S, Manginot-Durr C, Amine J, Petitdemange E, Petitdemange H (1995) Isolation and characterization of *Clostridium butyricum* DSM 5431 mutants with increased resistance to 1,3-propanediol and altered production of acids. Appl Environ Microbiol 61:4413–4417
78. Petitdemange E, Dürr C, Abbad-Andaloussi S, Raval G (1995) Fermentation of raw glycerol to 1,3-propanediol by new strains of *Clostridium butyricum*. J Ind Microbiol 15:498–502
79. Reimann A, Biebl H (1996) Production of 1, 3-propanediol by *Clostridium butyricum* DSM 5431 and product tolerant mutants in fedbatch culture: feeding strategy for glycerol and ammonium. Biotechnol Lett 18:827–832
80. Hartlep M, Zeng A-P (2002) Fedbatch-Verfahren für die mikrobielle Herstellung von 1, 3-Propandiol in *Klebsiella pneumoniae* und *Clostridium butyricum*. Chem Ing Tech 74:663–664
81. Szymanowska-Powalowska D (2014) 1,3-Propanediol production from crude glycerol by *Clostridium butyricum* DSP1 in repeated batch. Electron J Biotechnol 17:322–328
82. Himmi EH, Bories A, Barbirato F (1999) Nutrient requirements for glycerol conversion to 1,3-propanediol by *Clostridium butyricum*. Biores Technol 67:123–128
83. Gonzalez-Pajuelo M, Meynial-Salles I, Mendes F, Andrade JC, Vasconcelos I, Soucaille P (2005) Metabolic engineering of *Clostridium acetobutylicum* for the industrial production of 1,3-propanediol from glycerol. Metab Eng 7:329–336
84. Hirschmann S, Koschik I, Baganz K, Vorlop K (2005) Development of an integrated bioconversion process for the production of 1,3-propanediol from raw glycerol waters. Landbauforschung Völkenrode 55:261–267
85. Jensen TO, Kvist T, Mikkelsen MJ, Westermann P (2012) Production of 1,3-PDO and butanol by a mutant strain of *Clostridium pasteurianum* with increased tolerance towards crude glycerol. AMB Express 2:44
86. Metsoviti M, Zeng A-P, Koutinas AA, Papanikolaou S (2013) Enhanced 1,3-propanediol production by a newly isolated *Citrobacter freundii* strain cultivated on biodiesel-derived waste glycerol through sterile and non-sterile bioprocesses. J Biotechnol 163:408–418
87. Yang G, Tian J, Li J (2007) Fermentation of 1,3-propanediol by a lactate deficient mutant of *Klebsiella oxytoca* under microaerobic conditions. Appl Microbiol Biotechnol 73:1017–1024
88. Jolly J, Hitzmann B, Ramalingam S, Ramachandran KB (2014) Biosynthesis of 1,3-propanediol from glycerol with *Lactobacillus reuteri*: effect of operating variables. J Biosci Bioeng 118:188–194

89. Biebl H (1991) Glycerol fermentation of 1,3-propanediol by *Clostridium butyricum*. Measurement of product inhibition by use of a pH-auxostat. Appl Microbiol Biotechnol 35:701–705
90. Zeng A-P (1996) Pathway and kinetic analysis of 1,3-propanediol production from glycerol fermentation by *Clostridium butyricum*. Bioprocess Eng 14:169–175
91. Zeng AP, Biebl H (2002) Bulk chemicals from biotechnology: the case of 1,3-propanediol production and the new trends. Adv Biochem Eng Biotechnol 74:239–259
92. Clomburg JM, Gonzalez R (2013) Anaerobic fermentation of glycerol: a platform for renewable fuels and chemicals. Trends Biotechnol 31:20–28
93. Rieckenberg F, Ardoa I, Rujananon R, Zeng AP (2014) Cell-free synthesis of 1,3-propanediol from glycerol with a high yield. Eng Life Sci 14:380–386
94. Zhang YH (2014) Production of biofuels and biochemicals by in vitro synthetic biosystems: opportunities and challenges. Biotechnol Adv. doi:10.1016/j.biotechadv.2014.10.009
95. Nakamura CE, Whited GM (2003) Metabolic engineering for the microbial production of 1,3-propanediol. Curr Opin Biotechnol 14:454–459
96. Chen Z, Geng F, Zeng AP (2015) Protein design and engineering of a de novo pathway for microbial production of 1,3-propanediol from glucose. Biotechnol J 10:284–289
97. Celinska E (2015) Fully glycerol-independent microbial production of 1,3-propanediol via non-natural pathway: paving the way to success with synthetic tiles. Biotechnol J 10:242–243
98. Jung JY, Yun HS, Lee J, Oh MK (2011) Production of 1,2-propanediol from glycerol in *Saccharomyces cerevisiae*. J Microbiol Biotechnol 21:846–853
99. Clomburg JM, Gonzalez R (2011) Metabolic engineering of *Escherichia coli* for the production of 1,2-propanediol from glycerol. Biotechnol Bioeng 108:867–879
100. Rode CV, Ghalwadkar AA, Mane RB, Hengne AM, Jadkar ST, Biradar NS (2010) Selective hydrogenolysis of glycerol to 1,2-propanediol: comparison of batch and continuous process operations. Org Process Res Dev 14:1385–1392
101. Cameron DC, Cooney CL (1986) A novel fermentation: the production of (R)-1,2-propanediol and acetol by *Clostridium thermosaccharolyticum*. Biores Technol 4:651–654
102. Altaras N, Etzel M, Cameron DC (2001) Conversion of sugars to 1,2-propanediol by *Thermoanaerobacterium thermosaccharolyticum* HG-8. Biotechnol Prog 17:52–56
103. Turner KW, Roberton AM (1979) Xylose, arabinose, and rhamnose fermentation by *Bacteroides ruminicola*. Appl Environ Microbiol 38:7–12
104. Tran-Din K, Gottschalk G (1985) Formation of D (-)-1, 2-propanediol and D (-)-lactate from glucose by *Clostridium sphenoides* under phosphate limitation. Arch Microbiol 142:87–92
105. Oude Elferink SJ, Krooneman J, Gottschal JC, Spoelstra SF, Faber F, Driehuis F (2001) Anaerobic conversion of lactic acid to acetic acid and 1, 2-propanediol by *Lactobacillus buchneri*. Appl Environ Microbiol 67:125–132
106. Bennett GN, San KY (2001) Microbial formation, biotechnological production and applications of 1,2-propanediol. Appl Microbiol Biotechnol 55:1–9
107. Koch D, Meurer S-J, Eck J (2014) New means and methods for producing propanediol. BRAIN Biotechnology Research and Information Network AG, editor, Zwingenberg, Germany. US patent: 14/125,409(US 2014/0178953 A1), p 1–12
108. Li H, Liao JC (2013) Engineering a cyanobacterium as the catalyst for the photosynthetic conversion of CO_2 to 1,2-propanediol. Microb Cell Fact 12:4
109. Xiu ZL, Zeng A-P (2008) Present state and perspective of downstream processing of biologically produced 1,3-propanediol and 2,3-butanediol. Appl Microbiol Biotechnol 78:917–926
110. Willke T, Vorlop K (2008) Biotransformation of glycerol into 1,3 propanediol. Eur J Lipid Sci Technol 110:831–840
111. Anvari M, Khayati G (2009) In situ recovery of 2,3-butanediol from fermentation by liquid-liquid extraction. J Ind Microbiol Biotechnol 36:313–317
112. Li Z, Teng H, Xiu Z-L (2010) Aqueous two-phase extraction of 2,3-butanediol from fermentation broths using an ethanol/ammonium sulfate system. Proc Biochem 45:731–737

113. Sun LH, Jiang B, Xiu Z-L (2009) Aqueous two-phase extraction of 2,3-butanediol from fermentation broths by isopropanol/ammonium sulfate system. Biotechnol Lett 31:371–376
114. Birajdar SD, Rajagopalan S, Sawant JS, Padmanabhan S (2015) Continuous countercurrent liquid-liquid extraction method for the separation of 2,3-butanediol from fermentation broth using n-butanol and phosphate salt. Proc Biochem. doi:10.1016/j.procbio.2015.05.016
115. Qureshi N, Meagher MM, Hutkins RW (1994) Recovery of 2,3-Butanediol by vacuum membrane distillation. Sep Sci Technol 29:1733–1748
116. Li Y, Wu Y, Zhu J, Liu J, Shen Y (2013) Separating 2,3-butanediol from fermentation broth using n-butylaldehyde. J Saudi Chem Soc. doi:10.1016/j.jscs.2013.02.005
117. Malinowski J (1999) Evaluation of liquid extraction potentials for downstream separation of 1,3-propanediol. Biotechnol Tech 13:127–130
118. Li Z-G, Sun Y-Q, Zheng W-L, Teng H, Xiu Z-L (2013) A novel and environment-friendly bioprocess of 1,3-propanediol fermentation integrated with aqueous two-phase extraction by ethanol/sodium carbonate system. Biochem Eng J 80:68–75
119. Li Z, Teng H, Xiu Z (2011) Extraction of 1,3-propanediol from glycerol-based fermentation broths with methanol/phosphate aqueous two-phase system. Proc Biochem 46:586–591
120. Müller A, Gorak A (2012) Extraction of 1,3-propanediol from aqueous solutions using different ionic liquid-based aqueous two-phase systems. Sep Purif Technol 97:130–136
121. Malinowski J (2000) Reactive extraction for downstream separation of 1,3-Propanediol. Biotechnol Prog 16:76–79
122. Matsumoto M, Nagai K, Kondo K (2015) Reactive extraction of 1, 3-propanediol with aldehydes in the presence of a hydrophobic acidic ionic liquid as a catalyst. Solvent Extr Res Dev Jpn 22:209–213
123. Broekhuis RR, Lynn S, King CJ (1994) Recovery of propylene glycol from dilute aqueous solutions via reversible reaction with aldehydes. Ind Eng Chem Res 33:3230–3237
124. Hilaly A, Binder T (2002) Method of recovering 1,3-propanediol from fermentation broth. US Patent 6479716
125. Wang S, Dai H, Yan Z, Zhu C, Huang L, Fang B (2014) 1,3-Propanediol adsorption on a cation exchange resin: adsorption isotherm, thermodynamics and mechanistic studies. Eng Life Sci 14:485–492
126. Cho M-H, Im Joen S, Pyo S-H, Mun S, Kim J-H (2006) A novel separation and purification process for 1,3-propanediol. Proc Biochem 41:739–744
127. Wang Z, Wu Z, Tan T (2013) Sorption equilibrium, mechanism and thermodynamics studies of 1,3-propanediol on beta zeolite from an aqueous solution. Bioresour Technol 145:37–42

Adv Biochem Eng Biotechnol (2016) 155: 199–235
DOI 10.1007/10_2015_335

Published Online: 17 October 2015

Cell Factories of Higher Fungi for Useful Metabolite Production

Hao Qin, Jun-Wei Xu, Jian-Hui Xiao, Ya-Jie Tang, Han Xiao and Jian-Jiang Zhong

Abstract Higher fungi or called as macro-fungi, consisting of the divisions ascomycetes, basidiomycetes, and imperfect fungi, are receiving great interest around the world, because studies of higher fungi help us not only to find new edible and officinal resources but also to understand their complicated biology. In recent decades, a large number of useful substances from higher fungi have been isolated, identified, and characterized, which have important biological functions, such as reducing blood pressure, enhancing immunity, and possessing anti-cancer and anti-HIV and other pharmacological activities. This chapter will review the genetic manipulation tools for higher fungi, omics analysis of higher-fungus cell factories, and production of useful metabolites by higher fungi, including those of terpenoids, heterocyclics, polysaccharides, and polyketides. Trends in future development of cell factories of higher fungi for useful metabolite production will also be analyzed.

H. Qin · H. Xiao · J.-J. Zhong (✉)
State Key Laboratory of Microbial Metabolism, Joint International Research Laboratory of Metabolic & Developmental Sciences, and Laboratory of Molecular Biochemical Engineering, School of Life Sciences and Biotechnology, Shanghai Jiao Tong University, 800 Dong-Chuan Road, Shanghai 200240, China
e-mail: jjzhong@sjtu.edu.cn

J.-W. Xu
Faculty of Life Science and Technology, Kunming University of Science and Technology, Kunming 650500, China

J.-H. Xiao
Center for Translational Medicine, Affiliated Hospital of Zunyi Medical University, Zunyi 563000, China

Y.-J. Tang
Key Laboratory of Fermentation Engineering (Ministry of Education), Hubei Provincial Cooperative Innovation Center of Industrial Fermentation, Hubei University of Technology, Wuhan 430068, China

Graphical Abstract

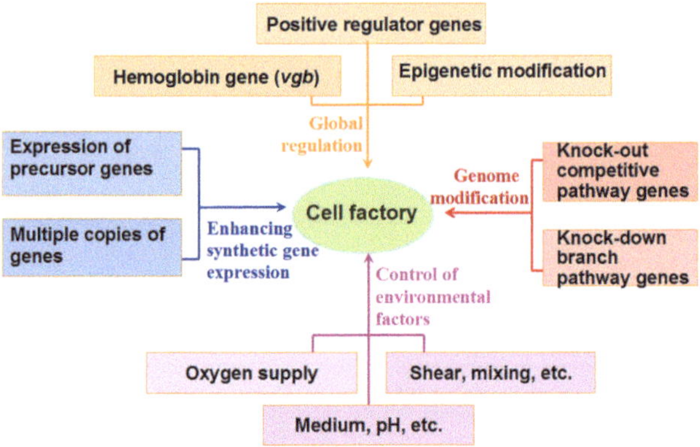

Strategies for improving cell factories of higher fungi for useful metabolite production

Keywords Genetic manipulation · Higher fungi · Omics analysis · Cell factory · Medicinal mushroom · Fermentation technology · Metabolic engineering · Secondary metabolite production

Contents

1 Introduction

Higher fungi or macrofungi, consisting of the divisions ascomycetes, basidiomycetes, and imperfect fungi, refer to the fungi with hyphae well-developed, septate, and usually at some stage of development interwoven into a compact tissue especially in the fruiting body. Studies of higher fungi help us not only to find new edible and officinal resources but also to understand their complicated biology. Several thousand years ago, people used higher fungi as medicinal and food sources, but at that time, their functional components were unknown. In recent decades, a large number of useful substances from higher fungi have been isolated, identified, and characterized, which could reduce blood pressure, enhance immunity, and possess anticancer and anti-HIV and other pharmacological activities. Some new bioactive compounds reported in recent years are shown in Table 1, and most of them are secondary metabolites, which are organic compounds not directly involved in the normal growth, development, or reproduction of an organism but often play an important role in plant defense against herbivory and other interspecies defenses. These compounds can be used as potential lead compounds for new drug development. However, bioactive secondary metabolites produced by higher fungi are generally of very low productivity and are thus unable to meet the requirement for (pre-)clinical study and large market supply. Furthermore, these metabolites are usually of complicated chemical structures and very difficult to synthesize chemically at high efficiency. Higher fungus cell factories, therefore, have received increasing attention to achieve large-scale industrial production of their unique and important natural compounds.

Microbial cell factories are the basis of biochemical conversion from low-cost raw materials and/or agro-industrial by-products into valuable medicine, chemicals and energy, or detoxifying harmful/toxic chemicals. When a beneficial compound from higher fungi is identified, we need to reveal its biosynthetic pathway and to improve its biosynthesis by genetic or metabolic engineering. Also, its large-scale efficient production using fermentation technology is necessary for industrial application. The framework in constructing higher fungus cell factories is shown in Fig. 1.

Cell factories of higher fungi have been more and more widely applied to produce value-added compounds. For example, the higher fungus *Ganoderma lucidum* is used for the production of ganoderic acid, a potential antitumor terpenoid; the key biosynthetic genes in its pathway upstream were overexpressed, and the fermentation process parameters were optimized, and a new two-stage fermentation strategy was also developed to achieve a high yield of the product [41–44]. In order to better understand the significance of cell factories of higher fungi, in the following, this article is outlined from the construction and analysis of high-fungus cell factories to their production of useful metabolites.

Table 1 New bioactive compounds recently reported in higher fungi

Higher fungi	Compounds	Bioactivities	References
Terpenoids			
Ganoderma boninense	Ganoboninketals A-C	Anticancer	[1]
Stereum hirsutum	Three hirsutane-type sesquiterpenoids	Antimicrobial and anticancer	[2]
Inonotus rickii	Inonotic acid A		[3]
	3-O-Formyl inonotic acid A		
	Inonotic acid B		
	3,6-Dihydroxycinnamolide	Anticancer	
Pleurotus cornucopiae		Antibacterial	[4]
Granulobasidium vellereum	2-Hydroxycoprinolone		[5]
	8-Deoxy-4a-hydroxytsugicoline		
	8-Deoxydihydrotsugicoline		
Naematoloma fasciculare	Four new lanostane triterpenoids	Anticancer	[6]
Flammulina velutipes	Enokipodins E-J	Antifungal activity	[7]
	Sterpurols A	Anticancer and antioxidant	
Sarcodon scabrosus	Secoscabronine M	Anticancer	[8]
	Scabronine M	Anticancer	[9]
Neonothopanus nambi	Nambinones A-C	Anticancer	[10]
	1-Epi-nambinone B		
	Nambinone D		
	Aurisin K	Antimalarial and anticancer	
Armillaria	Melleolide sesquiterpene aryl esters	Anticancer	[11]
Ganoderma lucidum	Lucidenic acids A-C, N	Anticancer	[12]
	7-O-Ethyl ganoderic acid O	Anticancer	[13]
	3α,22β-Diacetoxy-7α-hydroxyl-5α-lanost-8, 24E-dien-26-oic acid	Anticancer	[14]
	3-O-Acetylganoderic acid B	Antimicrobial, anti-HIV, antitumor, antioxidation	[15]
	8β,9α-Dihydroganoderic acid C		
	3-O-Acetylganoderic acid K		
	Ethyl 3-O-acetylganoderate B		
	Ethyl ganoderate J		

(continued)

Table 1 (continued)

Higher fungi	Compounds	Bioactivities	References
Heterocyclics			
Hericium erinaceus	Erinacerins C-L	Anticancer	[16]
	Isohericenone	Anticancer	[17]
	3-Hydroxyhericenone F, Hericenone I and J	Endoplasmic reticulum (ER) stress-suppressive activity	[18]
Sarcodon leucopus	Sarcoviolin beta and episarcoviolin beta	Antioxidant and anticancer	[19]
Phellinus ribis	Phelliribsin A	Anticancer	[20]
Trifolium nigrescens	4‴,5,5″,7,7″-Pentahydroxy-3′,3‴-dimethoxy-3-O-beta-D-glucosyl-3 ″,4′-O-biflavone	Antioxidant and tyrosinase inhibitory activities	[21]
Neolentinus lepideus	5-Methoxyisobenzofuran-4,7(1H,3H)-dione, 1,3-Dihydroisobenzofuran-4,6-diol	Antibacterial	[22]
Lasiosphaera fenzlii	4,6-Dihydroxy-1H-isoindole-1,3(2H)-dione, 4,6-Dihydroxy-2,3-dihydro-1H-isoindol-1-one	Anticancer	[23]
Macrolepiota neomastoidea	Macrolepiotin	Anticancer	[24]
Amanita exitialis	N-2-(1-Methoxycarbonylethyl)guanosine	Anticancer	[25]
Ganoderma colossum	Ganomycin 1	Anti-HIV	[26]
Xylaria sp. PSU-F100	Xylarisin 132	Antibacterial	[27]
Xylaria sp. (#2508)	Xylopyridine A	DNA-binding affinity	[28]
Phellinus linteus	Phellifuropyranone A	Anticancer	[29]
Cortinarius brunneus	N-Glucosyl-1H-indole derivatives		[30]
Cortinarius subtortus	(Iso)-Quinoline alkaloids	Antioxidant	[31]
Miscellaneous			
Cordyceps taii	Deacetylcytochalasin C Zygosporin D	Anticancer	[32]
Lentinus polychrous	6-Methylheptane-1,2,3,4,5-pentaol	Anticancer	[33]
Tuber indicum	Four novel cerebrosides	Antifatty liver, antitumor	[34]
Cordyceps jiangxiensis	Jiangxienone	Anticancer	[35]
Hericium erinaceums	Hericenone L	Anticancer	[36]
Cantharellus cibarius	(10E, 14Z)-9-Oxooctadeca-10,14-dien-12-ynoicacid	Anticancer	[37]

(continued)

Table 1 (continued)

Higher fungi	Compounds	Bioactivities	References
Thelephora aurantiotincta	Thelephantin O	Anticancer	[38]
Tuber indicum	5α-Androst-16-en-3α-ol	Increase the sexual arousal of human female, adjust moods, mediate human menstrual synchrony	[39]
Thelephora vialis	Vialinin A	Antioxidant and anticancer	[40]

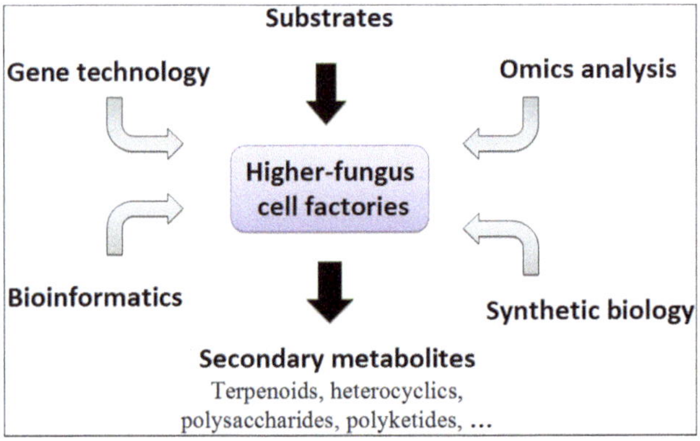

Fig. 1 Framework in constructing higher fungus cell factories for useful secondary metabolite production

2 Genetic Manipulations of Higher Fungi

Initially, the designing strategies of higher fungus cell factories by genetic engineering were mainly aimed at single gene modification and/or an individual pathway, which may not efficiently change the metabolic flux. With the advance of metabolic engineering and synthetic biology, the focus has gradually shifted to polygenic modifications and multimetabolic pathways. This has resulted in the shift in designing strategies of metabolic engineering from conventional deletion or overexpression of endogenous genes in individual metabolic pathways to current combinatorial approaches with manipulation of key gene expression of metabolic networks in an entire cell [45] (Fig. 2), while control of environmental factors (Fig. 2) could provide useful epigenetic information and cellular physiological and metabolic responses to cultivation conditions (especially in large-scale cultivation) with aid of omics analysis (Fig. 1). Recently, higher fungi as cell factories for the production of secondary metabolites have attracted extensive interests around the world. While compared with the achievement of the cell factories of *Escherichia*

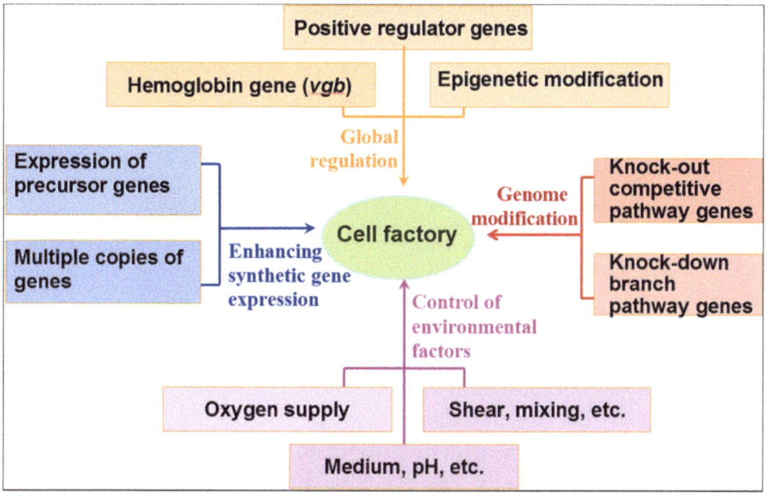

Fig. 2 Strategies for improving cell factories of higher fungi

coli and yeasts, the studies on higher fungi are obviously lagging behind, which requires more research inputs urgently. In this section, the tools for constructing high-fungus cell factories, including genetic transformation methods, gene over-expression, gene silence, and gene deletion, will be overviewed as follows.

2.1 *Genetic Transformation Methods*

The development of genetic transformation technology provides a powerful tool for the study of gene expression and gene function. Establishment of an efficient transformation system is an important premise for functional analysis of genes. The genetic transformation technology has been applied to various mushrooms, such as *Agaricus bisporus* [46], *Flammulina velutipes* [47], *Volvariella volvacea* [48], *Ganoderma weberianum* [49], *and Tremella fuciformis* [50]. Over the years, there have been great advances in transformation methods of higher fungi. For example, *Agrobacterium*-mediated transformation was successfully applied to higher fungi, which was originally used in plants. Here, the development of transformation technology in higher fungi is to be overviewed, including polyethylene glycol (PEG)-mediated transformation, electroporation transformation, *Agrobacterium tumefaciens*-mediated transformation, and restriction enzyme-mediated DNA integration. The comparison of those four methods is shown in Table 2.

Table 2 Comparison of four transformation methods

Method	Applicability	Efficiency	Operability	References
PEG-mediated transformation	Ordinary	Ordinary	Ordinary	[51]
Electroporation transformation	High	Ordinary	High	[52]
Agrobacterium tumefaciens-mediated transformation	Ordinary	Low	Low	[44]
Restriction enzyme-mediated DNA integration	Low	High	Ordinary	[53]

2.1.1 PEG-Mediated Transformation

The PEG protoplast transformation method has been applied to *G. lucidum* [51], *Pleurotus nebrodensis* [54], *Lentinus edodes* [55], etc. In the past, the efficiency of this transformation process was generally low and the exogenous gene was difficult to integrate into the genome by this method, and many transformants lost their resistance phenotype after several-week growth in the absence of selection pressure [56]. However, after many efforts, this method has been improved and become one of main ones in higher fungi.

Li et al. [51] developed an efficient PEG-mediated transformation method for *Pleurotus ostreatus*. Heparin, ATA, and spermidine were used to improve the transformation efficiency. As a result, 80–180 colonies could be obtained per μg of DNA per 10^7 protoplasts with the hygromycin B phosphotransferase gene (*hph*) as a selection marker, which was much higher than previously reported [57]. And 120–150 and 85–100 transformants per μg of DNA per 10^7 protoplasts were also obtained in *G. lucidum* and *L. edodes*, respectively. That means the PEG-mediated transformation could be a useful tool for genetic engineering in mushrooms.

Yu et al. [58] transformed a homogenous 3-hydroxy-3-methylglutaryl coenzyme A reductase (HMGR) gene into *G. lucidum* by PEG-mediated transformation method. The gene *sdhB* mutation was used as a selection marker. A total number of 15–20 transformants per μg plasmid DNA were obtained. What's more, they found that the transformants could maintain the resistance phenotype after five passages of cultivation on a non-selection medium. Southern blot analysis confirmed that the *sdhB* gene was stably integrated at multiple sites in the genome.

High efficiency, convenience, and genome integration are the merits of the PEG-mediated transformation method in the genetic engineering of higher fungi. However, the genes integrated into the genome are usually of multiple copies, and the protoplast preparation is also time-consuming.

2.1.2 Electroporation Transformation

Electroporation transformation has also been used in higher fungi, such as *T. fuciformis* [59], *L. edodes* [60], *F. velutipes* [61], and *G. lucidum* [62]. Compared with other transformation methods, it is simple, rapid, and of wide application range. In the past, electroporation transformation required protoplast preparation, which was tedious, but recently, scientists have developed a modified method for basidiospores or mycelial fragments. Kuo et al. [63] did transformation by electroporation of basidiospores or mycelial fragments in *F. velutipes*. While the basidiospores or mycelial fragments also need lysing enzymes to incubate for 2 h, it is much easier than protoplasting. The efficiency using basidiospores could be 30–150 transformants per μg DNA in *L. edodes* [60]. The resistance was stable for at least 6 months during subcultivation. Southern blotting analysis confirmed that the gene was stably integrated into the genome.

The modified electroporation transformation has many advantages. It is simple and cheap and does not require protoplasts. For some mushrooms, it is difficult to obtain a sufficient number of protoplasts and the regeneration efficiency from protoplasts was also very low. Thus, this method was more widely applicable than others under these circumstances. However, the method also has limitations [61]. During basidiospore isolation, contamination frequently happens. Moreover, multiple copies may be generated via random integration. It is not surprising that electroporation transformation is yet to be improved.

2.1.3 *Agrobacterium Tumefaciens*-Mediated Transformation

In 1995, Bundock used *A. tumefaciens*-mediated transformation (ATMT) in *Saccharomyces cerevisiae* for the first time [64]. De Groot et al. [65] applied ATMT to filamentous fungi—*Dictyostelium discoideum*. Recently, some higher fungi have been transformed by this method, including *A. bisporus* [66, 67], *F. velutipes* [68], *Heterobasidion annosum* [69], *P. nebrodensis* [70], *V. volvacea* [71], and *G. lucidum* [44]. The ATMT method has the following advantages: a wide range of transformation recipients, high degree of stability, and high proportion of single-copy transformants [44]. For the fungi lacking sexual stages, the exogenous gene single-copy transformation is very important. This made ATMT become one of the common higher fungal transformation methods.

Xu et al. [44] applied the *Agrobacterium*-mediated transformation to *G. lucidum* and obtained 10–15 transformants per 10^7 protoplasts. All tested transformants maintained resistance stably and the transformants showed 100 % mitotic stability for the *sdhB* selection marker. Most of the integrated DNA had a single copy in the genome.

The efficiency of ATMT was dependent on species, and this method was also reported to be unable to obtain transformants [72]. Many factors can affect the efficiency of this transformation, such as the conditions of protoplasts and *A. tumefaciens*, the ratio of the number of bacteria to protoplasts, and cocultivation temperature. Zhang et al. [73] tested different transformation parameters in *Hypsizygus marmoreus*. Their results showed that 25 °C and 2 days of coculture was best. Kemppainen et al. [74] studied the mechanism of ATMT and found no obvious sequence similarities between genomic sites and T-DNA (the transferred DNA of the tumor-inducing (Ti) plasmid of *A. tumefaciens*). About 75 % of the integrations took place at the predicted locus in the model mushroom *Laccaria bicolor*.

ATMT has the irreplaceable advantage in the single-copy DNA integration, so this method has been developed rapidly these years [75]. But it should also be improved due to limitations such as low applicability and fluctuant transformation efficiency. As a transformation method affects the fate of transformed DNA, it should be paid attention to during the design of metabolic engineering strategies. For example, the PEG method could be best in the case where multiple copies of a gene should be randomly integrated into the genome [76]. On the other hand, when targeted integration or gene deletion is expected, the ATMT method would be the choice [77].

2.1.4 Restriction Enzyme-Mediated DNA Integration (REMI)

The mechanism of REMI is that added restriction enzymes in the mixture enter recipient cells, recognize genome/plasmid enzyme cutting sites, and realize the cutting and integration, and then, incisions are connected by DNA ligase. The earliest REMI was established in yeasts [78], and it has a wide range of application [79]. Recently, with increasing studies on higher fungi, REMI has also been applied to *G. lucidum* [80], *C. cinereus* [81], *L. edodes* [82], *Pleurotus eryngii* [53], and others. For example, Noh et al. [53] transformed the enhanced cyan fluorescent protein (ECFP) gene in *P. eryngii* via REMI, resulting in 10–40 hygromycin-resistant transformants per μg of the Hind III-digested DNA and 10^6 protoplasts. Southern blot analysis revealed that the gene was integrated to the genome.

One of the advantages of REMI is that it can generate various mutants [83]. This advantage has been used to analyze the genes related to the mutant characteristics. For example, by using REMI, Nakazawa et al. [84, 85] obtained a lot of mutants and found that the *Cc.rmt1* gene encoded a putative arginine methyltransferase and the *Cc.ubc2* gene affected clamp cell morphogenesis and nuclear migration of *C. cinerea*.

REMI can also be used for the site of insertion and selection marker and provides great convenience for screening. But, it also needs protoplast preparation and requires the addition of PEG and $CaCl_2$ for higher transformation efficiency. Cutting sites of the restriction enzyme are random; at the same time, to obtain sufficient number of transformants, the REMI system needs to select the optimal transformation enzyme, plasmid, and recipient cells. Thus, this method has not been widely applied, compared to the ATMT and PEG methods described above.

As another transformation method, the particle bombardment method has also been used as a transformation tool in higher fungi, as reported by Sunagawa and Magae [86] and Sunagawa et al. [87]. But, because of requiring a special instrument —particle bombardment gun—this method is not the first choice.

2.2 Selectable Marker

A resistance marker is generally used for selection of positive transformants. The commonly used resistance markers in higher fungi are of drug resistance, so that the positive transformants could grow in the culture medium with the presence of drug (s). Until now, the most widely used marker is hygromycin B phosphotransferase (*hph*) gene derived from bacteria and the transformants would obtain hygromycin resistance. Interestingly, our recent work [44] found that the mutated *sdhB* gene encoding an iron-sulfur protein subunit of succinate dehydrogenase, which successfully conferred carboxin resistance upon transformation, was a suitable resistance marker, and the efficiency of transformants screening in carboxin resistance was within the range of other reported cases. But, the use of hygromycin resistance was found unsuccessful to obtain a positive transformant in the system.

2.3 Gene Overexpression

The exogenetic genes may exist stably in cells through transformation. However, the low gene expression level is a bottleneck to the construction of cell factories. Thus, in this case, the gene overexpression is critical to successful metabolic engineering. Upregulation expression of key genes can effectively improve the level of downstream metabolite synthesis. Recently, gene overexpression has been applied to gene engineering of higher fungi in developing useful cell factories.

Xu et al. [44] overexpressed the HMGR gene which is a key enzyme in the synthetic pathway of ganoderic acids in *G. lucidum*, by using a homogenous *gpd* as a promoter. As shown in Fig. 3, the ganoderic acid content reached twofold in the overexpressed cells compared to the wild type (control). The accumulation of intermediates, such as squalene and lanosterol, was also increased. The results showed the promising potential of metabolic engineering of the ganoderic acid pathway via the transgenic system.

Zhou et al. [88] improved the production of individual ganoderic acids by engineering the biosynthetic pathway of ganoderic acids in *G. lucidum* through overexpressing the gene of squalene synthase (SQS), which catalyzes the following reaction: 2 farnesyl diphosphate + NAD(P)H $\leftrightarrows$ squalene + 2 diphosphate + NAD(P) (+). The constructed SQS strain may be a suitable basis for further development of an industrial process for hyperproduction of the antitumor secondary metabolite (Fig. 4).

The direct expression of a target gene is one way of overexpression, while the promoter engineering is another way to optimize the expression of target genes. Chai et al. [89] adopted promoter engineering to enhance the biosynthesis of β-glucans in *P. ostreatus*, in which β-glucan synthase is its key enzyme. After changing the β-glucan synthase promoter into a strong one from *Aspergillus nidulans*, the product level was increased up to 32 %, which is the first report of successful swapping of promoters in higher fungi.

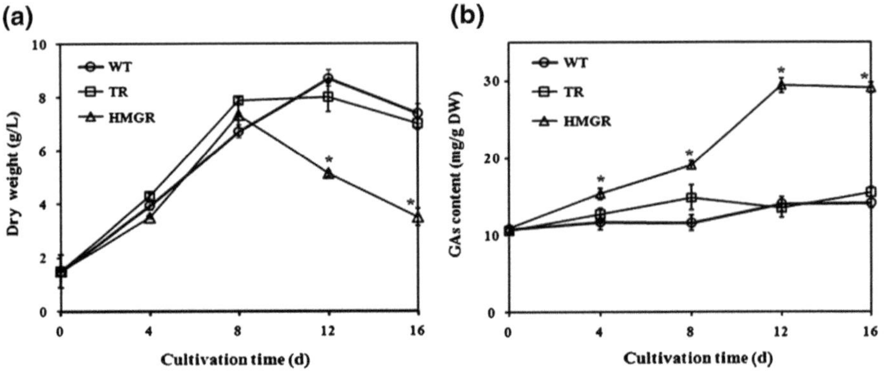

Fig. 3 Kinetic profiles of cell growth (**a**) and ganoderic acid content (**b**) in the wild-type strain (*open circle*), the strain transformed with a void plasmid (*open square*), and the tHMGR-overexpressed strain (*open triangle*). The error bars indicate the standard deviations from three independent samples. *Asterisks* statistical significance ($P < 0.05$) compared to the WT strain, *d* days [44]

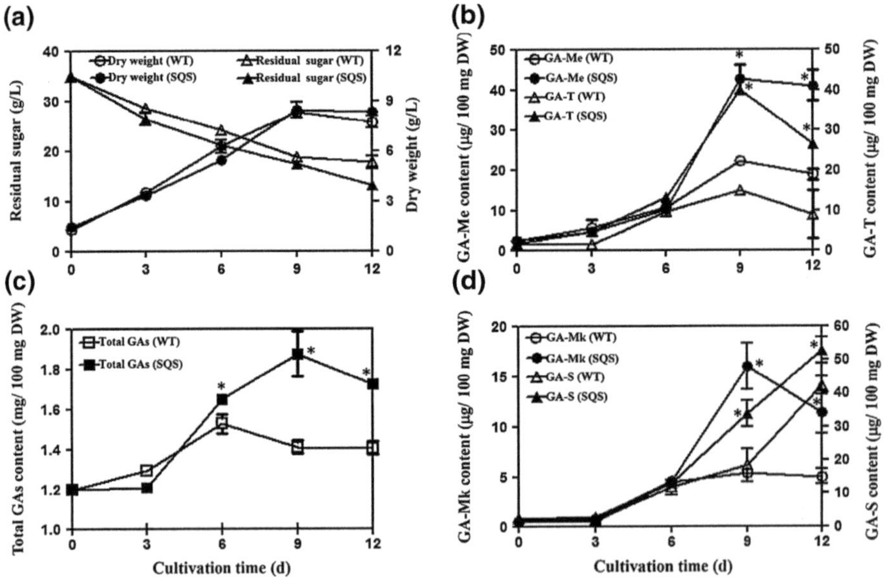

Fig. 4 Kinetic profiles of cell growth and residual sugar (**a**), content of total GAs (**c**), accumulation of GA-Me and GA-T (**b**), and GA-S and GA-Mk (**d**) in the WT strain and the SQS-overexpressed strain. The *error bars* indicate the standard deviations from three independent samples, *d* days [88]

The above-mentioned strategies are mainly related to changing promoters by introducing endogenous or heterogonous promoters to increase the expression of key genes in biosynthetic pathways. In another aspect, Lin et al. [90] reported that heterogonous protein expression in the enoki mushroom *F. velutipes* was notably enhanced by polycistronic strategy to express multiple copies of a single gene. This strategy can realize not only single gene copy but also coexpression of multiple genes.

Many efforts have been made to enhance the expression of targeted genes in higher fungi by promoter engineering, increasing gene copy number, etc. Coconi-Linares et al. [91] overexpressed three peroxidases in a single Phanerochaete chrysosporium strain, which increased the production yield of ligninolytic enzymes up to 4 times. However, the gene expression level is still difficult to control precisely, making the regulation of the metabolic flux of higher fungus cell factories not so easy.

2.4 Gene Silencing

Gene silencing, or gene knockdown, reduces the expression of a targeted gene [92]. By this, we can downregulate the metabolic branch pathway and decrease the flux to unfavorable metabolic pathway, and ultimately enhance the metabolic flux toward the targeted product biosynthesis. These years, gene silencing has been applied to *C. cinereus* [93], *A. bisporus* [94], *L. bicolor* [95, 96], *L. edodes* [97], and *G. lucidum*62. The results show that the gene silencing technology is important to efficient biosynthesis of useful metabolites and it also helps the characterization of gene functions in higher fungi [98].

RNA interference (RNAi) is an RNA-dependent gene silencing process that could inhibit the expression of specific genes at the posttranscriptional stage by introducing small double-stranded interfering RNAs [99]. Mu et al. [62] cosilenced the genes of orotidine 5′-monophosphate decarboxylase (URA3, as a reporter) and laccase in *G. lucidum*. Their results showed that the highest rate of URA3 silencing, reaching 81.9 %, was obtained by using the dual promoter vector.

RNAi is usually used for gene functional analysis. Godio et al. [100] proved that a squalene epoxidase gene (*erg1*) from basidiomycete *Hypholoma sublateritium* was involved in the biosynthesis of clavaric acid (an antitumor triterpenoid) and ergosterol. When *erg1* was cloned and expressed with the *gpdA* promoter from *A. bisporus*, the production of clavaric acid was increased, while silencing *erg1* gene by antisense RNA resulted in the reduction of clavaric acid production and appeared an ergosterol-dependent phenotype (Fig. 5).

Different from gene knockout, the efficiency of gene silencing cannot reach 100 %, but at least 70 % of the targeted RNA could be depleted. However, this 'limitation' also has an advantage over the gene knockout in cases where essential genes under investigation are still required [101].

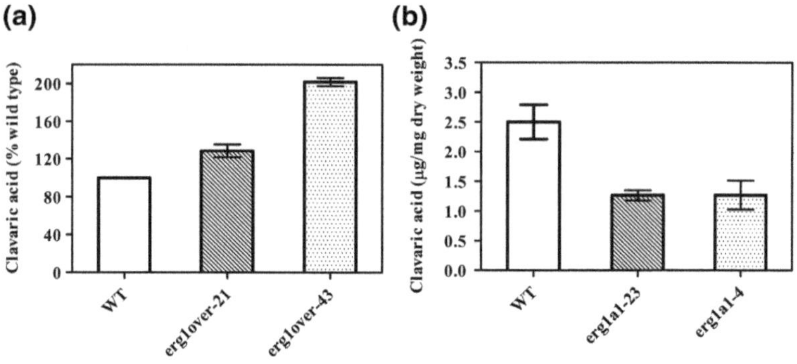

Fig. 5 Production (**a**) and content (**b**) of clavaric acid in cultures of *H. sublateritium* and the different transformants in minimal medium containing asparagine/glutamate as nitrogen source. WT, the wild-type strain; erg1over-21, the erg1-overexpressed strain 21; erg1over-43, the erg1-overexpressed strain 43; erg1a1-23, the erg1-silenced strain 23; erg1a1-4, the erg1-silenced strain 4 [100]

2.5 Gene Deletion

Gene deletion, the same as gene knockout or gene inactivation, is an important tool in learning about genes with unknown or incompletely known functions. This method usually occurs by homologous recombination. As a result, after recombination, no expression of targeted genes can be detected.

Gene deletion strategy has made a breakthrough in mushroom-forming fungus in recent years. In the past, gene inactivation was only reported in the mushroom *Schizophyllum commune* by homologous recombination, where the deletion efficiency was only 3.25 % for most targeted genes [102]. The low efficiency of homologous integration is the main obstacle to apply this method to other higher fungi. Recent progresses were made by the group of de Jong [103] and Ohm [104]. At first, they constructed a dedicated deletion vector pDelcas consisting of two antibiotic resistance cassettes. The nourseothricin resistance was located between the genes to be deleted, while the phleomycin resistance was an amplification of an ectopic integration. These transformants were screened out by using a fast colony PCR to confirm gene knockouts. Other scientists found that the inactivation of *ku80* and/or *ku70* which was related to the non-homologous end joining (NHEJ) could increase the frequency of the targeted gene knockout by homologous recombination [105]. Then, de Jong et al. [103] developed a constructive method to delete *ku80* gene and used the resulting strain for inactivation; finally, 7 out of 10 transformants were deleted, in which the efficiency of their gene disruption was greatly increased.

This improved system has been used in functional gene analysis in *S. commune*. van Peer et al. [106] deleted the *spc14* gene, which is related to the septal pore cap (SPC), and found that SPC was an organelle functioning in vegetative growth and mushroom formation. Ohm et al. [107] found that 5 transcription factor genes were related to the regulation of mushroom formation by deleting these genes in Δ*ku80* strain with pDelcas vector. In addition, Berends et al. [108] inactivated the *alg3* gene in *S. commune* by introducing pDelcas vector, and as a result, Man(3)GlcNAc (2) protein-linked N-glycans was predominantly produced.

Nakazawa et al. [109] also found the frequency of gene disruption was enhanced by inactivation of *ku70* gene in *C. cinerea*. Salame et al. [110] disrupted a targeted gene based on the Δ*ku80* strain and found the redundancy among genes of manganese peroxidases (MnPs) in *P. ostreatus* [111, 112].

3 Omics Analysis of Higher Fungus Cell Factories

Omics analysis, including genomics, transcriptomics, proteomics, and metabolomics, is an important approach to find new functional genes, proteins, metabolic networks, and metabolic products. Combining with bioinformatics, it has a great advantage to identify gene functions and to understand metabolite biosynthetic mechanisms. As we know, there are still many bioactive compounds of higher fungi, whose biosynthetic pathways are not yet identified. The development of omics analysis plays a significant role in elucidating biosynthetic pathways of metabolic products, and it is complementary to the traditional methods of drug candidate screening and identification. By omics technologies, scientists can further design, modify, and reconstruct higher fungus cell factories based on the obtained findings and implications, which can improve the rationality and validity of metabolic engineering and synthetic biology.

3.1 Genomic Analysis

More and more higher fungi genomes have been released due to technological progress. The development of genome sequencing in higher fungi provides opportunities for research and development of their metabolic products, as the genome information facilitates the discovery and biosynthetic study on bioactive compounds from higher fungi.

Recently, genomic information of a couple of higher fungi is obtained, including common edible mushroom such as *A. bisporus* [113], *F. velutipes* [114], *V. volvacea* [115], *Omphalotus olearius* [116], *S. commune* [117], *Lignosus rhinocerotis* [118], *Mycena chlorophos* [119], medicinal mushroom *G. sinese* [120], *G. lucidum* [121], and *Cordyceps militaris* [122].

F. velutipes, as an important edible mushroom, is also a rich source of secondary metabolites and enzymes which affect wood-degrading machinery and ethanol production. By sequencing and analyzing *F. velutipes* genome, 58 potential enzymes for ethanol production were identified, which provided new possibilities to use *F. velutipes* for ethanol fermentation [114].

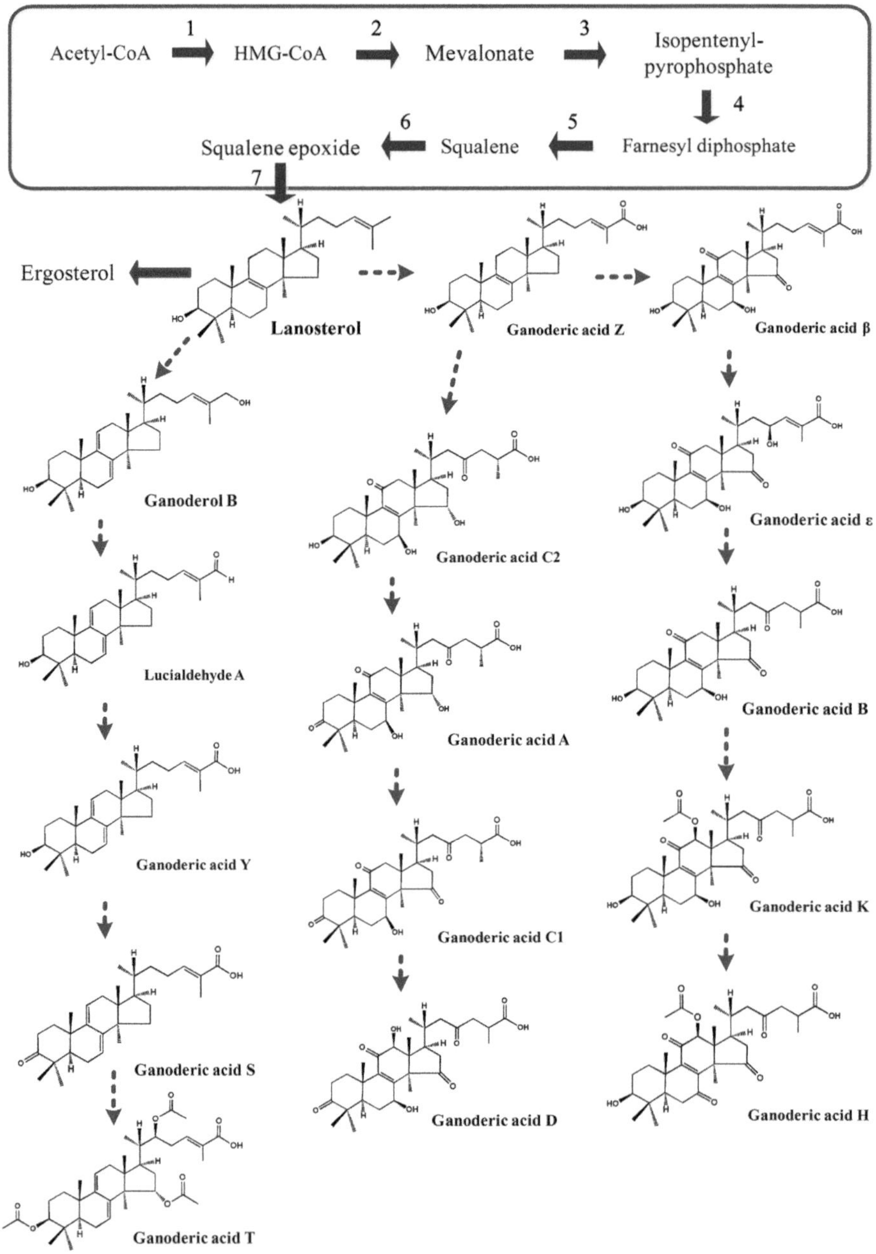

Fig. 6 Possible synthetic pathway of ganoderic acids [121]. *1* HMG-CoA synthase (*hmgs*); *2* HMG-CoA reductase (*hmgr*); *3* mevalonate-5-pyrophosphate decarboxylase (*mvd*); *4* farnesyl pyrophosphate synthase (*fps*); *5* squalene synthase (*sqs*); *6* squalene epoxidase (*se*); *7* lanosterol synthase (*ls*)

Zhu et al. reported the genome sequence of *Ganoderma sinense* and found that most of *G. sinense* gene clusters were silenced under common culture conditions. DNA methylation and small RNA-mediated reversible gene silencing tightly maintained this control, suggesting epigenetics may play critical roles in the regulation of *G. sinense* secondary metabolism [120].

Chen et al. [121] analyzed the genome of *G. lucidum*, which has been used as a drug with antitumor, antiaging, antiviral, and immunomodulatory activities since ancient times in East Asia [123]. Its triterpenoids—ganoderic acids and ganoderan—are known as main bioactive compounds [124]. Although the ganoderic acids are known to be synthesized via the mevalonate pathway (MVP) [125], it remains a mystery about the detailed modification of the lanosterol skeleton, which may be related to cytochrome P450 superfamily (CYPs). The predicted biosynthetic pathway of ganoderic acids is shown in Fig. 6. Sequencing analysis of *G. lucidum* genome revealed possible 16 CYPs involved in the terpenoid synthesis, which was favorable to the identification of ganoderic acid synthetic pathway and to the massive production of the triterpenoids as well as to the heterogonous expression via synthetic biotechnology. Then, Liu et al. [126] performed comprehensive annotation for these genes, which were analyzed from the genome of *G. lucidum*. Their work showed the genes related to triterpene biosynthesis and wood degradation. Recently, Qian et al. [127] identified simple sequence repeats (SSRs) or microsatellites in *G. lucidum* and analyzed their frequency and distribution in different genomic regions. Xu et al. [128] used qRT-PCR in *G. lucidum* gene analysis. Li et al. [129] identified and characterized long intergenic noncoding RNAs (lincRNA) in the mushroom. The study about the genome of *G. lucidum* will definitely promote the future R&D toward pharmacological and industrial applications.

3.2 Transcriptomic Analysis

Transcriptome is the set of all RNA molecules—mRNA, rRNA, tRNA, and other noncoding RNA transcribed in one cell or a cell population. It can be applied to the total set of transcripts in a given organism or to the specific subset of transcripts present in a particular cell type. Different from the genome which is roughly fixed for a given cell line, the transcriptome may vary significantly with external environmental conditions. As all mRNA transcripts in a cell are included, the transcriptome reflects the genes being actively expressed at any given time, except for mRNA degradation phenomena. The study of transcriptomics, also called as expression profiling, examines the mRNAs expression level in a given cell population, often using high-throughput techniques based on DNA microarray technology, or using next-generation sequencing technology known as RNA sequencing (RNA-Seq) to study the transcriptome at the nucleotide level. Transcriptomic analysis has been used in higher fungi *C. militaris* [130], *G. lucidum* [131], *L. edodes* [132], *V. volvacea* [133], and *P. ostreatus* [134].

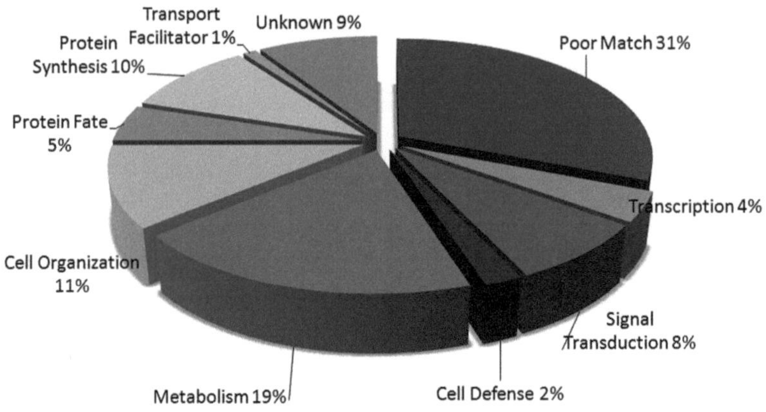

Fig. 7 Functional categories of 147 cDNA sequences in the *G. lucidum* SSH-cDNA library [135]

Xu et al. [135] did comparative transcriptome analysis of *G. lucidum* using suppression subtractive hybridization (SSH) technique to identify differentially expressed genes in liquid static culture versus shaking culture. As a result, 147 unigenes significantly expressed in static culture were identified, including those related to asexual sporulation and signal transduction (Fig. 7). Zhu et al. [136] screened for RNA editing sites at the genomic level in *G. lucidum* and revealed the role of transcriptional plasticity in the mushroom growth and development and in the regulation of secondary metabolic pathway.

RNA-Seq is a recently developed method for transcriptome profiling by deep-sequencing technologies [137], which has advantages: detected transcripts correspond to genomic sequences and it has a low background signal. Thus, RNA-Seq has become a main method for transcriptomic analysis. Yu et al. [138] analyzed the transcriptome of *G. lucidum* via Illumina high-throughput technology. Their studies performed the functional genes involved in the terpenoid biosynthesis pathway and wood degradation. Plaza et al. [139] reported that the exposure of different fungi tissues to different types of antagonists shaped the expression patterns of defense loci in a tissue-specific manner. Yang et al. [140] found enzymes related to saponin biosynthesis in the termite mushroom *Termitomyces albuminosus*, including 22 glycosyltransferase and 6 cytochrome P450 s genes by de novo sequencing and transcriptome analysis.

Ophiocordyceps sinensis, or called *Cordyceps sinensis*, has thousands of years of history as a traditional Chinese medicine because of its regulation function on human body [141]. In recent years, the main active ingredients of *O. sinensis* were identified. Transcriptome analysis by Xiang et al. [142] found that 121 genes might be involved in the regulation of signal transduction and transcription level of *O. sinensis*. They also analyzed the adenosine kinase, adenylate kinase, and 5′-nucleotidase probably related to the phosphorylation and dephosphorylation in the cordycepin biosynthesis, and the work provided useful information for identifying the cordycepin biosynthetic pathway.

3.3 Proteomic Analysis

Proteomic analysis is the systematic identification and quantification of the complete complement of proteins of a biological system such as a cell, tissue, or organism at a specific time point. Cell metabolic activities are directly/indirectly regulated by proteins. Therefore, proteomic analysis can help us better understand the cellular metabolism. But not all of the protein spots can be identified by proteomic analysis. Proteome study is still under development in higher fungi although some proteomic analyses were conducted in recent years in *Termitomyces heimii* [143], *A. bisporus* [144], *Pleurotus tuber-regium* [145], *Antrodia cinnamomea* [146], and *G. lucidum* [147].

Zhang et al. [148] investigated the mechanism of the effect of Tween 80 on the exopolysaccharide production by *P. tuber-regium* using proteomic analysis. They identified 32 differentially expressed proteins by one-dimensional gel electrophoresis, and the ATP:citrate lyase isoform 2 could stimulate exopolysaccharide production. Wang et al. [149] used high-throughput sequencing analysis to obtain the transcriptome and proteome of *Agrocybe aegerita* mycelia and fruiting bodies. *A. aegerita* possesses multiple pharmacological activities such as antitumor, antiaging, and reducing blood lipids [150]. The work helped in revealing the polysaccharide and sterol synthetic pathway, and it was also found that the polysaccharide was highly biosynthesized in fruiting bodies. The information provided important clues for establishing the mushroom cell factories toward future application.

3.4 Metabolomic Analysis

Metabolome refers to the complete set of small-molecule metabolites (usually less than 1 kDa in size), such as metabolic intermediates, hormones and other signaling molecules, and secondary metabolites, to be found in a biological sample. Like transcriptome and proteome, metabolome is dynamic and changing from second to second. Metabonomics, as a scientific study of chemical processes involving metabolites, is the quantitative measurement of dynamic multiparametric metabolic responses of living systems to pathophysiological stimuli or genetic modifications. Thus, metabolic profiling can give an instantaneous snapshot of the cell physiology. One of the challenges of systems biology and functional genomics is to integrate proteomic, transcriptomic, and metabolomic information to provide a better understanding of cellular biology.

The metabolomic study can not only find difference in external and internal environment disturbance response, but also distinguish different phenotypes; thus, it is an important technology in omics research. With technology development, such as sophisticated nuclear magnetic resonance (NMR), gas chromatography–mass spectrometry (GC/MS), and high-performance liquid chromatography–mass spectrometry (HPLC-MS), high-throughput metabolomics analysis has become possible on higher fungi metabolites.

Some scientists used metabolite profiles for the chemotaxonomy of higher fungi [151]. What's more, metabolomic study was also used for the analysis of different developmental stages or growth environments of higher fungi. For example, Park et al. [152] investigated the metabolic profiling of mycelia and fruiting bodies of *Cordyceps bassiana* by H-1 NMR spectroscopy and multivariate data analysis.

Metabolomic analysis is usually done by combining with other omics technologies such as transcriptomic and proteomic analyses to obtain more in-depth results. Matsuzaki et al. [153] performed the proteomic and metabolomic analyses of the benzoic acid metabolism of *Phanerochaete chrysosporium*. Bak et al. [154] used a polyomics-based analysis including metabolomics, proteomics, and transcriptomics of *P. chrysosporium* and tried to understand the metabolic and regulatory mechanisms of lignocellulose depolymerization.

4 Production of Useful Metabolites by Higher Fungus Cell Factories

As we know, during the long-term evolution, higher fungi have formed a special mechanism of metabolism in resisting unfavorable environments and becoming self-defense and survival during the entire life cycle. Diversified secondary metabolites with various bioactivities could be produced by higher fungi, such as terpenoids, heterocyclics, polysaccharides, polyketides, and polyphenols, which provide an abundant resource for drug discovery.

4.1 Terpenoids

Terpenes, as main bioactive compounds isolated from mushrooms, are known as an important category of naturally occurring bioactive metabolites produced by many higher fungi [116]. In particular, sesquiterpenoid, diterpenoid, and triterpenoid are typical representatives of terpenes with various interesting biological activities.

Ganoderic acids (GAs), a kind of highly oxygenated lanostane-type tetracyclic triterpene, are important secondary metabolites of *G. lucidum*. The production titer and productivity of GAs have reached great advancement in recent decade, as shown in Table 3. For example, Zhang and Tang [155] developed a novel three-stage light irradiation strategy for the efficient production of GAs and *Ganoderma* polysaccharides in submerged fermentation of *G. lucidum*. The maximal GA production was 69 % higher than the two-stage culture. Tang et al. [156] enhanced the GA production up to 754.6 mg/L through a pH-shift and DO-shift integrated fed-batch fermentation process. Liang et al. [42] found phenobarbital, a

Table 3 Typical studies on fermentation production of ganoderic acids

Strain	Ganoderic acid	Titer (mg/L)	Productivity [mg/(L h)]	Reference
G. lucidum CGMCC 5.616	Total GAs	210–1900	21–160	[158, 160–162]
G. lucidum CGMCC 5.616	Mk	0.71–260	0.088–22	[88, 158–160, 162]
G. lucidum CGMCC 5.616	T	0.71–240	0.089–20	[88, 158–160, 162]
G. lucidum CGMCC 5.616	S	0.81–270	0.10–17	[88, 158–160, 162]
G. lucidum CGMCC 5.616	Me	0.37–110	0.046–9.3	[88, 158–160, 162]
G. lucidum SB97	Total GAs	500	110	[163]
Ganoderma sinense SCIM 0701	Total GAs	260	37	[164]
Ganoderma applanatum ACCC-52297	Total GAs	293	65	[165]
G. lucidum SCIM 0006	Me	12	0.69	[166]
G. lucidum 5.26	Total GAs	680	75	[167]
G. lucidum 5.26	T	19	2.1	[167]

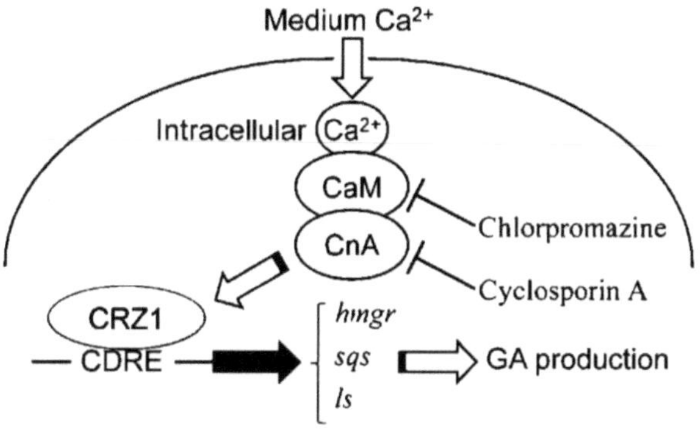

Fig. 8 Proposed Ca^{2+} induction mechanism in GA biosynthesis via calcineurin signaling [158]

P450 inducer, could enhance the production of total and individual GAs in two-stage cultivation. Zhao et al. [157] reported that nitrogen limitation led to more accumulation of GAs and upregulation of the biosynthetic genes.

Recently, signal transduction engineering of fungal secondary metabolism is receiving great interest for efficient production of valuable metabolites. Xu and Zhong [158] found that the calcineurin signal transduction was significant to GAs biosynthesis by *G. lucidum* (Fig. 8). Addition of calcium ion to static liquid cultures of *G. lucidum* resulted in the enhanced production of GAs, and the total GAs and individual GA-Mk, GA-S, GA-T, and GA-Me reached 3.7-, 2.6-, 3.2-, 4.5-, and 3.8-fold improvement compared to control, respectively. The group further reported that Na^+ addition [159] and Mn^{2+} addition [160] could both enhance the GAs accumulation.

Dou et al. [168] studied the oxygen supply effect on the biomass and helvolic acid production in submerged fermentation of *C. taii*. Helvolic acid belongs to a member of fusidane skeleton triterpenoid family, which has a significant bactericidal activity, but few studies were performed about its fermentation production. The results showed that the value of initial volumetric oxygen transfer coefficient (K_La) greatly affected the production of both biomass and helvolic acid.

4.2 Heterocyclics

Heterocyclic compounds, or heterocyclics, are those whose one or more of the ring carbon atoms are replaced with a different element such as oxygen, nitrogen, and sulfur. Many types of heterocyclics from higher fungi have been isolated, and their structures and biological activities have been analyzed. These include indoles, pyridines, cytochalasins, quinolines, flavonoids, and nucleosides, and their anticancer, anti-HIV, antibacterial, and other pharmacological activities have been reported [169].

Cordycepin, 3-deoxyadenosine, is a major bioactive compound of *C. militaris*, which has various pharmacological activities, including antitumor, immunomodulatory, anti-inflammatory, and antibacterial ones [170]. The biosynthetic pathway of cordycepin has not been completely elucidated; however, many efforts have been made to enhance its production. Mao et al. [171] found that NH_4^+ had a significant effect on cordycepin production. Das et al. [172] improved the productivity of cordycepin in *C. militaris* by mutation using high-energy ion beam irradiation. In the work by Fan et al. [173], the influence of ferrous sulfate on the production of cordycepin was studied in shake flask cultures. The results indicated that the highest cordycepin titer was about 70 % higher than that without ferrous sulfate addition. This work might also be useful for further understanding the cordycepin biosynthesis.

Lovastatin is a member of the drug class of statins found in oyster mushrooms [174], used in combination with diet, weight loss, and exercise for lowering cholesterol in those with hypercholesterolemia to reduce risk of cardiovascular disease. Statistical experimental designs were used to optimize the lovastatin production by submerged fermentation of *P. ostreatus* [175]. The maximum lovastatin production reached 114.82 mg/L, which was 50 times higher than that obtained under the conditions without optimization.

Flavonoids, a class of secondary metabolites which have antioxidant effects and inhibitory activities on melanin biosynthesis, mostly exist in plants but have also been reported in higher fungi [176]. The *Vitreoscilla* hemoglobin, an oxygen-binding protein, could enhance cell growth, and it was used to alleviate oxygen limitation during submerged fermentation. Zhu et al. [177] expressed the *Vitreoscilla* hemoglobin gene (*vgb*) by REMI in *Phellinus igniarius*, which resulted in the improved growth and production of both total flavones. The metabolites reached 11.43 and 1.33 g/L, respectively, in bioreactor cultivations.

4.3 Polysaccharides

Macrofungal polysaccharides have been well known as part of traditional diet and medicine. The polysaccharides comprise a variety of biopolymers, such as β-glucans, providing a mechanism for cell protection or attachment to others [178]. The fungal polysaccharides mainly include exopolysaccharides (EPS) and intracellular polysaccharides (IPS), which could be produced by submerged mycelial cultures with a variety of medical applications. Until now, many polysaccharides have been isolated from various mushrooms, such as *Morchella conica* [179], *G. applanatum* [180], *Laetiporus sulphureus* [181], and *C. taii* [182], which have many bioactivities including antitumor [183], antioxidant [184], immunomodulatory [185], cytostatic, and antibacterial properties. Large-scale production of polysaccharides is very important for application, and some reports have been published on how to enhance its production such as optimization of culture conditions [186], addition of metabolic inducers [189], and genetic modifications [89].

Hwang et al. [187] investigated the optimum culture conditions in submerged culture of *L. sulphureus* var. *miniatus*, an edible mushroom. Interestingly, the most suitable initial pH for the metabolite synthesis was 2.0, which is rare in submerged cultures of macrofungi. In addition, supplementation of deep seawater (DSW) was also used for cultivation of higher fungi, and DSW was found to efficiently increase the mycelial growth and EPS production. As a result, the maximum mycelial growth (4.1 g/L) and EPS production (0.6 g/L) were achieved. Their work also showed that the EPS of *L. sulphureus* could increase cell proliferation and promote insulin secretion. Cui and Zhang [188] found that Mg^{2+}, Mn^{2+}, sodium dodecyl sulfate (SDS), and Tween 80 significantly enhanced the EPS production during two-stage submerged cultivation of *C. militaris*. The highest EPS production reached 3.28 g/L under the optimal condition. The results showed that additions of metal ion and surfactant could be used for enhancing the EPS production by *C. militaris*.

Xu et al. [189] overexpressed the gene of phosphoglucomutase (PGM) which is a key enzyme in the biosynthetic pathway of nucleotide sugar precursors. This enzyme catalyzes the conversion of glucose-6-phosphate into glucose-1-phosphate representing a branch point in carbohydrate metabolism. The maximum IPS content and EPS production in *G. lucidum* overexpressing the PGM gene reached 23.67 mg/100 mg dry weight and 1.76 g/L, respectively, which was higher by 40.5 % and 44.3 % than by the wild-type strain (Fig. 9). Ji et al. [190] improved the polysaccharide production by engineering the biosynthetic pathway in *G. lucidum* by overexpressing the homologous UDP glucose pyrophosphorylase (UGP) gene. The maximum IPS content and EPS production in the strain were 24.32 mg/100 mg dry weight and 1.66 g/L, respectively (Fig. 10). Their results showed the feasibility to enhance polysaccharide production by altering the expression of genes involved in precursor supply.

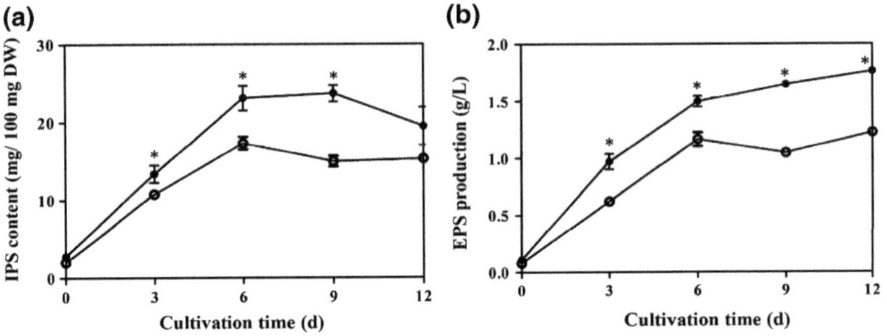

Fig. 9 Kinetic profiles of IPS content (**a**) and EPS production (**b**) in fermentation of the WT strain (*blank circles*) and the PGM strain (*dark circles*). The error bars indicate the standard deviations from three independent samples [189]

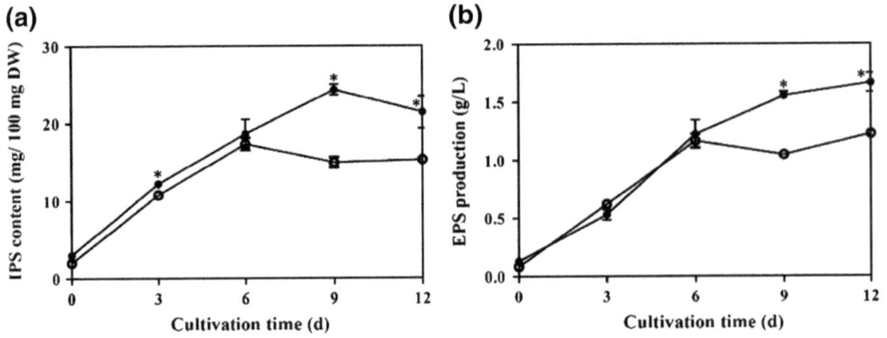

Fig. 10 Kinetic profiles of IPS content (**a**) and EPS production (**b**) in fermentation of the WT strain (*blank circles*) and the UGP strain (*dark circles*). The error bars indicate the standard deviations from three independent samples [190]

4.4 Polyketides

Polyketides (PKs), as a class of secondary metabolites, are structurally a very diverse family of complex organic compounds produced by certain living organisms in order to impart to them some survival advantages. They often possess biological activities and pharmacological properties such as antibacterial, anti-cancer, and antimalarial activities [191]. Polyketides are biosynthesized by polyketide synthases (PKSs) through the decarboxylative condensation of malonyl-CoA-derived extender units in a similar process to fatty acid synthesis [192]. In recent years, polyketides have been isolated from many higher fungi such as *Cordyceps* species and *Cortinarius purpurascens* [193].

Many fungal secondary metabolites are regulated by epigenetic modification, which not only affects metabolite titers, but also activates cryptic gene clusters

[194]. The application of epigenetic modification to higher fungi is becoming a new strategy for strain improvement and a powerful method to obtain novel natural products. For example, Strauss et al. [195] significantly enhanced the production of bisabolene-type sesquiterpenoids and xanthones analogs from *Aspergillus sydowii* by addition of 5-azacytidine, a chemical epigenetic modifier. As higher fungi can produce a variety of unique secondary metabolites including polyketides, the epigenetic modification could help to enhance the production titer and efficiency of those interesting metabolites.

Asai et al. [196] found that addition of epigenetic modification chemicals such as histone deacetylase (HDAC) inhibitor or DNA methyltransferase inhibitor could significantly enhance polyketide production by *Cordyceps annullata*, and a couple of new aromatic polyketides were isolated, such as indigotides C-F, 13-hydroxyindigotide A and 8-O-methylindigotide B [197].

5 Summary and Future Perspectives

Higher fungi have been used as both medicinal and edible materials for thousands of years in East Asia and many other regions around the world. Nowadays, with the quick development of biological science and related engineering fields, the research on higher fungi is deepening rapidly. As described above, many unique important bioactive components have been found, which is important to the development of higher fungus cell factories for industrial applications.

The technologies such as gene transformation, overexpression, silencing, and deletion are gradually applied to the studies of the construction of higher fungus cell factories. The combination of different gene editing methods will help to explore gene functions and promote the identification and optimization of metabolic pathways. However, some problems also need to be solved. The gene transformation methods have not yet mature, and their transformation efficiency and transformant stability still need improvement as well. The gene deletion and silencing have to be extended to more species. Meanwhile, the rapid development of bioinformatics will help to understand molecular characteristics of higher fungi and biosynthesis pathways of various secondary metabolites, which is important to large-scale commercial production.

Recently, omics technologies including genomics, transcriptomics, proteomics, and metabolomics play an important role in studies on behaviors and mechanisms of higher fungi. Whole genome transcription analysis will enable researchers to accurately evaluate the relationship between phenotype and expression of genes, helping understand the cellular metabolism. It also contributes to the identification of target genes for strain improvement and accelerates the rational design and construction of higher fungus cell factories. Because the regulation of all levels of cellular activity/metabolism is interacted with each other, single omic analysis technology has apparent limitations. Thus, the integrated utilization of omics technologies is important in obtaining complete information, which will deepen the understanding of complex biological systems and speed up the identification of target sites in metabolism. In order to have more rational and efficient improvement on cell/organism breeding, scientists of different professional backgrounds need to cooperate intimately with each other, especially with tools of bioinformatics and mathematics to simulate and design new biosynthesis systems, so as to enhance the existing bioproduct production and new bioproduct synthesis.

Without doubt, the establishment of cell factory platform has a big impact on large-scale production of useful metabolites by higher fungi. Through gene transformation, gene overexpression and other biological technologies, we can deepen our understanding on the synthetic mechanism and pathways of secondary metabolites, so as to realize the reconstruction of metabolic pathways and the massive accumulation of targeted metabolites. However, due to the short history of higher fungi research, their genetic technologies have yet to be improved to ensure exogenous genes or homologous genes expressed efficiently and stably in higher fungal cells; gene overexpression, silencing, and knockout technology also need to be applied widely in higher fungi. Furthermore, the construction of higher fungal cell factories needs the integration of genomics, gene technology platforms, and bioinformatics technologies, to have a better understanding about metabolic pathways and cellular metabolic mechanisms. Here, many new bioactive products may be found through these approaches, and the metabolic pathway can be improved to achieve efficient biomanufacturing of valuable metabolites. The expression of heterogonous genes in higher fungi, which resulted in the synthesis of other products, is also under development. Synthetic biology as an emerging discipline has become more and more popular due to its enormous potential, and how to better apply synthetic biology into higher fungi is becoming one of the targets in our future research. As the higher fungi are getting more and more global attention in recent years with their expanding markets, therefore research and development on higher fungal cell factories are anticipated to meet the growing demand of market by further promoting useful metabolites production.

Acknowledgements JJZ thanks the National 973 program (No. 2012CB721006) and 863 program (No. 2012AA021701) from MoST and the National Natural Science Foundation of China (No. 20776084) for financial support. JHX appreciates the Guizhou High-Level Innovative Talent Support Program (No. QKH-RC-20154028).

References

1. Ma K, Ren J, Han J, Bao L, Li L, Yao Y, Sun C, Zhou B, Liu H (2014) Ganoboninketals A-C, antiplasmodial 3,4-seco-27-norlanostane triterpenes from *Ganoderma boninense* Pat. J Nat Prod 77:1847–1852
2. Ma K, Bao L, Han J, Jin T, Yang X, Zhao F, Li S, Song F, Liu M, Liu H (2014) New benzoate derivatives and hirsutane type sesquiterpenoids with antimicrobial activity and cytotoxicity from the solid-state fermented rice by the medicinal mushroom *Stereum hirsutum*. Food Chem 143:239–245

3. Chen HP, Dong WB, Feng T, Yin X, Li ZH, Dong ZJ, Li Y, Liu JK (2014) Four new sesquiterpenoids from fruiting bodies of the fungus *Inonotus rickii*. J Asian Nat Prod Res 16:581–586
4. Wang S, Bao L, Zhao F, Wang Q, Li S, Ren J, Li L, Wen H, Guo L, Liu H (2013) Isolation, identification, and bioactivity of monoterpenoids and sesquiterpenoids from the mycelia of edible mushroom *Pleurotus cornucopiae*. J Agric Food Chem 61:5122–5129
5. Nord CL, Menkis A, Vasaitis R, Broberg A (2013) Protoilludane sesquiterpenes from the wood decomposing fungus *Granulobasidium vellereum* (Ellis & Cragin) Julich. Phytochemistry 90:128–134
6. Kim KH, Moon E, Choi SU, Kim SY, Lee KR (2013) Lanostane triterpenoids from the mushroom *Naematoloma fasciculare*. J Nat Prod 76:845–851
7. Wang YQ, Bao L, Yang XL, Li L, Li SF, Gao H, Yao XS, Wen HA, Lie HW (2012) Bioactive sesquiterpenoids from the solid culture of the edible mushroom *Flammulina velutipes* growing on cooked rice. Food Chem 132:1346–1353
8. Shi XW, Zhang AL, Pescitelli G, Gao JM (2012) Secoscabronine M, a novel diterpenoid from the Chinese bitter mushroom *Sarcodon scabrosus*. Chirality 24:386–390
9. Liu L, Shi XW, Zong SC, Tang JJ, Gao JM (2012) Scabronine M, a novel inhibitor of NGF-induced neurite outgrowth from PC12 cells from the fungus *Sarcodon scabrosus*. Bioorg Med Chem Lett 22:2401–2406
10. Kanokmedhakul S, Lekphrom R, Kanokmedhakul K, Hahnvajanawong C, Bua-art S, Saksirirat W, Prabpai S, Kongsaeree P (2012) Cytotoxic sesquiterpenes from luminescent mushroom *Neonothopanus nambi*. Tetrahedron 68:8261–8266
11. Bohnert M, Miethbauer S, Dahse HM, Ziemen J, Nett M, Hoffmeister D (2011) In vitro cytotoxicity of melleolide antibiotics: structural and mechanistic aspects. Bioorg Med Chem Lett 21:2003–2006
12. Weng CJ, Chau CF, Chen KD, Chen DH, Yen GC (2007) The anti-invasive effect of lucidenic acids isolated from a new *Ganoderma lucidum* strain. Mol Nutr Food Res 51:1472–1477
13. Wang JL, Li YB, Liu RM, Zhong JJ (2010) A new ganoderic acid from *Ganoderma lucidum* mycelia. J Asian Nat Prod Res 12:727–730
14. Li YB, Liu RM, Zhong JJ (2013) A new ganoderic acid from *Ganoderma lucidum* mycelia and its stability. Fitoterapia 84:115–122
15. Li YY, Mi ZY, Tang Y, Wang G, Li DS, Tang YJ (2009) Lanostanoids isolated from *Ganoderma lucidum* mycelium cultured by submerged fermentation. Helv Chim Acta 92:1586–1593
16. Wang K, Bao L, Qi Q, Zhao F, Ma K, Pei Y, Liu H (2015) Erinacerins C-L, isoindolin-1-ones with alpha-glucosidase inhibitory activity from cultures of the medicinal mushroom *Hericium erinaceus*. J Nat Prod 78:146–154
17. Kim KH, Noh HJ, Choi SU, Lee KR (2012) Isohericenone, a new cytotoxic isoindolinone alkaloid from *Hericium erinaceum*. J Antibiot (Tokyo) 65:575–577
18. Ueda K, Tsujimori M, Kodani S, Chiba A, Kubo M, Masuno K, Sekiya A, Nagai K, Kawagishi H (2008) An endoplasmic reticulum (ER) stress-suppressive compound and its analogues from the mushroom *Hericium erinaceum*. Bioorg Med Chem 16:9467–9470
19. Ma K, Han J, Bao L, Wei T, Liu H (2014) Two sarcoviolins with antioxidative and alpha-glucosidase inhibitory activity from the edible mushroom *Sarcodon leucopus* collected in Tibet. J Nat Prod 77:942–947
20. Kubo M, Liu Y, Ishida M, Harada K, Fukuyama Y (2014) A new spiroindene pigment from the medicinal fungus *Phellinus ribis*. Chem Pharm Bull (Tokyo) 62:122–124
21. Demirkiran O, Sabudak T, Ozturk M, Topcu G (2013) Antioxidant and tyrosinase inhibitory activities of flavonoids from *Trifolium nigrescens* Subsp. petrisavi. J Agric Food Chem 61:12598–12603
22. Li Y, Bao L, Song B, Han J, Li H, Zhao F, Liu H (2013) A new benzoquinone and a new benzofuran from the edible mushroom *Neolentinus lepideus* and their inhibitory activity in NO production inhibition assay. Food Chem 141:1614–1618

23. Lu WW, Gao YJ, Su MZ, Luo Z, Zhang W, Shi GB, Zhao QC (2013) Isoindolones from *Lasiosphaera fenzlii* Reich. and their bioactivities. Helv Chim Acta 96:109–113
24. Kim KH, Park KM, Choi SU, Lee KR (2009) Macrolepiotin, a new indole alkaloid from *Macrolepiota neomastoidea*. J Antibiot (Tokyo) 62:335–338
25. Chi YL, Zhang HY, Xue JH, Hao J, Liu MF, Wei XY (2009) N-2-(1-Methoxycarbonylethyl) guanosine, a new nucleoside coupled with an amino acid derivative from *Amanita exitialis*. Chinese Chem Lett 20:830–832
26. El Dine RS, El Halawany AM, Ma CM, Hattori M (2009) Inhibition of the dimerization and active site of HIV-1 protease by secondary metabolites from the Vietnamese mushroom *Ganoderma colossum*. J Nat Prod 72:2019–2023
27. Rukachaisirikul V, Khamthong N, Sukpondma Y, Pakawatchai C, Phongpaichit S, Sakayaroj J, Kirtikara K (2009) An [11]cytochalasin derivative from the marine-derived fungus *Xylaria* sp. PSU-F100. Chem Pharm Bull (Tokyo) 57:1409–1411
28. Xu F, Pang JY, Lu BT, Wang JJ, Zhang Y, She ZG, Vrijmoed LLP, Gareth JEB, Lin YC (2009) Two metabolites with DNA-binding affinity from the mangrove fungus *Xylaria* sp (#2508) from the south China sea coast. Chinese J Chem 27:365–368
29. Kojima K, Ohno T, Inoue M, Mizukami H, Nagatsu A (2008) Phellifuropyranone A: a new furopyranone compound isolated from fruit bodies of wild *Phellinus linteus*. Chem Pharm Bull (Tokyo) 56:173–175
30. Teichert A, Schmidt J, Porzel A, Arnold N, Wessjohann L (2008) N-glucosyl-1H-indole derivatives from *Cortinarius brunneus* (Basidiomycetes). Chem Biodivers 5:664–669
31. Teichert A, Schmidt J, Porzel A, Arnold N, Wessjohann L (2008) (Iso)-quinoline alkaloids from fungal fruiting bodies of *Cortinarius subtortus*. J Nat Prod 71:1092–1094
32. Li XG, Pan WD, Lou HY, Liu RM, Xiao JH, Zhong JJ (2015) New cytochalasins from medicinal macrofungus *Crodyceps taii* and their inhibitory activities against human cancer cells. Bioorg Med Chem Lett 25:1823–1826
33. Fangkrathok N, Sripanidkulchai B, Umehara K, Noguchi H (2013) Bioactive ergostanoids and a new polyhydroxyoctane from *Lentinus polychrous* mycelia and their inhibitory effects on E2-enhanced cell proliferation of T47D cells. Nat Prod Res 27:1611–1619
34. Li YY, Liu P, Tang Y, Li HM, Tang YL, Liang XH, Tang YJ (2013) Novel cerebrosides isolated from the fermentation mycelia of *Tuber indicum*. Helv Chim Acta 96:702–709
35. Xiao JH, Sun ZH, Pan WD, Lu YH, Chen DX, Zhong JJ (2012) Jiangxienone, a new compound with potent cytotoxicity against tumor cells from traditional Chinese medicinal mushroom *Cordyceps jiangxiensis*. Chem Biodivers 9:1349–1355
36. Ma BJ, Ma JC, Ruan Y (2012) Hericenone L, a new aromatic compound from the fruiting bodies of *Hericium erinaceums*. Chin J Nat Medicines 10:363–365
37. Hong SS, Lee JH, Jeong W, Kim N, Jin HZ, Hwang BY, Lee HJ, Lee SJ, Jang DS, Lee D (2012) Acetylenic acid analogues from the edible mushroom Chanterelle (*Cantharellus cibarius*) and their effects on the gene expression of peroxisome proliferator-activated receptor-gamma target genes. Bioorg Med Chem Lett 22:2347–2349
38. Norikura T, Fujiwara K, Narita T, Yamaguchi S, Morinaga Y, Iwai K, Matsue H (2011) Anticancer activities of thelephantin O and vialinin A isolated from *Thelephora aurantiotincta*. J Agric Food Chem 59:6974–6979
39. Wang G, Li YY, Li DS, Tang YJ (2008) Determination of 5 alpha-androst-16-en-3 alpha-ol in truffle fermentation broth by solid-phase extraction coupled with gas chromatography-flame ionization detector/electron impact mass spectrometry. J Chromatogr B 870:209–215
40. Onose JI, Xie C, Ye YQ, Sugaya K, Takahashi S, Koshino H, Yasunaga K, Abe N, Yoshikawa K (2008) Vialinin A, a novel potent inhibitor of TNF-alpha production from RBL-2H3 cells. Biol Pharm Bull 31:831–833
41. Xu JW, Xu YN, Zhong JJ (2010) Production of individual ganoderic acids and expression of biosynthetic genes in liquid static and shaking cultures of *Ganoderma lucidum*. Appl Microbiol Biotechnol 85:941–948

42. Liang CX, Li YB, Xu JW, Wang JL, Miao XL, Tang YJ, Gu T, Zhong JJ (2010) Enhanced biosynthetic gene expressions and production of ganoderic acids in static liquid culture of *Ganoderma lucidum* under phenobarbital induction. Appl Microbiol Biotechnol 86:1367–1374
43. Xu JW, Zhao W, Zhong JJ (2010) Biotechnological production and application of ganoderic acids. Appl Microbiol Biotechnol 87:457–466
44. Xu JW, Xu YN, Zhong JJ (2012) Enhancement of ganoderic acid accumulation by overexpression of an N-terminally truncated 3-hydroxy-3-methylglutaryl coenzyme A reductase gene in the basidiomycete *Ganoderma lucidum*. Appl Environ Microbiol 78:7968–7976
45. Yadav VG, De Mey M, Lim CG, Ajikumar PK, Stephanopoulos G (2012) The future of metabolic engineering and synthetic biology: towards a systematic practice. Metab Eng 14:233–241
46. van de Rhee MD, Graca PM, Huizing HJ, Mooibroek H (1996) Transformation of the cultivated mushroom, *Agaricus bisporus*, to hygromycin B resistance. Mol Gen Genet 250:252–258
47. Cho JH, Lee SE, Chang WB, Cha JS (2006) *Agrobacterium*-mediated transformation of the winter mushroom. Flammulina velutipes. Mycobiology 34:104–107
48. Guo LQ, Lin JF, Xiong S, Chen SC (2005) Transformation of *Volvariella volvacea* with a thermal hysteresis protein gene by particle bombardment. Wei Sheng Wu Xue Bao 45:39–43
49. Zhou YP, Chen MH, Lu JJ, Kang X, Chen QH, Huang XL, Tian CE (2015) A simple and efficient genetic transformation method of *Ganoderma weberianum*. Folia Microbiol (Praha) 60:417–423
50. Shin DI, Park HS (2013) Mechanical wounding of yeast-like conidium cells of *Tremella fuciformis* makes them susceptible to *Agrobacterium*-mediated transformation. Biosci Biotechnol Biochem 77:2157–2159
51. Li G, Li R, Liu Q, Wang Q, Chen M, Li B (2006) A highly efficient polyethylene glycol-mediated transformation method for mushrooms. FEMS Microbiol Lett 256:203–208
52. Sun L, Cai HQ, Xu WH, Hu YL, Gao Y, Lin ZP (2001) Efficient transformation of the medicinal mushroom *Ganoderma lucidum*. Plant Mol Biol Rep 19:383
53. Noh W, Kim SW, Dong-Won B, Kim JY, Ro HS (2010) Genetic introduction of foreign genes to *Pleurotus eryngii* by restriction enzyme-mediated integration. J Microbiol 48:253–256
54. Lin JF, Zheng MY, Wang J, Shu W, Guo LQ (2008) Efficient transformation and expression of *gfp* gene in the edible mushroom *Pleurotus nebrodensis*. Prog Nat Sci 18:819–824
55. Sun L, Xu WH, Cai HQ, Hu YL, Lin ZP (2001) PEG-mediated transformation of *Lentinus edodes*. Acta Bot Sin 43:1089–1092
56. Peng M, Singh NK, Lemke PA (1992) Recovery of recombinant plasmids from *Pleurotus ostreatus* transformants. Curr Genet 22:53–59
57. Irie T, Honda Y, Hirano T, Sato T, Enei H, Watanabe T, Kuwahara M (2001) Stable transformation of *Pleurotus ostreatus* to hygromycin B resistance using *Lentinus edodes* GPD expression signals. Appl Microbiol Biotechnol 56:707–709
58. Yu XY, Ji SL, He YL, Ren MF, Xu JW (2014) Development of an expression plasmid and its use in genetic manipulation of lingzhi or reishi medicinal mushroom, *Ganoderma lucidum* (higher basidiomycetes). Int J Med Mushrooms 16:161–168
59. Guo L, Liu Y, Zhao S, Liu E, Lin J (2008) Highly efficient transformation of intact yeast-like conidium cells of *Tremella fuciformis* by electroporation. Sci China C Life Sci 51:932–940
60. Kuo CY, Huang CT (2008) A reliable transformation method and heterologous expression of beta-glucuronidase in *Lentinula edodes*. J Microbiol Methods 72:111–115
61. Kuo CY, Chou SY, Hseu RS, Huang CT (2010) Heterologous expression of EGFP in enoki mushroom *Flammulina velutipes*. Bot Stud 51:303–309
62. Mu D, Shi L, Ren A, Li M, Wu F, Jiang A, Zhao M (2012) The development and application of a multiple gene co-silencing system using endogenous *URA3* as a reporter gene in *Ganoderma lucidum*. PLoS ONE 7:e43737

63. Kuo CY, Chou SY, Huang CT (2004) Cloning of glyceraldehyde-3-phosphate dehydrogenase gene and use of the *gpd* promoter for transformation in *Flammulina velutipes*. Appl Microbiol Biotechnol 65:593–599
64. Bundock P, den Dulk-Ras A, Beijersbergen A, Hooykaas PJ (1995) Trans-kingdom T-DNA transfer from *Agrobacterium tumefaciens* to *Saccharomyces cerevisiae*. EMBO J 14:3206–3214
65. Hirth KP, Edwards CA, Firtel RA (1982) A DNA-mediated transformation system for *Dictyostelium discoideum*. Proc Natl Acad Sci US 79:7356–7360
66. Chen X, Stone M, Schlagnhaufer C, Romaine CP (2000) A fruiting body tissue method for efficient *Agrobacterium*-mediated transformation of *Agaricus bisporus*. Appl Environ Microbiol 66:4510–4513
67. Burns C, Leach KM, Elliott TJ, Challen MP, Foster GD, Bailey A (2006) Evaluation of *Agrobacterium*-mediated transformation of *Agaricus bisporus* using a range of promoters linked to hygromycin resistance. Mol Biotechnol 32:129–138
68. Okamoto T, Yamada M, Sekiya S, Okuhara T, Taguchi G, Inatomi S, Shimosaka M (2010) *Agrobacterium tumefaciens*-mediated transformation of the vegetative dikaryotic mycelium of the cultivated mushroom *Flammulina velutipes*. Biosci Biotechnol Biochem 74:2327–2329
69. Samils N, Elfstrand M, Czederpiltz DLL, Fahleson J, Olson A, Dixelius C, Stenlid J (2006) Development of a rapid and simple *Agrobacterium tumefaciens*-mediated transformation system for the fungal pathogen *Heterobasidion annosum*. FEMS Microbiol Lett 255:82–88
70. Lin JF, Zheng MY, Wang J, Shu W, Guo LQ (2008) Efficient transformation and expression of *gfp* gene in the edible mushroom *Pleurotus nebrodensis*. Prog Nat Sci 18:819–824
71. Wang J, Guo L, Zhang K, Wu Q, Lin J (2008) Highly efficient *Agrobacterium*-mediated transformation of *Volvariella volvacea*. Bioresour Technol 99:8524–8527
72. Michielse CB, Hooykaas PJJ (2005) van den Hondel CAMJJ, Ram AFJ: *Agrobacterium*-mediated transformation as a tool for functional genomics in fungi. Curr Genet 48:1–17
73. Zhang J, Shi L, Chen H, Sun Y, Zhao M, Ren A, Chen M, Wang H, Feng Z (2014) An efficient *Agrobacterium*-mediated transformation method for the edible mushroom *Hypsizygus marmoreus*. Microbiol Res 169:741–748
74. Kemppainen M, Duplessis S, Martin F, Pardo AG (2008) T-DNA insertion, plasmid rescue and integration analysis in the model mycorrhizal fungus *Laccaria bicolor*. Microb Biotechnol 1:258–269
75. Park HS, Choi JW (2014) Functional expression of bovine growth hormone gene in *Pleurotus eryngii*. Biotechnol Bioproc E 19:33–42
76. Kilaru S, Collins CM, Hartley AJ, Bailey AM, Foster GD (2009) Establishing molecular tools for genetic manipulation of the pleuromutilin-producing fungus *Clitopilus passeckerianus*. Appl Environ Microbiol 75:7196–7204
77. Meyer V (2008) Genetic engineering of filamentous fungi–progress, obstacles and future trends. Biotechnol Adv 26:177–185
78. Schiestl RH, Petes TD (1991) Integration of DNA fragments by illegitimate recombination in *Saccharomyces cerevisiae*. Proc Natl Acad Sci USA 88:7585–7589
79. Granado JD, Kertesz-Chaloupkova K, Aebi M, Kues U (1997) Restriction enzyme-mediated DNA integration in *Coprinus cinereus*. Mol Gen Genet 256:28–36
80. Kim S, Song J, Choi HT (2004) Genetic transformation and mutant isolation in *Ganoderma lucidum* by restriction enzyme-mediated integration. FEMS Microbiol Lett 233:201–204
81. Han F, Liu Y, Guo LQ, Zeng XL, Liu ZM, Lin JF (2010) Heterologous expression of the immunomodulatory protein gene from *Ganoderma sinense* in the basidiomycete *Coprinopsis cinerea*. J Appl Microbiol 109:1838–1844
82. Irie T, Sato T, Saito K, Honda Y, Watanabe T, Kuwahara M, Enei H (2003) Construction of a homologous selectable marker gene for *Lentinula edodes* transformation. Biosci Biotechnol Biochem 67:2006–2009
83. Riggle PJ, Kumamoto CA (1998) Genetic analysis in fungi using restriction-enzyme-mediated integration. Curr Opin Microbiol 1:395–399

84. Nakazawa T, Tatsuta Y, Fujita T, Nakahori K, Kamada T (2010) Mutations in the *Cc.rmt1* gene encoding a putative protein arginine methyltransferase alter developmental programs in the basidiomycete *Coprinopsis cinerea*. Curr Genet 56:361–367
85. Nakazawa T, Kondo H, Nakahori K, Kamada T (2011) A mutation in the *Cc.ubc2* gene affects clamp cell morphogenesis as well as nuclear migration for dikaryosis in *Coprinopsis cinerea*. Fungal Genet Biol 48:519–525
86. Sunagawa M, Magae Y (2002) Transformation of the edible mushroom *Pleurotus ostreatus* by particle bombardment. FEMS Microbiol Lett 211:143–146
87. Sunagawa M, Murata H, Miyazaki Y, Nakamura M (2007) Transformation of the mycorrhizal basidiomycetes, *Suilllus grevillei* and *S. bovinus*, by particle bombardment. Biosci Biotech Bioch 71:47–50
88. Zhou JS, Ji SL, Ren MF, He YL, Jing XR, Xu JW (2014) Enhanced accumulation of individual ganoderic acids in a submerged culture of *Ganoderma lucidum* by the overexpression of squalene synthase gene. Biochem Eng J 90:178–183
89. Chai R, Qiu C, Liu D, Qi Y, Gao Y, Shen J, Qiu L (2013) beta-Glucan synthase gene overexpression and beta-glucans overproduction in *Pleurotus ostreatus* using promoter swapping. PLoS ONE 8:e61693
90. Lin YJ, Huang LH, Huang CT (2013) Enhancement of heterologous gene expression in *Flammulina velutipes* using polycistronic vectors containing a viral 2A cleavage sequence. PLoS One 8:e59099
91. Coconi-Linares N, Magana-Ortiz D, Guzman-Ortiz DA, Fernandez F, Loske AM, Gomez-Lim MA (2014) High-yield production of manganese peroxidase, lignin peroxidase, and versatile peroxidase in Phanerochaete chrysosporium. Appl Microbiol Biotechnol 98:9283–9294
92. Hood E (2004) RNAi: what's all the noise about gene silencing? Environ Health Perspect 112:A224–A229
93. Namekawa SH, Iwabata K, Sugawara H, Hamada FN, Koshiyama A, Chiku H, Kamada T, Sakaguchi K (2005) Knockdown of *LIM15/DMC1* in the mushroom *Coprinus cinereus* by double-stranded RNA-mediated gene silencing. Microbiology 151:3669–3678
94. Eastwood DC, Challen MP, Zhang C, Jenkins H, Henderson J, Burton KS (2008) Hairpin-mediated down-regulation of the urea cycle enzyme argininosuccinate lyase in *Agaricus bisporus*. Mycol Res 112:708–716
95. Kemppainen M, Duplessis S, Martin F, Pardo AG (2009) RNA silencing in the model mycorrhizal fungus *Laccaria bicolor*: gene knock-down of nitrate reductase results in inhibition of symbiosis with *Populus*. Environ Microbiol 11:1878–1896
96. Kemppainen MJ, Pardo AG (2010) Gene knockdown by ihpRNA-triggering in the ectomycorrhizal basidiomycete fungus *Laccaria bicolor*. Bioeng Bugs 1:354–358
97. Nakade K, Watanabe H, Sakamoto Y, Sato T (2011) Gene silencing of the *Lentinula edodes lcc1* gene by expression of a homologous inverted repeat sequence. Microbiol Res 166:484–493
98. Dang Y, Yang Q, Xue Z, Liu Y (2011) RNA interference in fungi: pathways, functions, and applications. Eukaryot Cell 10:1148–1155
99. Zhong S, Leng Y, Bolton MD (2012) Construction of hairpin RNA-expressing vectors for RNA-mediated gene silencing in fungi. Methods Mol Biol 835:623–633
100. Godio RP, Fouces R, Martin JF (2007) A squalene epoxidase is involved in biosynthesis of both the antitumor compound clavaric acid and sterols in the basidiomycete *H. sublateritium*. Chem Biol 14:1334–1346
101. Chang SS, Zhang Z, Liu Y (2012) RNA interference pathways in fungi: mechanisms and functions. Annu Rev Microbiol 66:305–323
102. Lugones LG, de Jong JF, de Vries OM, Jalving R, Dijksterhuis J, Wosten HA (2004) The SC15 protein of *Schizophyllum commune* mediates formation of aerial hyphae and attachment in the absence of the SC3 hydrophobin. Mol Microbiol 53:707–716

103. de Jong JF, Ohm RA, de Bekker C, Wosten HA, Lugones LG (2010) Inactivation of *ku80* in the mushroom-forming fungus *Schizophyllum commune* increases the relative incidence of homologous recombination. FEMS Microbiol Lett 310:91–95
104. Ohm RA, de Jong JF, Berends E, Wang F, Wosten HA, Lugones LG (2010) An efficient gene deletion procedure for the mushroom-forming basidiomycete *Schizophyllum commune*. World J Microbiol Biotechnol 26:1919–1923
105. Ninomiya Y, Suzuki K, Ishii C, Inoue H (2004) Highly efficient gene replacements in *Neurospora* strains deficient for nonhomologous end-joining. Proc Natl Acad Sci USA 101:12248–12253
106. van Peer AF, Wang FF, van Driel KGA, de Jong JF, van Donselaar EG, Mueller WH, Boekhout T, Lugones LG, Wosten HAB (2010) The septal pore cap is an organelle that functions in vegetative growth and mushroom formation of the wood-rot fungus *Schizophyllum commune*. Environ Microbiol 12:833–844
107. Ohm RA, de Jong JF, de Bekker C, Wosten HA, Lugones LG (2011) Transcription factor genes of *Schizophyllum commune* involved in regulation of mushroom formation. Mol Microbiol 81:1433–1445
108. Berends E, Lehle L, Henquet M, Hesselink T, Wosten HA, Lugones LG, Bosch D (2013) Identification of *alg3* in the mushroom-forming fungus *Schizophyllum commune* and analysis of the Delt*aalg3* knockout mutant. Glycobiology 23:147–154
109. Nakazawa T, Ando Y, Kitaaki K, Nakahori K, Kamada T (2011) Efficient gene targeting in Delta Cc.*ku70* or Delta *Cc.lig4* mutants of the agaricomycete *Coprinopsis cinerea*. Fungal Genet Biol 48:939–946
110. Salame TM, Knop D, Tal D, Levinson D, Yarden O, Hadar Y (2012) Predominance of a versatile-peroxidase-encoding gene, *mnp4*, as demonstrated by gene replacement via a gene targeting system for *Pleurotus ostreatus*. Appl Environ Microbiol 78:5341–5352
111. Salame TM, Knop D, Levinson D, Yarden O, Hadar Y (2013) Redundancy among manganese peroxidases in *Pleurotus ostreatus*. Appl Environ Microbiol 79:2405–2415
112. Salame TM, Knop D, Levinson D, Mabjeesh SJ, Yarden O, Hadar Y (2014) Inactivation of a *Pleurotus ostreatus* versatile peroxidase-encoding gene (*mnp2*) results in reduced lignin degradation. Environ Microbiol 16:265–277
113. Morin E, Kohler A, Baker AR, Foulongne-Oriol M, Lombard V, Nagy LG, Ohm RA, Patyshakuliyeva A, Brun A, Aerts AL et al (2012) Genome sequence of the button mushroom *Agaricus bisporus* reveals mechanisms governing adaptation to a humic-rich ecological niche. Proc Natl Acad Sci USA 109:17501–17506
114. Park YJ, Baek JH, Lee S, Kim C, Rhee H, Kim H, Seo JS, Park HR, Yoon DE, Nam JY et al (2014) Whole genome and global gene expression analyses of the model mushroom *Flammulina velutipes* reveal a high capacity for lignocellulose degradation. PLoS ONE 9: e93560
115. Bao D, Gong M, Zheng H, Chen M, Zhang L, Wang H, Jiang J, Wu L, Zhu Y, Zhu G et al (2013) Sequencing and comparative analysis of the straw mushroom (*Volvariella volvacea*) genome. PLoS ONE 8:e58294
116. Wawrzyn GT, Quin MB, Choudhary S, Lopez-Gallego F, Schmidt-Dannert C (2012) Draft genome of *Omphalotus olearius* provides a predictive framework for sesquiterpenoid natural product biosynthesis in Basidiomycota. Chem Biol 19:772–783
117. Ohm RA, de Jong JF, Lugones LG, Aerts A, Kothe E, Stajich JE, de Vries RP, Record E, Levasseur A, Baker SE et al (2010) Genome sequence of the model mushroom *Schizophyllum commune*. Nat Biotechnol 28:957–963
118. Yap HY, Chooi YH, Firdaus-Raih M, Fung SY, Ng ST, Tan CS, Tan NH (2014) The genome of the Tiger Milk mushroom, *Lignosus rhinocerotis*, provides insights into the genetic basis of its medicinal properties. BMC Genom 15:635
119. Tanaka Y, Kasuga D, Oba Y, Hase S, Sato K, Oba Y, Sakakibara Y (2014) Genome sequence of the luminous mushroom *Mycena chlorophos* for searching fungal bioluminescence genes. Luminescence 29:47–48

120. Zhu YJ, Xu J, Sun C, Zhou SG, Xu HB, Nelson DR, Qian J, Song JY, Luo HM, Xiang L et al (2015) Chromosome-level genome map provides insights into diverse defense mechanisms in the medicinal fungus *Ganoderma sinense*. Sci Rep 5:11087
121. Chen S, Xu J, Liu C, Zhu Y, Nelson DR, Zhou S, Li C, Wang L, Guo X, Sun Y et al (2012) Genome sequence of the model medicinal mushroom *Ganoderma lucidum*. Nat Commun 3:913
122. Zheng P, Xia Y, Xiao G, Xiong C, Hu X, Zhang S, Zheng H, Huang Y, Zhou Y, Wang S et al (2011) Genome sequence of the insect pathogenic fungus *Cordyceps militaris*, a valued traditional Chinese medicine. Genome Biol 12:R116
123. Boh B, Berovic M, Zhang J, Zhi-Bin L (2007) *Ganoderma lucidum* and its pharmaceutically active compounds. Biotechnol Annu Rev 13:265–301
124. Shiao MS (2003) Natural products of the medicinal fungus *Ganoderma lucidum*: occurrence, biological activities, and pharmacological functions. Chem Rec 3:172–180
125. Shi L, Ren A, Mu D, Zhao M (2010) Current progress in the study on biosynthesis and regulation of ganoderic acids. Appl Microbiol Biotechnol 88:1243–1251
126. Liu DB, Gong J, Dai WK, Kang XC, Huang Z, Zhang HM, Liu W, Liu L, Ma JP, Xia ZL et al (2012) The genome of *Ganderma lucidum* provide insights into triterpense biosynthesis and wood degradation. PLoS One 7:e36146
127. Qian J, Xu H, Song J, Xu J, Zhu Y, Chen S (2013) Genome-wide analysis of simple sequence repeats in the model medicinal mushroom *Ganoderma lucidum*. Gene 512:331–336
128. Xu J, Xu Z, Zhu Y, Luo H, Qian J, Ji A, Hu Y, Sun W, Wang B, Song J et al (2014) Identification and evaluation of reference genes for qRT-PCR normalization in *Ganoderma lucidum*. Curr Microbiol 68:120–126
129. Li J, Wu B, Xu J, Liu C (2014) Genome-wide identification and characterization of long intergenic non-coding RNAs in *Ganoderma lucidum*. PLoS ONE 9:e99442
130. Yin Y, Yu G, Chen Y, Jiang S, Wang M, Jin Y, Lan X, Liang Y, Sun H (2012) Genome-wide transcriptome and proteome analysis on different developmental stages of *Cordyceps militaris*. PLoS ONE 7:e51853
131. Li J, Zhang J, Chen H, Chen X, Lan J, Liu C (2013) Complete mitochondrial genome of the medicinal mushroom *Ganoderma lucidum*. PLoS ONE 8:e72038
132. Tang LH, Jian HH, Song CY, Bao DP, Shang XD, Wu DQ, Tan Q, Zhang XH (2013) Transcriptome analysis of candidate genes and signaling pathways associated with light-induced brown film formation in *Lentinula edodes*. Appl Microbiol Biotechnol 97:4977–4989
133. Tao Y, van Peer AF, Chen B, Chen Z, Zhu J, Deng Y, Jiang Y, Li S, Wu T, Xie B (2014) Gene expression profiling reveals large regulatory switches between succeeding stipe stages in *Volvariella volvacea*. PLoS ONE 9:e97789
134. Ramirez L, Oguiza JA, Perez G, Lavin JL, Omarini A, Santoyo F, Alfaro M, Castanera R, Parenti A, Muguerza E, Pisabarro AG (2011) Genomics and transcriptomics characterization of genes expressed during postharvest at 4 & #xB0;C by the edible basidiomycete Pleurotus ostreatus. Int Microbiol 14:111–120
135. Xu JW, Zhao W, Xu YN, Zhong JJ (2012) Isolation and analysis of differentially expressed genes during asexual sporulation in liquid static culture of *Ganoderma lucidum* by suppression subtractive hybridization. Mol Biol Rep 39:3603–3610
136. Zhu Y, Luo H, Zhang X, Song J, Sun C, Ji A, Xu J, Chen S (2014) Abundant and selective RNA-editing events in the medicinal mushroom *Ganoderma lucidum*. Genetics 196:1047–1057
137. Wang Z, Gerstein M, Snyder M (2009) RNA-Seq: a revolutionary tool for transcriptomics. Nat Rev Genet 10:57–63
138. Yu GJ, Wang M, Huang J, Yin YL, Chen YJ, Jiang S, Jin YX, Lan XQ, Wong BHC, Liang Y, Sun H (2012) Deep insight into the *Ganoderma lucidum* by comprehensive analysis of its transcriptome. PLoS One 7:e44031

139. Plaza DF, Lin CW, van der Velden NS, Aebi M, Kunzler M (2014) Comparative transcriptomics of the model mushroom *Coprinopsis cinerea* reveals tissue-specific armories and a conserved circuitry for sexual development. BMC Genom 15:492
140. Yang F, Xu B, Zhao S, Li J, Yang Y, Tang X, Wang F, Peng M, Huang Z (2012) De novo sequencing and analysis of the termite mushroom (*Termitomyces albuminosus*) transcriptome to discover putative genes involved in bioactive component biosynthesis. J Biosci Bioeng 114:228–231
141. Shashidhar MG, Giridhar P, Sankar KU, Manohar B (2013) Bioactive principles from *Cordyceps sinensis*: a potent food supplement—a review. J Funct Foods 5:1013–1030
142. Xiang L, Li Y, Zhu YJ, Luo HM, Li CF, Xu XL, Sun C, Song JY, Shi LC, He L et al (2014) Transcriptome analysis of the *Ophiocordyceps sinensis* fruiting body reveals putative genes involved in fruiting body development and cordycepin biosynthesis. Genomics 103:154–159
143. Rahmad N, Al-Obaidi JR, Al-Obaidi JR, Nor Rashid NM, Zean NB, Mohd Yusoff MH, Shaharuddin NS, Mohd Jamil NA, Mohd Saleh N (2014) Comparative proteomic analysis of different developmental stages of the edible mushroom *Termitomyces heimii*. Biol Res 47:30
144. Lau CC, Abdullah N, Shuib AS, Aminudin N (2012) Proteomic analysis of antihypertensive proteins in edible mushrooms. J Agric Food Chem 60:12341–12348
145. Chen L, Zhang BB, Cheung PC (2012) Comparative proteomic analysis of mushroom cell wall proteins among the different developmental stages of *Pleurotus tuber-regium*. J Agric Food Chem 60:6173–6182
146. Lin YL, Wen TN, Chang ST, Chu FH (2011) Proteomic analysis of differently cultured endemic medicinal mushroom *Antrodia cinnamomea* T.T. Chang et W.N. Chou from Taiwan. Int J Med Mushrooms 13:473–481
147. Yu GJ, Yin YL, Yu WH, Liu W, Jin YX, Shrestha A, Yang Q, Ye XD, Sun H (2015) Proteome exploration to provide a resource for the investigation of *Ganoderma lucidum*. PLoS ONE 10:e0119439
148. Zhang BB, Chen L, Cheung PCK (2012) Proteomic insights into the stimulatory effect of Tween 80 on mycelial growth and exopolysaccharide production of an edible mushroom *Pleurotus tuber-regium*. Biotechnol Lett 34:1863–1867
149. Wang M, Gu B, Huang J, Jiang S, Chen Y, Yin Y, Pan Y, Yu G, Li Y, Wong BH et al (2013) Transcriptome and proteome exploration to provide a resource for the study of *Agrocybe aegerita*. PLoS ONE 8:e56686
150. Petrovic J, Glamoclija J, Stojkovic D, Nikolic M, Ciric A, Fernandes A, Ferreira IC, Sokovic M (2014) Bioactive composition, antimicrobial activities and the influence of *Agrocybe aegerita* (Brig.) Sing on certain quorum-sensing-regulated functions and biofilm formation by *Pseudomonas aeruginosa*. Food Funct 5:3296–3303
151. Heinke R, Schone P, Arnold N, Wessjohann L, Schmidt J, Schmidt J (2014) Metabolite profiling and fingerprinting of *Suillus* species (Basidiomycetes) by electrospray mass spectrometry. Eur J Mass Spectrom (Chichester, Eng) 20:85–97
152. Park SJ, Hyun SH, Suh HW, Lee SY, Sung GH, Kim SH, Choi HK (2013) Biochemical characterization of cultivated *Cordyceps bassiana* mycelia and fruiting bodies by H-1 nuclear magnetic resonance spectroscopy. Metabolomics 9:236–246
153. Matsuzaki F, Shimizu M, Wariishi H (2008) Proteomic and metabolomic analyses of the white-rot fungus *Phanerochaete chrysosporium* exposed to exogenous benzoic acid. J Proteome Res 7:2342–2350
154. Bak JS (2015) Lignocellulose depolymerization occurs via an environmentally adapted metabolic cascades in the wood-rotting basidiomycete *Phanerochaete chrysosporium*. Microbiologyopen 4:151–166
155. Zhang W, Tang YJ (2008) A novel three-stage light irradiation strategy in the submerged fermentation of medicinal mushroom *Ganoderma lucidum* for the efficient production of ganoderic acid and *Ganoderma* polysaccharides. Biotechnol Prog 24:1249–1261
156. Tang YJ, Zhang W, Zhong JJ (2009) Performance analyses of a pH-shift and DOT-shift integrated fed-batch fermentation process for the production of ganoderic acid and

Ganoderma polysaccharides by medicinal mushroom *Ganoderma lucidum*. Bioresour Technol 100:1852–1859
157. Zhao W, Xu JW, Zhong JJ (2011) Enhanced production of ganoderic acids in static liquid culture of *Ganoderma lucidum* under nitrogen-limiting conditions. Bioresour Technol 102:8185–8190
158. Xu YN, Zhong JJ (2012) Impacts of calcium signal transduction on the fermentation production of antitumor ganoderic acids by medicinal mushroom *Ganoderma lucidum*. Biotechnol Adv 30:1301–1308
159. Xu YN, Xia XX, Zhong JJ (2013) Induced effect of Na(+) on ganoderic acid biosynthesis in static liquid culture of *Ganoderma lucidum* via calcineurin signal transduction. Biotechnol Bioeng 110:1913–1923
160. Xu YN, Xia XX, Zhong JJ (2014) Induction of ganoderic acid biosynthesis by Mn^{2+} in static liquid cultivation of *Ganoderma lucidum*. Biotechnol Bioeng 111:2358–2365
161. Fang QH, Zhong JJ (2002) Two-stage culture process for improved production of ganoderic acid by liquid fermentation of higher fungus *Ganoderma lucidum*. Biotechnol Prog 18:51–54
162. Zhang JM, Zhong JJ, Geng AL (2014) Improvement of ganoderic acid production by fermentation of *Ganoderma lucidum* with cellulase as an elicitor. Process Biochem 49:1580–1586
163. Xu P, Ding ZY, Qian Z, Zhao CX, Zhang KC (2008) Improved production of mycelial biomass and ganoderic acid by submerged culture of *Ganoderma lucidum* SB97 using complex media. Enzyme Microb Technol 42:325–331
164. Liu GQ, Ren GP, Wang XL, Zhao Y (2011) Statistical optimization of the key medium components by response surface methodology to promote ganoderic acid formation by medicinal mushroom *Ganoderma sinense* in submerged culture. J Med Plants Res 5:425–431
165. Zhang ZC, Shen WL, Liu D, Li JS (2011) Enhanced production of mycelial biomass and ganoderic acid in submerged culture of *Ganoderma applanatum* ACCC-52297 elicited by feeding rutin. Afr J Microbiol Res 5:3452–3461
166. Liu GQ, Wang XL, Han WJ, Lin QL (2012) Improving the fermentation production of the individual key triterpene ganoderic acid Me by the medicinal fungus *Ganoderma lucidum* in submerged culture. Molecules 17:12575–12586
167. Wei ZH, Duan YY, Qian YQ, Guo XF, Li YJ, Jin SH, Zhou ZX, Shan SY, Wang CR, Chen XJ et al (2014) Screening of *Ganoderma* strains with high polysaccharides and ganoderic acid contents and optimization of the fermentation medium by statistical methods. Bioprocess Biosyst Eng 37:1789–1797
168. Dou Y, Xiao JH, Xia XX, Zhong JJ (2013) Effect of oxygen supply on biomass and helvolic acid production in submerged fermentation of *Cordyceps taii*. Biochem Eng J 81:73–79
169. Jiang MY, Feng T, Liu JK (2011) N-containing compounds of macromycetes. Nat Prod Rep 28:783–808
170. Nakamura K, Yoshikawa N, Yamaguchi Y, Kagota S, Shinozuka K, Kunitomo M (2006) Antitumor effect of cordycepin (3′-deoxyadenosine) on mouse melanoma and lung carcinoma cells involves adenosine A3 receptor stimulation. Anticancer Res 26:43–47
171. Mao XB, Zhong JJ (2006) Significant effect of NH_4^+ on cordycepin production by submerged cultivation of medicinal mushroom *Cordyceps militaris*. Enzyme Microb Technol 38:343–350
172. Das SK, Masuda M, Hatashita M, Sakurai A, Sakakibara M (2008) A new approach for improving cordycepin productivity in surface liquid culture of *Cordyceps militaris* using high-energy ion beam irradiation. Lett Appl Microbiol 47:534–538
173. Fan DD, Wang W, Zhong JJ (2012) Enhancement of cordycepin production in submerged cultures of *Cordyceps militaris* by addition of ferrous sulfate. Biochem Eng J 60:30–35
174. Gunde-Cimerman N, Cimerman A (1995) *Pleurotus* fruiting bodies contain the inhibitor of 3-hydroxy-3-methylglutaryl-coenzyme A reductase-lovastatin. Exp Mycol 19:1–6
175. Atli B, Yamac M, Yildiz Z (2013) Optimization of submerged fermentation conditions for lovastatin production by the culinary-medicinal oyster mushroom, *Pleurotus ostreatus* (higher basidiomycetes). Int J Med Mushrooms 15:487–495

176. Lan WC, Tzeng CW, Lin CC, Yen FL, Ko HH (2013) Prenylated flavonoids from *Artocarpus altilis*: Antioxidant activities and inhibitory effects on melanin production. Phytochemistry 89:78–88
177. Zhu H, Sun S, Zhang S (2011) Enhanced production of total flavones and exopolysaccharides via *Vitreoscilla* hemoglobin biosynthesis in *Phellinus igniarius*. Bioresour Technol 102:1747–1751
178. Giavasis I (2014) Bioactive fungal polysaccharides as potential functional ingredients in food and nutraceuticals. Curr Opin Biotechnol 26:162–173
179. Su CA, Xu XY, Liu DY, Wu M, Zeng FQ, Zeng MY, Wei W, Jiang N, Luo X (2013) Isolation and characterization of exopolysaccharide with immunomodulatory activity from fermentation broth of *Morchella conica*. Daru J Pharm Sci 21:5
180. Osinska-Jaroszuk M, Jaszek M, Mizerska-Dudka M, Blachowicz A, Rejczak TP, Janusz G, Wydrych J, Polak J, Jarosz-WilkoBazka A, Kandefer-Szerszen M (2014) Exopolysaccharide from *Ganoderma applanatum* as a promising bioactive compound with cytostatic and antibacterial properties. Biomed Res Int :743812
181. Seo MJ, Kang BW, Park JU, Kim MJ, Lee HH, Choi YH, Jeong YK (2011) Biochemical characterization of the exopolysaccharide purified from *Laetiporus sulphureus* mycelia. J Microbiol Biotechnol 21:1287–1293
182. Xiao JH, Xiao DM, Chen DX, Xiao Y, Liang ZQ, Zhong JJ (2012) Polysaccharides from the medicinal mushroom *Cordyceps taii* show antioxidant and immunoenhancing activities in a D-galactose-induced aging mouse model. Evid-Based Compl Alt :273435
183. Ren L, Perera C, Hemar Y (2012) Antitumor activity of mushroom polysaccharides: a review. Food Funct 3:1118–1130
184. Xiang Y, Xu X, Li J (2012) Chemical properties and antioxidant activity of exopolysaccharides fractions from mycelial culture of *Inonotus obliquus* in a ground corn stover medium. Food Chem 134:1899–1905
185. Im SA, Wang W, Lee CK, Lee YN (2010) Activation of macrophages by exopolysaccharide produced by MK1 bacterial strain isolated from neungee mushroom, *Sarcodon aspratus*. Immune Netw 10:230–238
186. Zhu H, Cao CX, Zhang SS, Zhang Y, Zou WS (2011) pH-control modes in a 5-L stirred-tank bioreactor for cell biomass and exopolysaccharide production by *Tremella fuciformis* spore. Bioresour Technol 102:9175–9178
187. Hwang HS, Lee SH, Baek YM, Kim SW, Jeong YK, Yun JW (2008) Production of extracellular polysaccharides by submerged mycelial culture of *Laetiporus sulphureus* var. *miniatus* and their insulinotropic properties. Appl Microbiol Biotechnol 78:419–429
188. Cui JD, Zhang YN (2012) Evaluation of metal ions and surfactants effect on cell growth and exopolysaccharide production in two-stage submerged culture of *Cordyceps militaris*. Appl Biochem Biotechnol 168:1394–1404
189. Xu JW, Ji SL, Li HJ, Zhou JS, Duan YQ, Dang LZ, Mo MH (2015) Increased polysaccharide production and biosynthetic gene expressions in a submerged culture of *Ganoderma lucidum* by the overexpression of the homologous alpha-phosphoglucomutase gene. Bioprocess Biosyst Eng 38:399–405
190. Ji SL, Liu R, Ren MF, Li HJ, Xu JW (2015) Enhanced production of polysaccharide through the overexpression of the homologous UDP glucose pyrophosphorylase gene in a submerged culture of Lingzhi or Reishi medicinal mushroom, *Ganoderma lucidum* (higher basidiomycetes). Int J Med Mushrooms 17:435–442
191. Molnar I, Gibson DM, Krasnoff SB (2010) Secondary metabolites from entomopathogenic Hypocrealean fungi. Nat Prod Rep 27:1241–1275
192. Robinson JA (1991) Polyketide synthase complexes—their structure and function in antibiotic biosynthesis. Philos T Roy Soc B 332:107–114
193. Bai MS, Wang C, Zong SC, Lei M, Gao JM (2013) Antioxidant polyketide phenolic metabolites from the edible mushroom *Cortinarius purpurascens*. Food Chem 141:3424–3427

194. Strauss J, Reyes-Dominguez Y (2011) Regulation of secondary metabolism by chromatin structure and epigenetic codes. Fungal Genet Biol 48:62–69
195. Chung YM, Wei CK, Chuang DW, El-Shazly M, Hsieh CT, Asai T, Oshima Y, Hsieh TJ, Hwang TL, Wu YC, Chang FR (2013) An epigenetic modifier enhances the production of anti-diabetic and anti-inflammatory sesquiterpenoids from *Aspergillus sydowii*. Bioorg Med Chem 21:3866–3872
196. Asai T, Luo D, Obara Y, Taniguchi T, Monde K, Yamashita K, Oshima Y (2012) Dihydrobenzofurans as cannabinoid receptor ligands from *Cordyceps annullata*, an entomopathogenic fungus cultivated in the presence of an HDAC inhibitor. Tetrahedron Lett 53:2239–2243
197. Asai T, Yamamoto T, Oshima Y (2012) Aromatic polyketide production in *Cordyceps indigotica*, an entomopathogenic fungus, induced by exposure to a histone deacetylase inhibitor. Org Lett 14:2006–2009

Adv Biochem Eng Biotechnol (2016) 155: 237–241
DOI 10.1007/978-3-662-49161-4

Index

MIX
Papier aus verantwortungsvollen Quellen
Paper from responsible sources
FSC® C105338

If you have any concerns about our products,
you can contact us on
ProductSafety@springernature.com

In case Publisher is established outside the EU,
the EU authorized representative is:
Springer Nature Customer Service Center GmbH
Europaplatz 3, 69115 Heidelberg, Germany

Printed by Libri Plureos GmbH
in Hamburg, Germany